Building services and equipment

Volume 2

Building services and equipment

Volume 2
Second edition

Frederick E Hall
H.N.C., C.Ed., C.G.F.T.C., Dip C.I.O.B.

Addison Wesley Longman Limited
Edinburgh Gate, Harlow,
Essex CM20 2JE, England

and Associated Companies throughout the world

First published 1977
Second edition 1994
Second impression 1995
Third impression 1997

British Library Cataloguing in Publication Data
A catalogue entry for this title is available from the British Library.

ISBN 0-582-22968-5

Set by 18 in 10/12pt IBM Journal
Produced through Longman Malaysia, VVP

Contents

Preface

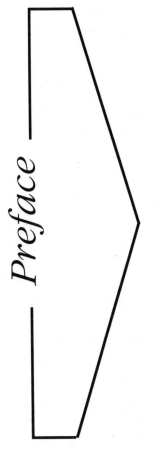

Because of the extensive amount of services being installed in modern buildings it was found necessary to write this second volume. The book, along with volume 1, should complete most of the subjects in services and equipment required for the various examinations, including the Business and Technology Education Council (BTEC) higher diplomas and certificates, City and Guilds of London Institute supplementary studies certificate and the Chartered Institute of Building examination. Students studying services for the various Royal Institution of Chartered Surveyors examinations should also find both volumes helpful.

Although this volume, as with volume 1, has been written primarily for students attending courses for various examinations, practising technicians in building, architecture and surveying should find it useful as a source of reference and students in structural engineering will find it helpful, as it provides an overall picture of services which is essential for a member of the design team.

The subjects covered in this volume are different from those covered in volume 1 and may be studied separately, but it will, of course, be necessary to study both volumes in order to cover most of the requirements of the various examinations.

SI units are used throughout the book and a description of the units has been included in the appendix. The emphasis has been on the illustrations of the installation of the various services by detailed diagrams which are described in the text. Where appropriate, tables and calculations have been included as an introduction to the design of services.

In this second edition the chapters on gas and artificial lighting have been revised and extended to comply with recent developments.

I should like to thank the publishers, C. R. Bassett for his encouragement and helpful constructive criticisms and my late wife for typing the entire manuscript without whose help this book would never have been written.

F. Hall
1994

Acknowledgements

We are grateful to the following for permission to reproduce copyright material:

Fire Officers' Committee for extracts from the *Committee's Rules for Automatic Sprinkler Installations* 29th edition; The Institute of Plumbing for charts and tables from *Pipe Sizing for Hot and Cold Water Installations*; Lighting Industry Federation Ltd for extracts from *Interior Lighting Design*.

We are also grateful to the following for permission to reproduce diagrams:

British Gas; Chubb Fire Security Ltd; The Walter Kidde Co. Ltd; Mather & Platt Ltd.

Chapter 1

Pipe sizing for hot- and cold-water installations

Design principles

In designing a hot- and cold-water supply installation, an assessment must first be made of the probable maximum water flow. In most buildings it seldom happens that the total number of appliances installed are ever in use at the same time, and therefore, for economic reasons, it is usual for a system to be designed for a peak usage which is less than the possible maximum usage. The probable demand will depend upon the type of sanitary appliances, the type of building in which they are installed, and the frequency of usage.

H. A. Howick has devised a method of assessing the probable maximum demand based upon the theory of probability. With this method a 'loading unit' rating has been devised for each type of sanitary appliance, based on its rate of water delivery, the time the taps are open during usage, and the simultaneous demand for the particular type of appliance.

Table 1.1 gives the 'loading unit' rating for various appliances.

Table 1.1

	Loading unit rating
Dwellings and flats	
W.C. flushing cistern	2
Wash basin	1½
Bath	10
Sink	3–5
Offices	
W.C. flushing cistern	2
Wash basin (distributed use)	1½
Wash basin (concentrated use)	3
Schools and industrial buildings	
W.C. flushing cistern	2
Wash basin	3
Shower (with nozzle)	3
Public bath	22

Note: Certain sanitary appliances require a continuous flow of water throughout the whole of the time that they are being used. These include: ablution appliances fitted with spray taps, umbrella sprays, shower nozzles or similar fittings.

In buildings where high peak demands occur, a loading unit rating for such appliances is not applicable and 100 per cent of the flow rate for these appliances is required as shown in Table 1.2. The same applies to automatic flushing cisterns and for urinals.

Table 1.2 Recommended minimum rate of flow at various appliances

Type of appliance	Rate of flow (litre/s)
W.C. flushing cistern	0.12
Wash basin	0.15
Wash basin with spray taps	0.04
Bath (private)	0.30
Bath (public)	0.60
Shower (with nozzle)	0.12
Sink with 13 mm taps	0.20
Sink with 19 mm taps	0.30
Sink with 25 mm taps	0.60

Example 1.1 *Determine the design flow rate for a hot or cold water distributing pipe supplying 8 W.C.s, and 12 wash basins in an office.*

8 W.C.s × 2 = 16 loading units

12 wash basins × 1½ = 18 loading units

= 34 loading units

From Fig. 1.1 the flow rate required for thirty-four loading units would be about 0.6 litre/s (litres per second).

Example 1.2. *Determine the design flow rate for a cold-water distributing pipe supplying 20 W.C.s, 24 wash basins, 10 urinals, 6 showers, and 4 cleaners' sinks, in a factory where there is a high peak demand for the use of showers.*

20 W.C.s x 2 = 40 loading units

24 wash basins x 3 = 72 loading units

4 sinks x 4 = 16 loading units

= 128 loading units

From Fig. 1.1 the flow rate required for 128 loading units would be about 1.6 litre/s and to this must be added the water required for urinal flushing and continuous use of showers. The urinals would require flushing every 20 min, and each urinal would require a 4.50 litre flush.

10 x 4.50 = 45 litre every 20 min

or

$$\frac{45}{20 \times 60} = 0.0375 \text{ litre/s}$$

The six showers would require

6 x 0.12 = 0.72 litre/s

The total flow rate required would be

1.6 + 0.0375 + 0.72 = 2.3575 litre/s

or

2.4 litre/s (approx.)

Effective length of pipe

The diameter of the pipe necessary to give a required flow rate will depend upon the head of water available, the smoothness of the internal bore of the pipe and the effective length of the pipe. An allowance for the frictional resistance set up by fittings such as elbows, tees, taps and valves must be added to the actual length of the pipe.

Table 1.3 gives the allowance for fittings expressed in equivalent pipe lengths.

Table 1.3 Frictional resistances of fittings expressed in equivalent pipe lengths

Copper			Galvanised steel			
Nominal outside diameter (mm)	Metre run of pipe		Nominal outside diameter (mm)	Metre run of pipe		
	Elbow	Tee		Elbow	Bend	Tee
15	0.5	0.6	15	0.5	0.4	1.2
22	0.8	1.0	20	0.6	0.5	1.4
28	1.0	1.5	25	0.7	0.6	1.8
35	1.4	2.0	32	1.0	0.7	2.3
42	1.7	2.5	40	1.2	1.0	2.7
54	2.3	3.5	50	1.4	1.2	3.4
62	3.0	4.5	65	1.7	1.3	4.2
76	3.4	5.8	80	2.0	1.6	5.3
108	4.5	8.0	100	2.7	2.0	6.8

In calculating the diameter of a pipe to supply individual fittings, the loss of head through the draw-off tap should also be taken into account. Table 1.4 gives the allowances for draw-off taps expressed in equivalent pipe lengths.

Table 1.4 Frictional resistances of draw-off taps expressed as equivalent pipe lengths

Fitting (BS 1010)	Discharge rate tap fully open (litre/s)	Equivalent length of pipe of same diameter as tap (m)	
		Copper	Galvanised steel
15 mm diameter bib-tap or pillar tap	0.20	2.70	4.00
20 mm diameter bib-tap or pillar tap	0.30	8.50	5.75
25 mm diameter bib-tap or pillar tap	0.60	20.00	13.00

Determination of pipe diameter

To determine the diameter of a pipe for a given flow rate, the allowable loss of head per metre run of effective pipe length must first be calculated.

Example 1.3. *Calculate the loss of head per metre run of pipe, when a copper hot- or cold-water distributing pipe having an actual length of 15 m, with six elbows in the run, is required to discharge 2 litre/s under a constant head of water of 6 m.*

Assuming a pipe diameter of 32 mm, the effective length of pipe would be

$15 + (6 \times 1.4) = 23.4$ m

As the available head is 6 m, the permissible loss of head per metre would be

$$\frac{\text{head}}{\text{effective length}} = \frac{6}{23.4}$$
$$= 0.2564 \text{ m/m run}$$
$$= 0.3 \text{ (approx.)}$$

To find the diameter of pipe see Fig. 1.2 which shows that a 32 mm diameter copper pipe will discharge about 2.5 litre/s when the loss of head per metre, per metre run of pipe, is 0.3.

Example 1.4. *Determine the diameter of a copper cold-water rising main capable of discharging 2 litre/s through a 20 mm orifice, when the pressure on the main is 500 kPa (kilopascal) (51 m head), the height of the ball valve above the main is 10 m and the actual length of pipe from the main to the ball valve is 40 m with four elbows and two stop valves in the run.*

Assuming a 28 mm (outside diameter) copper pipe, then the effective length would be (see Table 1.3)

$40 + (4 \times 1) = 44$ m

By reference to Fig. 1.3 at a flow rate of 2 litre/s, the loss of head through a ball valve with a 20 mm orifice is about 3 m and by reference to Fig. 1.4 the loss of head through a 25 mm stop valve would be about 4.5 m.

The available head will be

$51 - (10 + 3 + 4.5) = 33.5$ m

The permissible loss of head per metre run of effective length of pipe would be

$$\frac{\text{head}}{\text{length}} = \frac{33.5}{44} = 0.76 \text{ m/m run}$$

By reference to Fig. 1.2 a 28 mm o.d. (outside diameter) copper pipe under these conditions would convey about 2.4 litre/s so that this size of pipe would be satisfactory.

Pipe sizing for multi-storey buildings

The same principles of pipe sizing can be applied to pipe sizing of multi-storey buildings. The design factors that must be taken into account are as follows:

1. *Loading units*

Determine from the loading unit Table 1.1:

(*a*) the total number of loading units for each branch pipe;

(*b*) the total number of loading units required at the commencement of the main distributing pipe and the reduced number of loading units required at each branch.

2. *Rate of flow*

Using the loading ratings calculated, determine the flow rate in litres per second at each branch from Fig. 1.1.

3. *Percentage demand*

From a comparison between the probable demand and the estimated maximum demand in litres per second, for which frictional resistances of the pipework should be taken into account, can be found.

Example 1.5. *Determine the diameters of the main pipes for the two-storey office block shown in Fig. 1.5. Copper pipes are to be used and it may be assumed that the short branch pipes to the fittings will be 15 mm (o.d.). For clarity, it will be assumed that the main branch pipes will not be reduced in diameter along the run.*

Probable demand:

16 W.C.s × 2 = 32 loading units
16 wash basins × 1½ = 24 loading units
= 56 loading units

From Fig. 1.1 the probable demand will be about 0.85 litre/s. The estimated maximum demand is 1.6 litre/s (0.8 litre/s per floor) and the ratio of the probable and the estimated maximum demand expressed as a percentage is

$$\frac{0.85}{1.6} \times \frac{100}{1} = 53 \text{ per cent}$$

When considering the resistances it is only necessary to take into account the fittings, valves and taps for the upper floors. Below the upper floors, the frictional resistances of the main vertical distributing pipe only, need be taken into account.

Note: If it is the intention of the designer to allow for 100 per cent demand Table 1.2 would be used and the pipework must therefore supply 2.16 litre/s to each floor. However, this method is over-generous and for large installations this would result in oversizing and increased costs.

Tabulation

In order to be able to see clearly the determination of the pipe size, a form of tabulation is required. The distributing pipe in Example 1.5 can be divided into six parts, the part serving each floor being given a number, with the top floor used as the point of commencement (see Table 1.5).

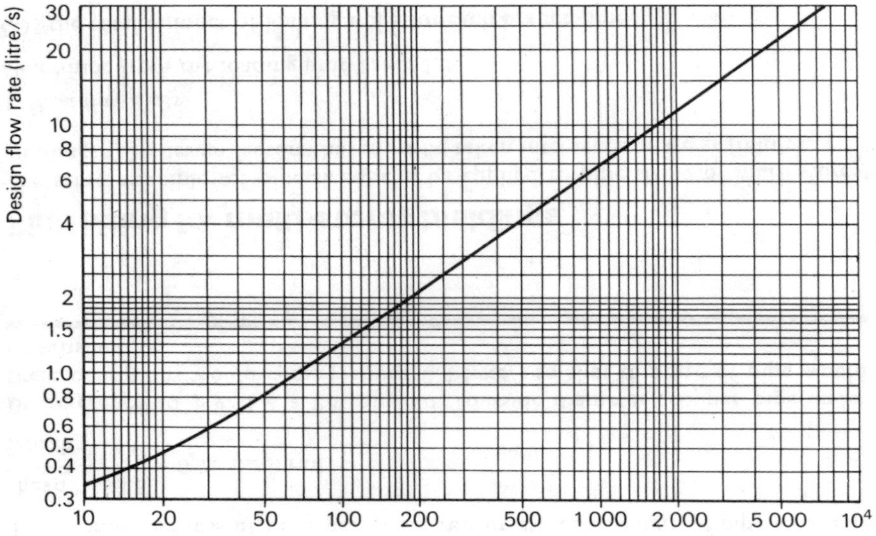
Design flow rate (litre/s)
30 20 10 8 6 4 2 1.5 1.0 0.8 0.6 0.5 0.4 0.3
10 20 50 100 200 500 1 000 2 000 5 000 10^4

Fig.1.1 Loading units

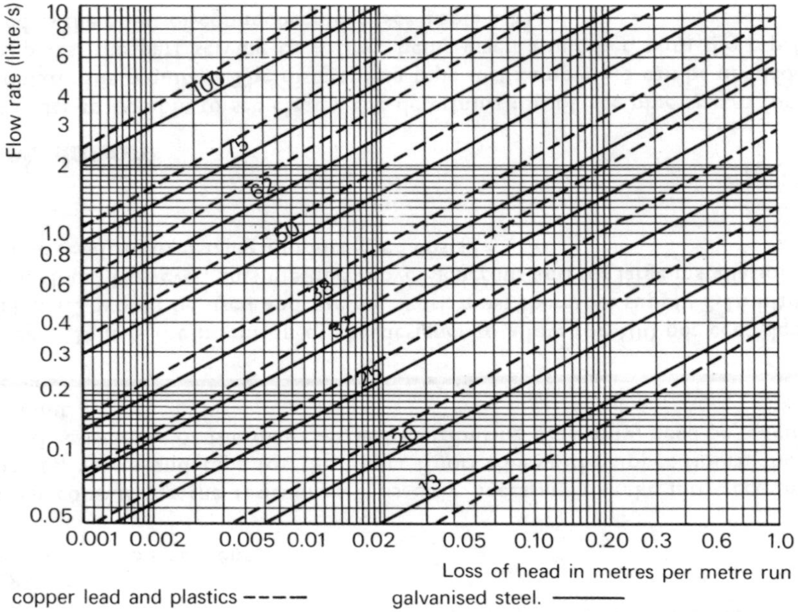
Flow rate (litre/s)
10 8 6 4 3 2 1.0 0.8 0.6 0.4 0.3 0.2 0.1 0.05
100
75
62
50
38
32
25
20
13
0.001 0.002 0.005 0.01 0.02 0.05 0.10 0.20 0.3 0.6 1.0
Loss of head in metres per metre run
copper lead and plastics ----
galvanised steel. ———

Fig.1.2 Pipe-sizing chart

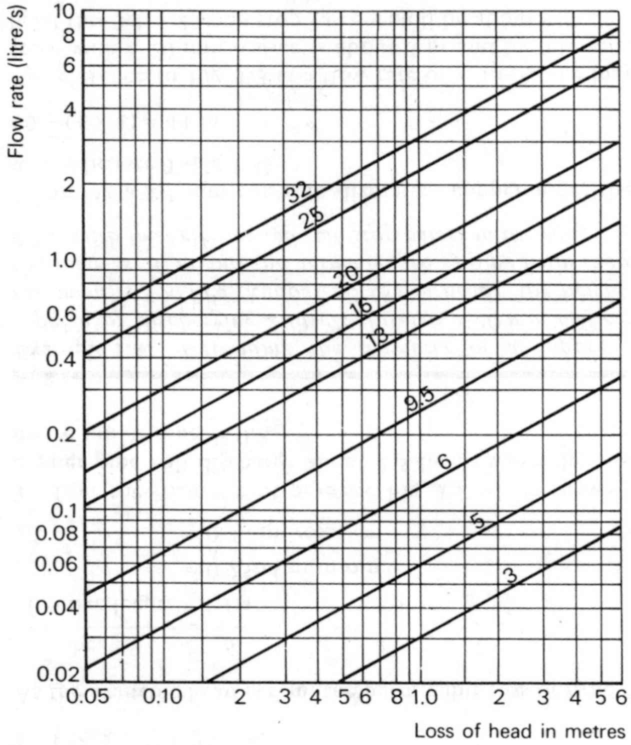
Flow rate (litre/s)
10 8 6 4 2 1.0 0.6 0.4 0.2 0.1 0.08 0.06 0.04 0.02
32
25
20
16
13
9.5
6
5
3
0.05 0.10 2 3 4 5 6 8 1.0 2 3 4 5 6
Loss of head in metres

Fig.1.3 Loss of head through ball-valve orifices

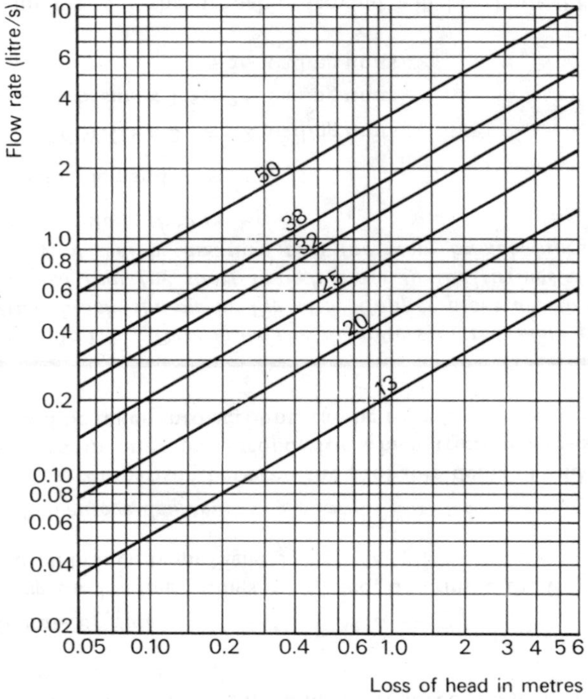
Flow rate (litre/s)
10 6 4 2 1.0 0.8 0.6 0.4 0.2 0.10 0.08 0.06 0.04 0.02
50
38
32
25
20
13
0.05 0.10 0.2 0.4 0.6 1.0 2 3 4 5 6
Loss of head in metres

Fig.1.4 Loss of head through stop valves

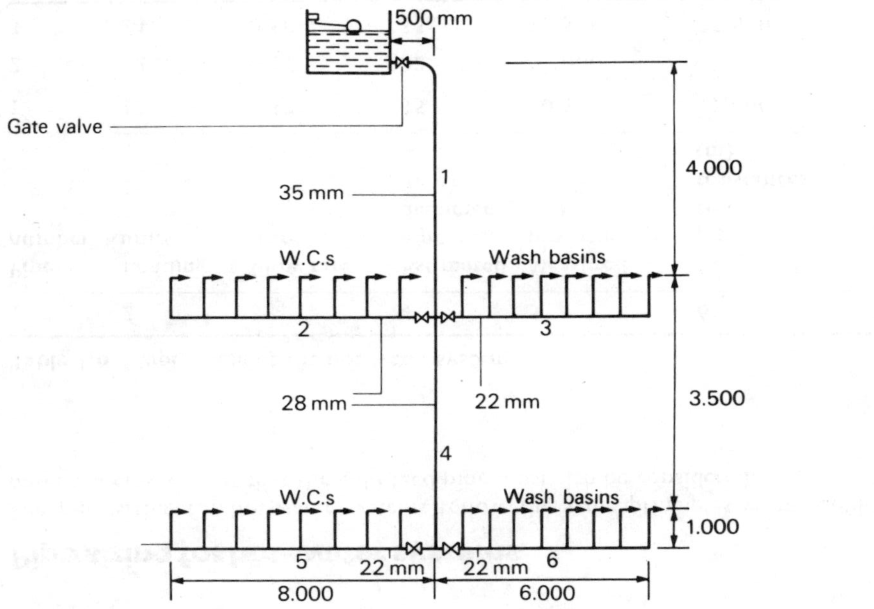
500 mm
Gate valve
35 mm
1
4.000
W.C.s
Wash basins
2
3
28 mm
22 mm
3.500
4
W.C.s
Wash basins
1.000
5
22 mm
22 mm
6
8.000
6.000
Fig.1.5 Pipe sizing of cold-water services to a two-storey office

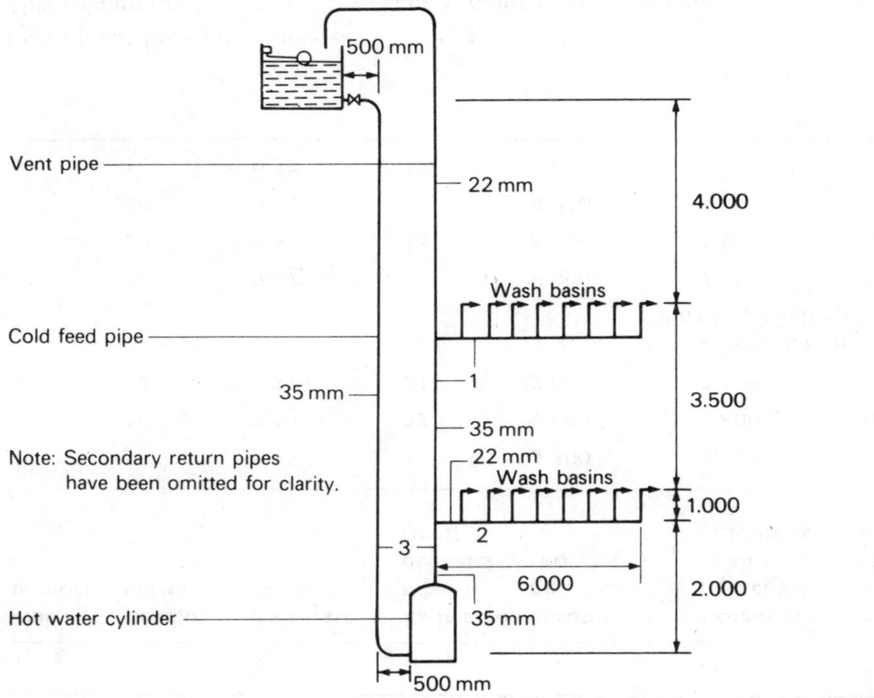
500 mm
Vent pipe
22 mm
4.000
Wash basins
Cold feed pipe
35 mm
1
3.500
35 mm
22 mm
Note: Secondary return pipes
have been omitted for clarity.
Wash basins
1.000
2
3
6.000
2.000
Hot water cylinder
35 mm
500 mm
Fig.1.6 Pipe sizing of hot-water supply service to a two-storey office

Explanation of Table 1.5

Column 1: Pipe number

The number allotted to each pipe section is entered in this column.

Column 2: Loading units

The sum of the loading units required for the appliances supplied by each pipe is found from Table 1.1 and tabulated.

Column 3: Flow rate

The flow rate required in litres per second at each floor can be read from Fig. 1.1.

Column 4: Assumed pipe diameter

To enable the frictional resistance equal to the equivalent length of the pipe to be obtained, a preliminary estimate of the required pipe diameter should be entered in this column. Where the estimated pipe diameter is too small, the excessive head loss shown in column 9 will indicate that an increase in pipe diameter is required. Alternatively, if the head loss is considerably less than the head available, a smaller pipe may be used.

Column 5: Measured pipe run

The actual measured pipe length in metres, ignoring any allowances for frictional losses in fittings, should be entered in this column.

Column 6: Equivalent pipe length of fittings

In this column is entered the sum of the equivalent pipe lengths of all fittings.

Column 7: Effective pipe length

The effective length of pipe in each section will be the sum of the measured length and the equivalent resistance length. For example: column 5 + column 6 = column 7.

Column 8: Loss of head in metres per metre run of pipe

Loss of head in metres, per metre run of pipe should be read from the pipe-sizing chart (see Fig. 1.2).

Column 9: Head consumed in metres

Head consumed in metres for any pipe section is obtained by multiplying the effective length of pipe in metres, by the loss of head in metres, per metre run. For example: column 7 x column 8 = column 9.

Column 10: Total head consumed in metres

Total head consumed is calculated by adding together the individual heads consumed in column 9. The total head consumed for parts of the pipework at each floor should be slightly less than the actual head available. If the difference is excessive, some change of pipe diameter in certain sections is required.

Column 11: Actual head available

The actual head available at any floor is the vertical height measured from the outlet, from the cistern above the draw-off points.

Table 1.5 Pipe sizing of the cold-water system

1	2	3	4	5	6	7	8	9	10	11	12
Pipe number	**Loading units**	**Flow rate (litre/s)**	**Estimated pipe diameter (mm)**	**Measured pipe run (m)**	**Length of pipe equal to all resistances (m)**	**Effective pipe length (m) (columns 5 + 6)**	**Loss of head (m/m run)**	**Head consumed (m)**	**Total head consumed (m)**	**Head available at point of discharge (m)**	**Final pipe size (o.d.) (mm)**
1	56	0.85	35	5.500	3.400	8.900	0.029	0.258	0.258	4.000	35
2	16	0.42	22	9.000	7.800	16.800	0.150	2.520	2.778	4.000	—
3	12	0.37	22	7.000	7.800	14.800	0.120	1.776	4.554	4.000	22
					Total head consumed in pipe No. 3 is too high. Pipe No. 2 is therefore increased to 28 mm						
2	16	0.42	28	9.000	7.800	16.800	0.040	0.672	2.706	4.000	28
4	28	0.58	28	3.500	1.500	5.000	0.070	0.350	3.056	7.500	28
5	16	0.42	22	9.000	—	9.000	0.150	1.35	4.406	7.500	22
6	12	0.37	22	7.000	—	7.000	0.120	0.84	5.046	7.500	22

Column 12: Final pipe diameter

This column can be used for any changes found necessary due to over- or under-sizing at the preliminary estimate.

Pipe sizing for hot-water systems

The pipe sizing for hot-water systems follow the same principles as the cold-water systems, except that the cold feed pipe must also be considered.

Example 1.6. *Determine the diameters of the main pipes for the two-storey office block shown in Fig. 1.6. Copper pipes are to be used and it may be assumed that the short branch pipes to the fittings will be 15 mm (o.d.). For clarity, it will again be assumed that the main branch pipes will not be reduced in diameter along the pipe run.*

Table 1.6 shows the method of determining the pipe diameters. The vent pipe may be assumed to be 22 mm (o.d.).

Table 1.6 Pipe sizing of the hot-water system

1	2	3	4	5	6	7	8	9	10	11	12
Pipe number	**Loading units**	**Flow rate (litre/s)**	**Estimated pipe diameter (mm)**	**Measured pipe run (m)**	**Length of pipe equal to all resistances (m)**	**Effective pipe length (m) (columns 5 + 6)**	**Loss of head (m/m run)**	**Head consumed (m)**	**Total head consumed (m)**	**Head available at point of discharge (m)**	**Final pipe size (o.d.) (mm)**
1	12	0.37	35	10.5	17.400	28.000	0.120	3.36	3.36	4.000	35
2	12	0.37	22	7.000	—	7.000	0.120	0.84	4.200	7.500	22
3	24	0.50	35	13.5	5.800	19.300	0.019	0.37	4.570	7.500	35

Relative discharging power of pipes

Relative discharging power of pipes are as the square root of the fifth power of their diameters.

$$N = \sqrt{\left(\frac{D}{d}\right)^5}$$

where

N = number of branch pipes
D = diameter of main pipe
d = diameter of branch pipes

Table 1.7 has been based on the formula.

Table 1.7 Relative discharging power of pipes

Main pipe	Branch pipe								Internal diameter
Nominal diameter of pipe size	Nominal diameter of pipe size (mm)								
Internal diameter (mm)	100	75	65	50	40	32	25	20	15
100	1	2	3	6	10	17	32	56	115
75		1	2	3	5	9	16	28	56
65			1	2	4	6	11	19	39
50				1	2	3	6	10	21
40					1	2	4	6	12
32						1	2	4	7
25							1	2	4
20								1	2
15									1
	Number of branch pipes								

Example 1.7. *Determine the number of 20 mm diameter branch pipes that may be supplied by a 50 mm diameter main pipe.*

From Table 1.7 ten branch pipes may be supplied by a 50 mm diameter main pipe.

Example 1.8. *Determine the diameter of a main pipe to supply thirty-nine 15 mm diameter branch pipes.*

From Table 1.7 the diameter of main pipe is 65 mm.

Thomas Box formula

This is a well-known practical formula for pipe sizing given in the following expression:

$$q = \sqrt{\frac{d^5 \times H}{25 \times L \times 10^5}}$$

where

q = discharge through pipe in litres per second
d = diameter of pipe in millimetres
H = head of water in metres
L = total length of pipe in metres

Example 1.9. *Calculate the diameter of a pipe to discharge 1.25 litre/s when the head is 4 m and the effective length is 45.5 m.*

By transposition of formula:

$$q^2 = \frac{d^5 \times H}{25 \times L \times 10^5}$$

$$d = \sqrt[5]{\frac{q^2 \times 25 \times L \times 10^5}{H}}$$

$$d = \sqrt[5]{\frac{1.25^2 \times 25 \times 45.5 \times 10^5}{4}}$$

$$d = 33.85$$

The nearest pipe size is 32 mm which is approximately 2 mm undersize, the next size is 38 mm which is approximately 4 mm oversize and it would be left to the designer which size to choose, depending upon the circumstances.

Example 1.10. *Calculate the discharge in litre/s through a 50 mm bore pipe when the head of water is 8 m and the effective length of pipe is 15 m.*

$$q = \sqrt{\frac{50^5 \times 8}{25 \times 15 \times 10^5}}$$

q = 8.2 litre/s approx.

Chapter 2

Electrical terms and calculations

Definition of terms

1. **Ampere, A (amp)** The unit of electric current. The ampere is a basic SI unit and is defined as that constant current which, if maintained in two straight parallel conductors of infinite length, of negligible circular cross-section, and placed 1 m apart in a vacuum, would produce between these conductors a force equal to 2×10^{-7} N/m (newton per metre) of length.

2. **Alternating current (a.c.)** A current which may start at zero, increase to reach maximum, fall away to zero, and then increase to an equal but opposite maximum and fall away again to zero. This series is known as one cycle and the frequency of an alternating current is the number of cycles completed in 1 s. The normal supply in the British Isles is an alternating current at a frequency of 50 cycles per second (50 Hz (hertz)). Figure 2.1 shows a single-phase alternating current wave form.

3. **Capacitor** An arrangement of conductors in the form of metal sheets, or foil separated by a thin dielectric of paper, or other insulating material. It provides for what is known as electrical capacitance and alternating current will flow through it. Capacitors are widely used in radio apparatus, radio interference suppression and for power factor correction purposes.

4. **Capacitance** The property of a system of conductors and insulators, which allows them to store an electrical charge when a potential difference exists between the conductors. The unit of capacitance the farad.

5. **Conductor** A substance that allows an electrical current to pass through it relatively freely; for example, has a high conductivity. Copper and silver are good conductors.

6. **Current** The movement or passage of electricity along a conductor (generally measured in amperes).

7. **Dielectric** The insulation between two electrically charged bodies. Paper, insulating oil, ebonite and mica, are common dielectric materials.

8. **Direct current (d.c.)** The form of electric current that flows continuously in the same direction.

9. **Discharge lamp** A lamp in which the current passing through a mixture of gas and metallic vapour forms a luminous electric discharge. The main advantage of fluorescent lamps is their high efficacy. In general, the range for tungsten lamps is about 10–15 lm/W (lumens per watt) (depending upon life and on wattage) and for fluorescent lamps 20–60 lm/W (depending mainly on colour).

10. **Distribution board** A board from which connections are taken for the distribution of electrical circuits, generally through circuit-breakers or fuses.

11. **Distribution main** The low-voltage main used for the supply to the consumers. The main is usually fed from a high-voltage ring main through a transformer in a sub-station, and is either buried underground or carried overhead.

12. **Diversity factor** The relation between the total connected load and the maximum demand. In some electrical installations it is extremely unlikely that all the lights and other equipment will be in use simultaneously. If say 100 appliances, each loaded to 1 kW (kilowatt), produce a statistical maximum demand of 25 kW, the diversity factor would be 1 in 4, or 25 per cent. The Electrical Regulations permit a reduction to be made in the size of conductors and switchgear of circuits, other than final sub-circuits, based upon the diversity factor.

13. **Earth leakage circuit-breaker** A circuit-breaker which disconnects the supply when the voltage or current with respect to earth, reaches a predetermined limit.

14. **Earth and earthing** A connection with the ground made in such a manner that an immediate and safe discharge of electrical energy is ensured.

15. **Extra low voltage** Normally not exceeding 50 V (volts) between conductors and not exceeding 30 V a.c. or 50 V d.c. between any conductor and earth.

16. **Fluorescent lamp** A tubular discharge lamp internally coated with a powder that fluoresces under the action of an electrical discharge, producing a white or coloured light.

17. **Four-wire distribution** The system of distribution on three-phase a.c. which consists of three phase wires and one neutral.

18. **Fuse** A safety device consisting of a short length of relatively fine wire in a suitable holder. The device is inserted in the live conductor at a suitable point and if the current exceeds the design value, the fuse wire melts and thus prevents damage to the circuit it protects.

19. **Grid** The high-voltage transmission system operated by the National Grid Company plc (N.G.C.). The 'grid' owes its name to the fact that a map of the system resembles a grid iron.

20. **High voltage** A voltage normally exceeding 650.

21. **Impedance** The total reactance of an electric circuit to the flow of an alternating current.

22. **Insulator** A non-conducting material such as rubber, plastic, porcelain, glass or mica, surrounding or supporting a conductor.

23. **Kilovolt (kV)** A unit equal to 1000 V.

24. **Kilovolt-ampere (kVA)** The unit of apparent power equal to 1000 VA (volt-amperes). With some types of a.c. electrical equipment such as generators, the voltage and current waves are displaced to each other and the power is less than the product of voltage and amperage. The load is therefore rated in kilovolt-amperes rather than kilowatts.
25. **Kilowatt (kW)** The unit of true power equal to 1000 W (watts).
26. **Kilowatt-hours (kWh)** The quantity of electrical power consumed, measured by kilowatts multiplied by hours. A 3 kW electric heater which operates for 3 h (hours) would consume 9 kWh. The kilowatt-hour is commonly known as the unit of electricity and is equal to 1 kW used in 1 h.
27. **Load** The power supplied to a building or a piece of apparatus or the power delivered by a generator.
28. **Load factor** The proportion expressed as a percentage of the number of units used or generated, to the number of units that would have been used or generated if the maximum load had been maintained steadily and continuously e.g. if a motor runs fully loaded for 44 h per week and 50 weeks in a year, its load factor would be

$$\frac{44 \times 50}{24 \times 365} = 25 \text{ per cent}$$

29. **Low voltage** Not exceeding 250 V between any two conductors, or between one conductor and earth.
30. **Maximum demand** The greatest rate of consumption of electricity.
31. **Medium voltage** A voltage exceeding 250 V, but not exceeding 650 V between any two conductors or between one conductor and earth.
32. **Megavolt-ampere (MVA)** A unit equal to 1 million VA.
33. **Megawatt (MW)** A unit equal to 1 million W.
34. **Milliampere (mA)** A thousandth part of an ampere.
35. **Millivolt (mV)** A thousandth part of a volt.
36. **Neutral conductor** A conductor connected to the neutral point of a star connected three-phase system.
37. **Ohm (Ω)** The unit of electrical resistance. A resistance of 1 Ω will pass a current of 1 A, when a potential difference of 1 V (d.c.) is applied between its ends. This relationship between current and voltage is known as Ohm's law.
38. **Power factor** Applies to alternating current only. The ratio between true power and the product of the voltage and current is called the power factor.

$$\frac{\text{kW}}{\text{kVA}} = \text{power factor}$$

The power factor is the percentage of current which can be used as energy. A power factor of 0.8 means that 80 per cent of the current supplied is used as energy and the remaining 20 per cent is idle. In an inductive circuit such as produced by a motor or solenoid, the electromagnetic force opposes the applied voltage which causes the current wave to lag behind the voltage wave. In a capacitive circuit, the current leads the voltage because the capacitor stores energy as the current rises and discharges energy as the current falls. Most electrical equipment however is inductive and it is therefore usual for the current to lag the voltage and the whole current available is not used. Electricity boards do not allow equipment to be installed that has a power factor of less than a certain figure, e.g. 0.85.

Note: An electric meter measures true power.

39. **Root mean square value (r.m.s.)** (see Fig. 2.1) Alternating current and voltage values fluctuate between zero and maximum potential while direct current values remain about some steady point. In order to obtain the same values for a.c. as are present in d.c. the r.m.s. value is used. The r.m.s. value of 1 A a.c. gives the same effective value in heat production as 1 A d.c., and a.c. voltage follows the same reasoning.

The relationship of r.m.s. value to maximum value a.c. is

$$\frac{\text{r.m.s. value}}{\text{maximum value}} = 0.707$$

and

$$\frac{\text{maximum value}}{\text{r.m.s. value}} = 1.414$$

40. **Three-phase** The electricity generated and transmitted in Great Britain is known as the three-phase alternating current supply. This means that there are three separate single-phase supplies (as shown in Fig. 2.2) out of step with each other by 120° produced by an alternating current generator known as an alternator. It is possible to take three pairs of conductors from one alternator, thus providing three separate three-phase supplies, but it is more economical to join the windings together (as shown in Fig. 2.3) and use three live or line conductors and one common return conductor or neutral.

The voltage between any line conductor and neutral (phase voltage) is 240 and between any two line conductors (line voltage) 1.73 times the phase voltage, e.g. 415 V. The relationship between phase and line voltages in a three-phase, four-wire supply is shown in Fig. 2.4, or may be expressed as follows:

$$\text{line voltage} = \text{phase voltage} \times \sqrt{3}$$
$$= \text{phase voltage} \times 1.73$$

and

$$\text{phase voltage} = \frac{\text{line voltage}}{1.73}$$

A three-phase supply is required for all types of buildings where electric motors larger than those required for domestic buildings are used. Electricity boards usually require a three-phase supply when a motor requires 3730 W or over, of power. Three-phase supply is more efficient than single-phase.

Example 2.1. *Calculate the gain in power, expressed as a percentage, of three-phase over single-phase supply, assuming unity power factor and with 240 V and 10 A, respectively.*

Single-phase power = VA
Single-phase power = 2400 W

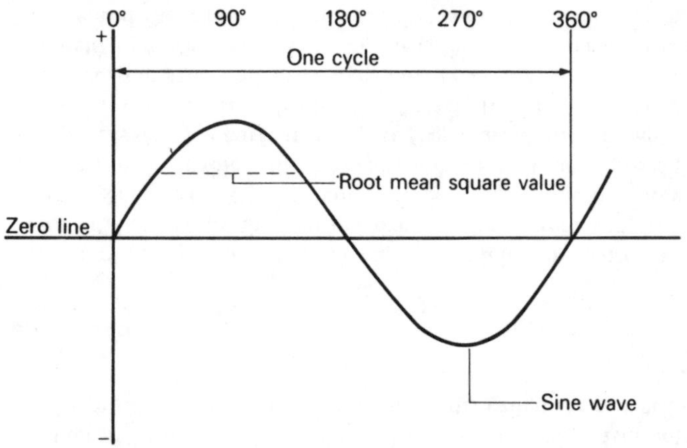
0°
90°
180°
270°
360°
+
One cycle
Root mean square value
Zero line
Sine wave
–

Fig.2.1 Single-phase alternating current

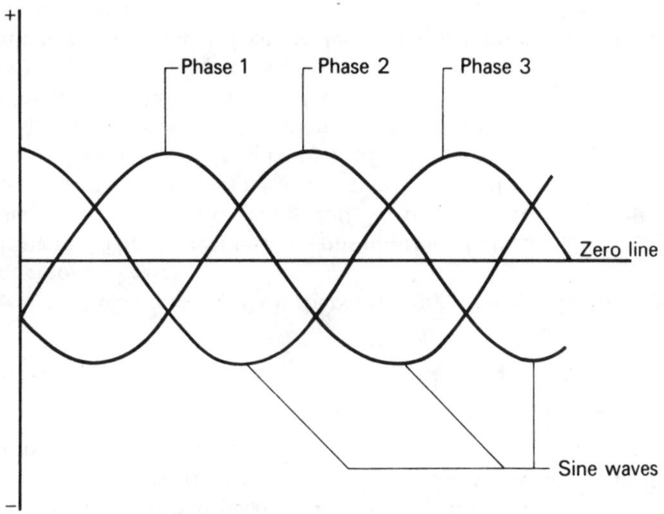
+
Phase 1
Phase 2
Phase 3
Zero line
Sine waves
–

Fig.2.2 Three-phase alternating current

$$\text{Three-phase power} = VA\sqrt{3}$$
$$\text{Three-phase power} = 240 \times 10 \times 1.73$$
$$= 4152\ \text{W}$$
$$\text{Gain in power of three-phase} = 4152 - 2400$$
$$= 1752\ \text{W}$$
$$\text{Gain in power of three-phase expressed as a percentage} = \frac{1752}{2400} \times \frac{100}{1}$$
$$= 73\ \text{per cent}$$

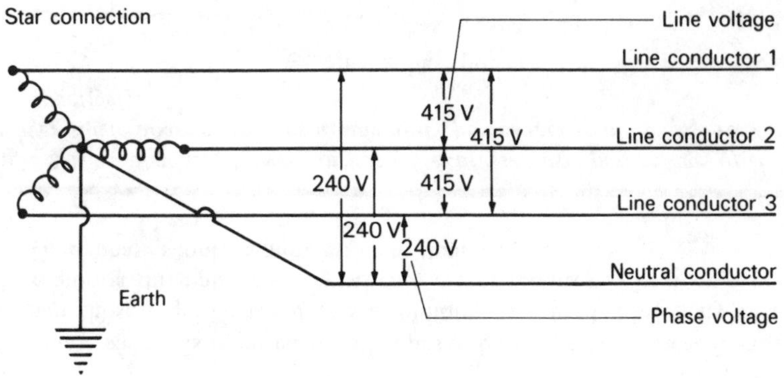
Star connection
Line voltage
Line conductor 1
415 V
415 V
Line conductor 2
240 V
415 V
Line conductor 3
240 V
240 V
Neutral conductor
Earth
Phase voltage

Fig.2.3 Line and phase voltages of a three-phase, four-wire supply

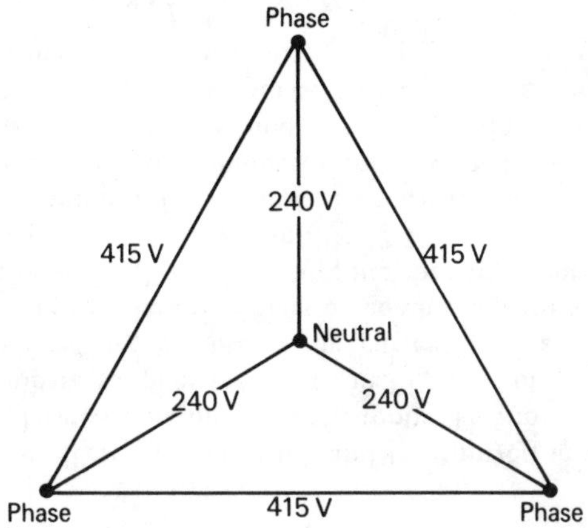
Phase
240 V
415 V
415 V
Neutral
240 V
240 V
Phase
415 V
Phase

Fig.2.4 Relationship between voltages in a three-phase, four-wire 415/240 V supply

Therefore for 50 per cent more wire, a three-phase supply provides 73 per cent more power.

41. Volt (V) The unit used for measuring the force which causes an electric current to flow along a conductor.

42. Voltage drop The portion of the total voltage in a circuit which is used to overcome resistance or impedance to the flow of electricity.

43. Watt (W)

1. With direct current, volts x amps = watts.

2. With single-phase alternating current, volts x amps x power factor = watts.
With three-phase alternating current, volts x amps x power factor x 1.73 = watts.

Electrical calculations

Power

When the potential difference between two points of a conductor is 1 V, when carrying a constant current of 1 A, the power dissipated between the two points is equal to 1 W.

Watts = electrical potential (volts) x current

1 W = 1 V x 1 A

or

watts = volts x amps

or

$$W = VA \qquad [2.1]$$

Potential and resistance

If the resistance between two points of a conductor is 1 Ω, when carrying a constant current of 1 A, the difference of potential of 1 V must be applied between the two points.

Electrical potential (volts) = current in amperes x resistance in ohms

or

volts = amps x ohms

$$V = AR \qquad [2.2]$$

or

$$\frac{V}{A} = R$$

and

$$A = \frac{V}{R}$$

The formulas [2.1] and [2.2] may be combined as follows:

[2.1] $W = VA$ [2.2] $V = AR$

Substituting [2.2] in [2.1]

$$W = AR \times A$$

$$W = A^2 \times R \qquad [2.3]$$

also from [2.2]

$$A = \frac{V}{R}$$

and substituting in [2.1]

$$W = V \times \frac{V}{R}$$

$$W = \frac{V^2}{R} \qquad [2.4]$$

Note: From formulas [2.3] and [2.4] it can be seen that the power in watts varies either with the square of the current, or the square of the voltage.

Example 2.2. *Calculate the current flowing in and the resistance of a 2 kW immersion heater connected to a 250 V supply.*

$$W = VA$$

$$A = \frac{W}{V}$$

$$\text{current} = \frac{2 \times 1000}{250}$$

$$= 8\ \text{A}$$

$$R = \frac{V}{A} = \frac{250}{8} = 31.25$$

resistance = 31.25 Ω

Check using formula [2.3]

$$W = A^2R$$

$$W = 8^2 \times 31.25$$

$$W = 2000 \text{ or } 2\ \text{kW}$$

Example 2.3. *Calculate the current flowing in and the resistance of a lamp rated at 240 V and 60 W.*

$$W = VA$$

$$A = \frac{W}{V}$$

$$\text{current} = \frac{60}{240} = 0.25\ \text{A}$$

$$R = \frac{V}{A}$$

$$R = \frac{240}{0.25} = 960\ \Omega$$

The heat energy or the specific heat capacity (s.h.c.) of water is equal to 4.2 kJ/kg °C. In other words, it requires 4.2 kJ of heat energy to raise the temperature of 1 kg of water through 1 °C. Also

$$\text{kW} = \frac{\text{kJ}}{\text{seconds}} \quad \text{or} \quad \frac{\text{s.h.c.}}{\text{seconds}}$$

Example 2.4. *Calculate the power in kW required to raise the temperature of 136 kg of water from 10 °C to 60 °C in 2 h, when the heat losses are 20 per cent.*

$$\text{kW} = \frac{\text{s.h.c.} \times \text{temperature rise °C} \times \text{kg} \times 100}{\text{heating time in seconds} \times \text{efficiency}}$$

$$\text{kW} = \frac{4.2 \times (10 - 60) \times 136 \times 100}{2 \times 3600 \times 80}$$

$$\text{kW} = 4.95$$

A 5 kW electric immersion heater would be required.

Example 2.5. *Calculate the time taken in hours, and the cost in pence, of heating 227 kg of water from 10 °C to 60 °C by means of a 5 kW immersion heater when the heat losses are 20 per cent and 1 kWh costs 8.5p.*

$$\text{kW} = \frac{\text{s.h.c.} \times \text{kg} \times \text{temperature rise °C} \times 100}{\text{heating time in seconds} \times \text{efficiency}}$$

By transposition:

$$\text{Time in seconds} = \frac{\text{s.h.c.} \times \text{kg} \times \text{°C} \times 100}{\text{kW} \times \text{efficiency}}$$

$$\text{Time in seconds} = \frac{4.2 \times 227 \times 50 \times 100}{5 \times 80}$$

$$\text{Time in seconds} = 11\,917.5$$

$$\text{Time in hours} = \frac{11\,917.5}{3600}$$

$$\text{Time in hours} = 3.31 \text{ h}$$

$$\begin{aligned}\text{Cost} &= \text{kWh} \times 8.5\\ &= 5 \times 3.31 \times 8.5\\ &= £1.40 \text{ approx}\end{aligned}$$

Example 2.6. *An immersion heater has a resistance of 20 Ω and has a current passing through of 12 A. If the heat losses are 20 per cent, calculate the mass of water in kilograms that can be heated from 5 °C to 60 °C in 1 h.*

Volts = current x resistance

$V = AR$

$V = 12 \times 20$

$V = 240$

Watts = volts x amps

$W = VA$

$W = 240 \times 12$

$W = 2880$

At 80 per cent efficiency:

$$W = 2880 \times \frac{80}{100}$$

$W = 2304$

$\therefore kW = 2.3$ (approx.)

kJ = kW x seconds

kJ = 2.3 x 3600

kJ = 8280

Heat in water in kilojoules = 8280. Therefore

8280 = kg x 4.2 x temperature rise °C

8280 = kg x 4.2 x 55

$$\text{kg} = \frac{8280}{4.2 \times 55}$$

= 35.84 kg, say 36 kg

Chapter 3

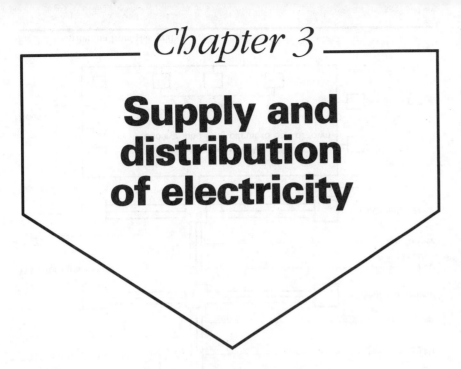

Supply and distribution of electricity

The public supply of electricity

In 1831, Michael Faraday plunged a bar magnet into a coil of wire and thus generated electricity. Faraday's original experiment of holding a coil of wire stationary while varying the magnetic field, is used today for the generation of electricity, but the coils are arranged so that their windings are cut by the magnetic field as the magnet rotates. The mechanical energy required to turn the magnets is converted into electrical energy in the stator windings. These stator windings have an angular spacing of 120° and the electromotive forces or voltages are out of phase by this angle for each revolution of the magnets, thus producing a three-phase alternating current supply (see Fig. 3.1). In a power station the magnets in each generator are turned by engines, usually steam turbines, called prime movers. The steam required for the turbines is produced from water heated by burning oil, gas or coal, or by nuclear fission. Hot gases from jet engines and water are sometimes used instead of steam to drive the turbines and at a few smaller power stations, diesel engines provide the motive force.

The National 'Grid' system

Having produced the electricity in the generating stations, it is necessary to distribute it to the consumer and this can be achieved by having small power stations located near the consumers, or by large power stations with a transmission system, known as the 'grid' to connect them to the main distributing points. Larger stations are more economical than smaller ones and the interconnecting transmission system also means that each station can rely on the others during maintenance periods, thus the amount of spare plant is kept to a minimum.

The electricity produced in the stator windings of the generators is at 25 kV and is fed through a transformer which steps up the voltage to 132, 275 or 400 kV. The original 'grid' established in the 1930s was designed to work at 132 kV and was strengthened by the 275 kV system in 1953. In 1960, the CEGB decided to use the 400 kV system for major new lines in order to increase the carrying capacity and to reduce the number of new lines needed: one 400 kV line has three times the carrying capacity of one 275 kV line and eighteen times the carrying capacity of a 132 kV line. A 160 MW direct current cross-Channel link with France, enables some exchange of supply between the two countries which is particularly useful during peak demands.

Regional Electricity Companies

There are twelve Regional Electricity Companies in England and Wales, and two in Scotland. The Companies are responsible to their consumers for the efficient and economical supply of electricity. They are alsp permitted to sell electrical appliances and to carry out electrical installations in buildings. Under reasonable circumstances the Companies are obliged to supply electricity to a consumer who is prepared to provide space in his building for the Companies equipment, pay towards the cost incurred, and to provide an electrical installation of a satisfactory standard.

Application for a supply should be made at the earliest opportunity and for new premises, the Company will want to know the approximate installed load, the nature of the development, and the date on which the supply is required. The Companies obtain their supply from the 132, 275 or 400 kV transmission systems operated by private generating businesses. The electricity required by an Electricity Company for use by its consumers, is taken and metered at points of supply which may be located at 'grid' stations of the transmission system, or at generating stations. Most Companies distribute electricity through four networks operating at 132, 33 or 11 kV and at 415 and 240 V, depending upon the load to be supplied. The supply is carried overhead in rural areas and underground in urban areas.

Figure 3.2 shows a public electricity distribution system from the 'grid' to an area covered by an Electricity Company. Figure 3.3 shows the supply from the underground ring circuit to the buildings.

Supply to small buildings

The supply cable to small buildings such as domestic premises, is brought into the building through a trench and left in a position near the entrance ready for the installation of meters and fuse gear. If the meter and fuse gear is to be sited on the inside wall a 100 mm diameter duct is required for the cable, as shown in Fig. 3.4. The meter and fuse gear may, however, be sited on the outside wall, as shown in Fig. 3.5, in which case a shorter duct will be required. A more recent method for the supply intake to a small building has been the use of an external metal meter cabinet, as shown in Fig. 3.6. This method allows the meter to be

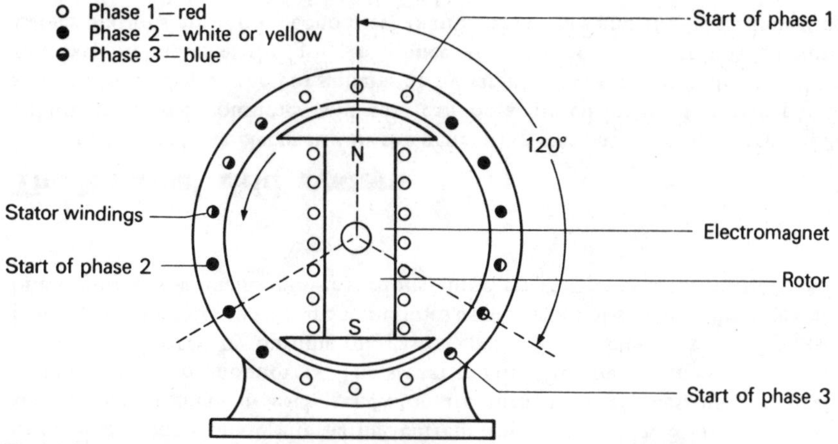
Phase 1—red
Phase 2—white or yellow
Phase 3—blue
Start of phase 1
120°
N
S
Stator windings
Start of phase 2
Electromagnet
Rotor
Start of phase 3

Fig.3.1 Simplified detail of three-phase alternating current generator or alternator

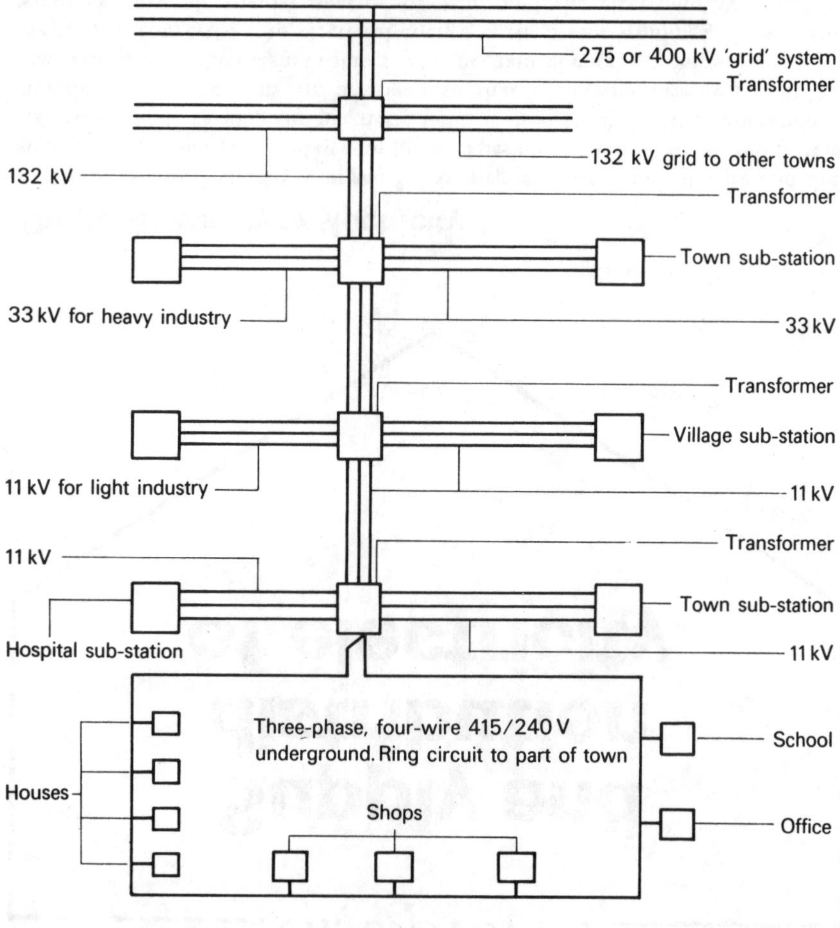
275 or 400 kV 'grid' system
Transformer
132 kV grid to other towns
132 kV
Transformer
Town sub-station
33 kV for heavy industry
33 kV
Transformer
Village sub-station
11 kV for light industry
11 kV
11 kV
Transformer
Town sub-station
Hospital sub-station
11 kV
Three-phase, four-wire 415/240 V
underground. Ring circuit to part of town
Houses
School
Shops
Office

Fig.3.2 Distribution from the 'grid' to various buildings

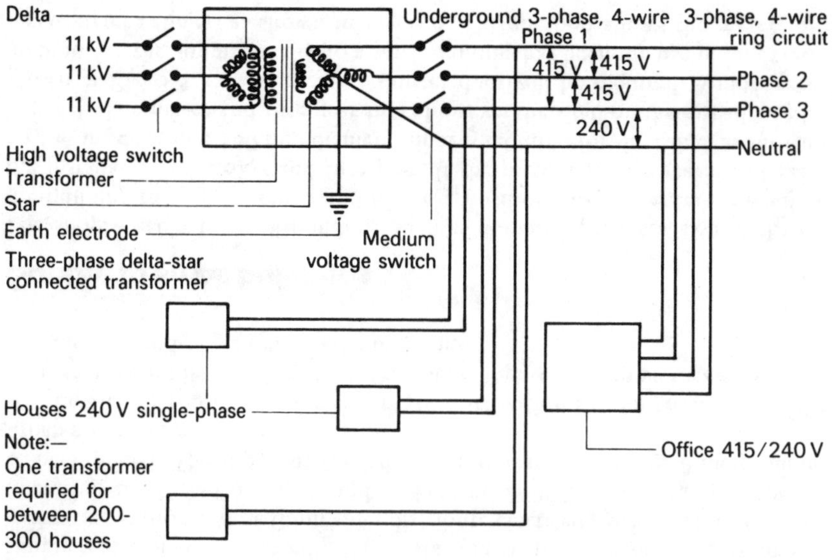
Delta
11 kV
11 kV
11 kV
Underground 3-phase, 4-wire
Phase 1
3-phase, 4-wire
ring circuit
415 V
415 V
415 V
Phase 2
Phase 3
240 V
Neutral
High voltage switch
Transformer
Star
Earth electrode
Medium
voltage switch
Three-phase delta-star
connected transformer
Houses 240 V single-phase
Office 415/240 V
Note:—
One transformer
required for
between 200-
300 houses

Fig.3.3 Supply from underground ring circuit to various buildings

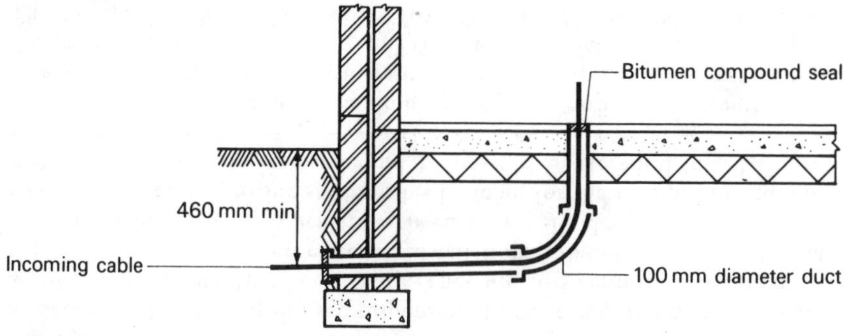
Bitumen compound seal
460 mm min
Incoming cable
100 mm diameter duct

Fig.3.4 Service cable to meter cupboard fixing on the inside wall

Minimum depths of cable below ground		
	In open ground	Under roads
Extra low voltage	460 mm	760 mm
Low voltage	460 mm	760 mm
Medium voltage	460 mm	760 mm

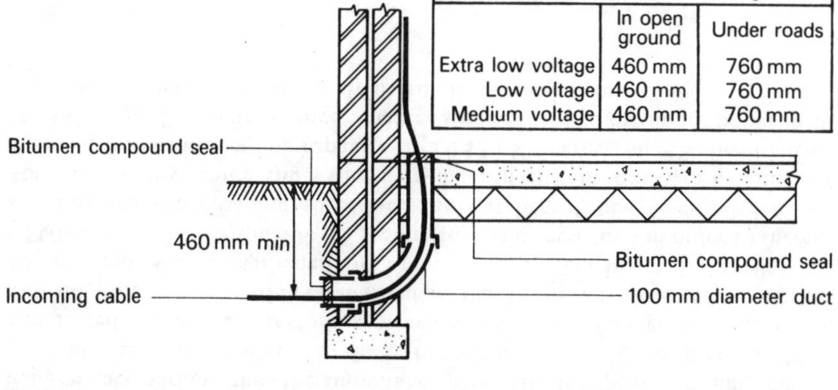
Bitumen compound seal
460 mm min
Incoming cable
Bitumen compound seal
100 mm diameter duct

Fig.3.5 Service cable to meter cupboard fixing on the outside wall

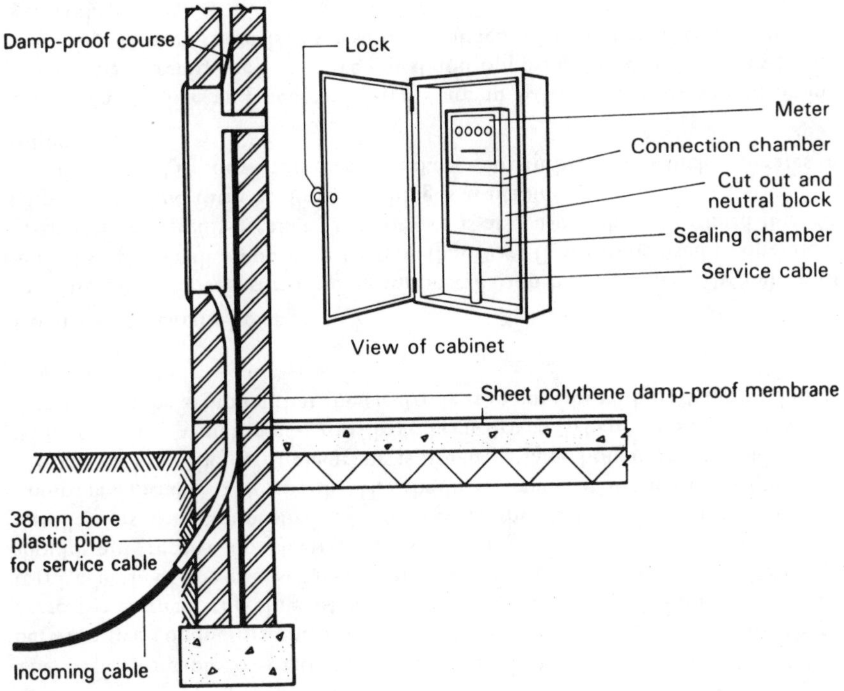
Damp-proof course
Lock
Meter
Connection chamber
Cut out and
neutral block
Sealing chamber
Service cable
View of cabinet
Sheet polythene damp-proof membrane
38 mm bore
plastic pipe
for service cable
Incoming cable

Fig.3.6 External cabinet for easy meter reading

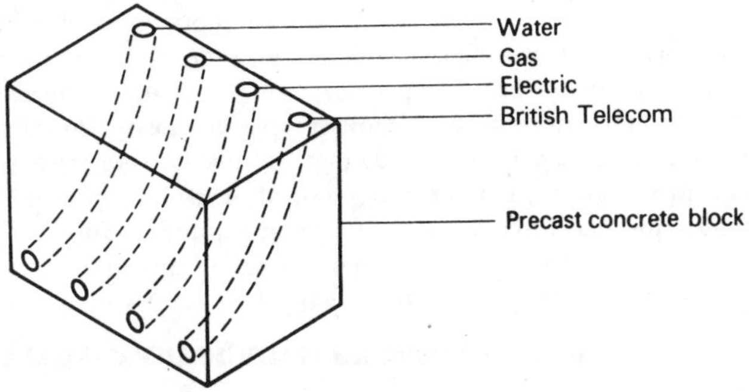
Water
Gas
Electric
British Telecom
Precast concrete block

Fig.3.7 The BRE precast concrete services intake component

read from outside the premises and thus prevents the difficulty of obtaining access to meters when the building is unoccupied.

The Building Research Establishment (BRE) has developed a combined service intake component for use in domestic premises. The unit consists of a precast concrete block which has provision for the entry of electric, water, gas, telephone and television services (see Fig. 3.7). If the electric meter and fuses are housed in the same cupboard as the gas meter, a fire-resisting partition should separate the two meters, as shown in Fig. 3.8.

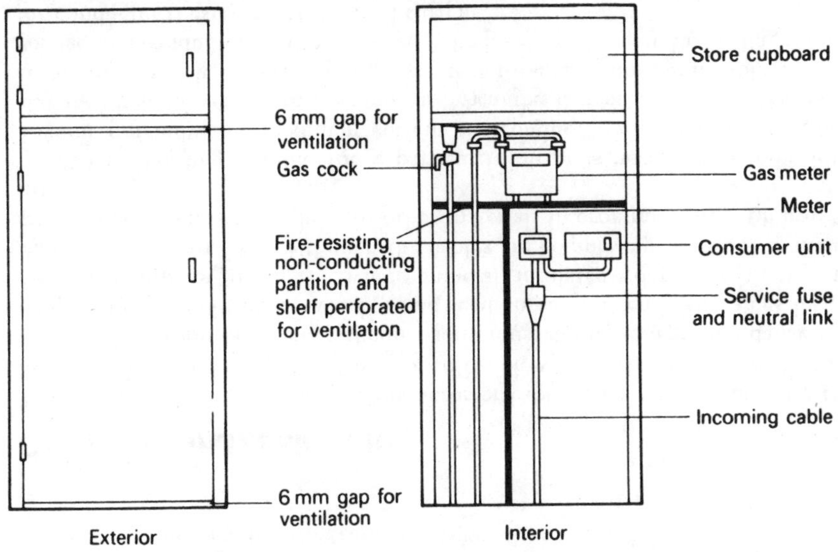
6 mm gap for
ventilation
Gas cock
Fire-resisting
non-conducting
partition and
shelf perforated
for ventilation
6 mm gap for
ventilation
Exterior
Interior
Store cupboard
Gas meter
Meter
Consumer unit
Service fuse
and neutral link
Incoming cable

Fig.3.8 Gas and electric meters in same cupboard

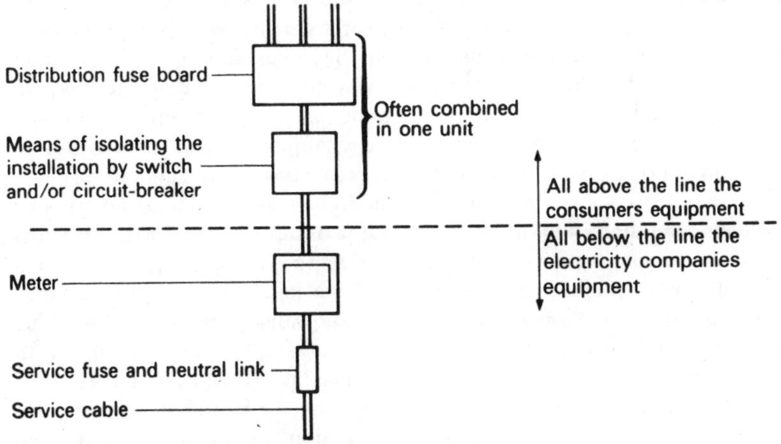
Distribution fuse board
Means of isolating the
installation by switch
and/or circuit-breaker
Often combined
in one unit
All above the line the
consumers equipment
All below the line the
electricity companies
equipment
Meter
Service fuse and neutral link
Service cable

Fig.3.9 Sequence of supply controls

Supply controls

The service cable usually terminates inside the building in a main cut-out, fitted as near as possible to the service cable entry. A meter is fitted after the main cut-out and everything up to and including the meter, are the property and responsibility of the Electricity Company. A switch or circuit-breaker is fitted after the meter and a distribution board follows the switch. Everything from the switch to the furthest outlet point, is the property and responsibility of the building owner. Figure 3.9 shows the sequence of supply controls.

The service intake and the control unit is 240 V single-phase for a domestic or similar small building, and for larger buildings a 415 V three-phase supply may be required, depending upon the load. The control unit must be sited so as to fulfil the following conditions:

1. It must allow the supply cable to be brought in without undue difficulty.
2. It must be reasonably accessible for meter reading and general maintenance.
3. It must be separated from any gas meter by a fire-resisting partition.
4. It should be placed in a position where heavy condensation is unlikely.
5. There should be an easy and accessible route for the outgoing cables.

Protection against excess current

All circuits must be protected against excess current and three devices are available, miniature circuit-breakers, rewireable fuses and cartridge fuses. Since miniature circuit-breakers are virtually tamper-proof, their use is to be recommended for circuit protection. The Institution of Electrical Engineers (IEE) Regulations recognises that miniature circuit-breakers and certain types of cartridge fuses, provide a more reliable protection against excess current than rewireable fuses and are classified as 'close' circuit protection. Wireable fuses and certain types of slow-acting cartridge fuses, are classified as 'coarse' excess current protection.

Earthing

The basic principle of earthing is that of limiting the difference in potential between live conductors and earth. If a person touches a live conductor that is correctly earthed, the flow of electricity through the earth conductor should form a path of lower resistance than that of a person's body and the person should not receive an electric shock (see Fig. 3.10). The earthing of most installations is accomplished by connecting the earth wire to the Electricity Companies metallic cable sheath. The sheath is connected to an earth electrode at the sub-station. The earth electrode is usually a large hole in the ground packed with coke, into which metal rods have been driven. Where the soil is reasonably damp, it can be assumed that the electrode is in contact with underground metal pipes.

Earth leakage circuit-breakers

In some cases the soil surrounding the installation has a particularly high resistance, but the small amount of current that does flow can be used to operate an earth leakage circuit-breaker. This device is also used where overhead lines are employed and no form of direct earthing is available.

Figure 3.11 shows an earth leakage circuit-breaker which operates as follows:

1. When the live and neutral currents are in balance, as they should be in a normal circuit, they produce equal and opposing fluxes in the transformer core. Hence, there is no resultant voltage generated in the fault-detector coil.
2. If, however, the earthed metalwork has a flow of current passing through it to earth, more current will flow in the live side than on the neutral side and an out-of-balance flux will be detected by the fault-detector coil.
3. The fault-detector coil is arranged to trip the main switch through the solenoid.

Note: The main switch will open well under 0.1 s, which is the British Standard requirement and excellent shock protection will therefore be obtained.

Bonding of services

The IEE Regulations require that the metal sheaths and armour of all cables operating at low or medium voltage, along with all metal conduits, ducts, trunking and bare earth continuity conductors which might come into contact with other fixed metalwork, shall either be effectively segregated or bonded to it. The purpose of this bonding, is to prevent appreciable voltage differences at any point of contact. Many fatalities have been caused by a person coming into contact with exposed metal made live as a result of a fault. The exposed metalwork, other than that of other services, includes baths, radiators, sinks, tanks, structural steelwork and the steel framework of mobile equipment such as cranes and lifts on which electrical equipment is mounted.

The bonding of the services shall be as close as possible to the point of entry of the services in the building. This is to ensure that the maximum amount of pipework in the building will be at the same potential. Figure 3.12 shows the method of earth bonding of services. If the pipework includes an insulation insert such as jointing material, the bonding connection must be made to the metalwork on the consumer's side of the section.

At one time, the sole means of earthing was by connecting an earth lead to the metal water pipe. The use of plastic water pipe resulted in an inefficient earthing and this practice is no longer accepted. The purpose of earth bonding to the water and gas pipes is to ensure a common potential, and the connection to the electric cable sheath is the true earth connection.

Conductor and cable rating

The amount of current which a conductor or cable can carry is limited by the heating effect caused by the resistance to the flow of electricity. The maximum permissible current under normal conditions, must not be so high that dangerous temperatures are attained, which could lead to fires. Even with cables inside metal conduits or ducts, or where mineral insulated copper or aluminium sheathed cables are used, although the cables are completely fireproof in themselves, the transmission of heat to other materials in proximity may still lead to fires.

When choosing a cable for a particular job, it is necessary to take into account not only the maximum current the conductor will have to carry, but also the drop in voltage that will occur when the current is carried. The 16th edition of the IEE Regulations stipulate that the maximum permissible drop in voltage in a conductor shall not exceed 2.5 per cent of the nominal voltage when the conductor is carrying its full-load current.

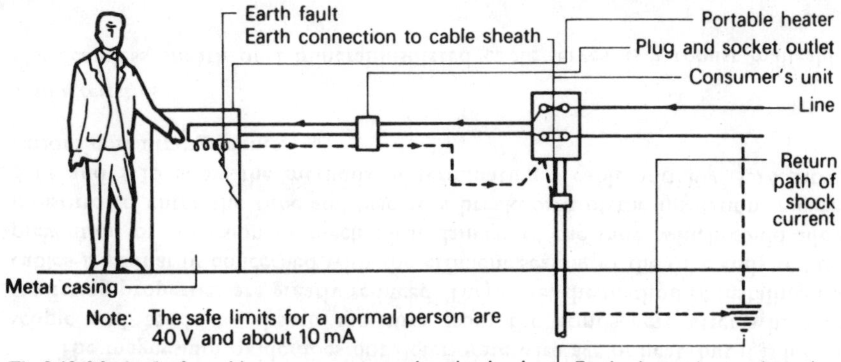
Earth fault
Earth connection to cable sheath
Portable heater
Plug and socket outlet
Consumer's unit
Line
Return path of shock current
Metal casing
Note: The safe limits for a normal person are 40 V and about 10 mA
Earth electrode at sub-station

Fig. 3.10 Method of earthing to prevent an electric shock

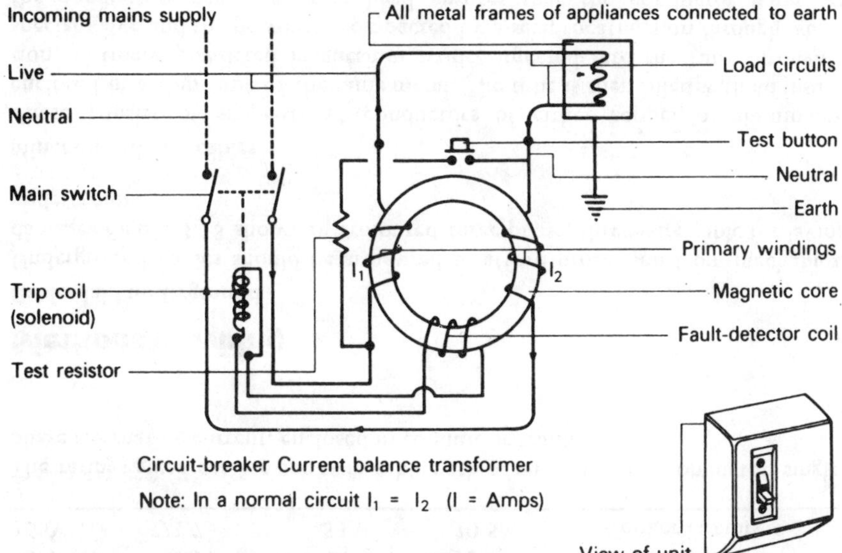
Incoming mains supply
All metal frames of appliances connected to earth
Live
Neutral
Load circuits
Test button
Neutral
Main switch
Earth
Primary windings
I_1
I_2
Trip coil (solenoid)
Magnetic core
Fault-detector coil
Test resistor
Circuit-breaker Current balance transformer
Note: In a normal circuit $I_1 = I_2$ (I = Amps)
View of unit

Fig. 3.11 Current-operated earth leakage circuit-breaker

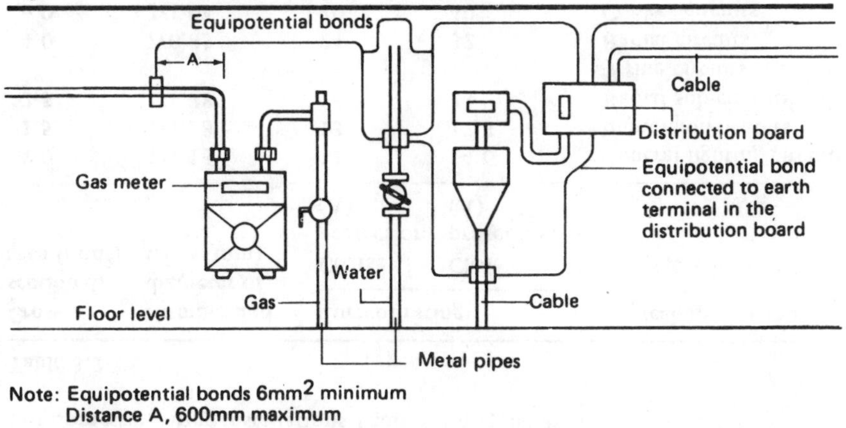
Equipotential bonds
A
Cable
Distribution board
Equipotential bond connected to earth terminal in the distribution board
Gas meter
Water
Gas
Cable
Floor level
Metal pipes
Note: Equipotential bonds 6mm^2 minimum
Distance A, 600mm maximum

Fig. 3.12 Bonding of services

The temperature reached by a cable is also affected by the following operating conditions:

1. Whether the cable is surrounded by the room air, or is enclosed in a conduit or duct.
2. The closeness to other cables which may cause heat to build up, due to induced currents.
3. The temperature of the ambient or surrounding air.

The British Standards provide tables of cable sizes for various operating conditions.

Table 3.1 (Part tables of B.S. 6004 and 6346). Single-circuit current ratings and associated voltage drops for single-core polyvinyl-chloride (PVC) insulated cables, non-armoured with or without sheath (copper conductors). The ratings tabulated apply where the cable is provided with coarse excess current protection. Conductor operating temperature 70°C

Conductor		Enclosed in conduit or trunking	
		Two core cable single phase a.c. or d.c.	
Nominal cross-sectional area (mm^2)	Number and diameter of wires (mm)	Current rating (A)	Voltage drop (mV) per ampere, per metre
1.0	1/1.13	13	44
1.5	1/1.38	16.5	29
2.5	1/1.78	23	18
4	7/0.85	30	11
6	7/1.04	38	7.3
10	7/1.35	52	4.4

Example 3.1. *One PVC insulated two core cable with copper conductors, non armoured and enclosed in conduit is 14 m in total length and is required to supply a 5 kW electric heater. If the nominal voltage of the supply is 240 find the size of cable required, coarse excess current protection being provided.*

$$\text{Current} = \frac{5000}{240} = 20.8 \text{ A}$$

From part of Tables BS 6004 and 6346, a 1/1.78 cable will allow a current of 23 A to flow with a drop of 18 mV (0.018 V) per ampere, per metre run.

Voltage drop in cable = 20.8 x 0.018 x 14 = 5.242 approx

Check percentage of nominal voltage:

$$\frac{5.242}{240} \times \frac{100}{1} = 2.184 \text{ per cent.}$$

4 per cent is the maximum voltage drop and the cable is therefore suitable.

Domestic cable sizing

For domestic wiring installations Table 3.2 may be used.

Table 3.2

Cross sectional area (mm^2)	Number and diameter of wires (mm)	Current rating: Coarse protection (A)	Current rating: Close protection (A)	Typical applications
1.0	1/1.13	11	14.6	General lighting circuits
1.5	1/1.38	13	17.3	Immersion heaters
2.5	1/1.78	18	24	Radial sub-circuits and ring circuits
4.0	7/0.85	24	32	Radial circuits
6.0	7/1.04	30	40	Cooker circuits
10.0	7/1.35	40	53	Cooker circuits
16.0	7/1.7	53	70.5	Cooker circuits

The ratings given are for one twin cable with or without earth conductor, single-phase alternating current, enclosed in conduit or trunking.

Methods of wiring

Cables laid underground

Underground cables should be armoured to afford protection from mechanical damage. Figure 3.13 shows an armoured three-phase, three-wire cable for laying underground.

Mineral-insulated cables

These consist of single-strand conductors of either copper or aluminium, enclosed in a thin tube of the same metal. The tube is then filled with an insulation of finely powdered magnesium oxide, injected into the tube through a special valve and the powder is compacted by a reciprocating ram through which the conductors pass. The ram head ensures that the conductor wires are accurately placed within the tube in accordance with the specification.

The magnesium oxide does not deteriorate with age or heat, but it is hygroscopic and tends to absorb moisture from the atmosphere after which its insulating properties are greatly reduced. Therefore, the method of installing the cables is primarily concerned with the efficient sealing of the tube ends and the prevention of corrosion or mechanical damage of the tube, which could allow moisture to enter the tube and lead to a breakdown of the insulation. Figures 3.14 and 3.15 show the methods of terminating a cable and Fig. 3.16 shows various core arrangements.

Characteristics

The seamless sheath of a mineral-insulated cable serves as a robust malleable conduit, obviating the threading and drawing through of cables required with some conduit systems. The malleability of the cable permits the bending of all but the largest cable by hand. Bending levers are available for large cables. Although the cable is easily worked, it still possesses firmness and does not require close-centred support. It can be fixed neatly and quickly to all types of surfaces and does not normally require protection from mechanical damage. The cables are unaffected by extremes of temperature, moisture or oil, but some acids in contact with the cable may lead to corrosion.

Earthing

When used in conjunction with close-tolerance brass terminations and glands, the metal sheathing provides an excellent earth conductor and it is one of the safest installations for earthing purposes.

Bending

All bends on the cable should be restricted to a minimum of six times the cable outside diameter. This allows for subsequent straightening and rebending of the cable.

Expansion bends

In some circumstances, expansion bends or loops have to be introduced into the run of the cable. These circumstances may be encountered where a cable has to pass expansion joints in structure, or where connections must be made to units subject to movement, such as compressors.

Note: Since magnesium oxide insulation has an affinity for moisture, some form of temporary seal to the tube ends must be applied to prevent moisture from penetrating at the ends of the cable. A temporary seal is used for this purpose during dispatch and storage, and a similar temporary seal must be applied to the open end of the cable after cutting if the cable is not to be used immediately.

Conduit systems

A conduit means a metal or non-metallic tube into which PVC insulated cables are drawn. The function of the conduit is to provide mechanical protection of the cables, permit rewiring and, if the conduit is of metal, to provide an earth conductor.

Heavy-gauge steel conduit (Class B)

This is mostly used in industry, its main function being for flame-proof installations. It is not usually used for low-voltage installations. The conduit is available in three grades:

1. Solid drawn, which is seamless.
2. Welded, where the conduit is formed from steel strip with the joint welded.
3. Brazed, where the conduit is formed from strip steel with the joint brazed.

The conduit is protected from corrosion both inside and outside by black bituminous enamel paint, or by galvanising or sheradising. The steel is thick enough for taper threads to be cut on the conduit by means of stocks and dies.

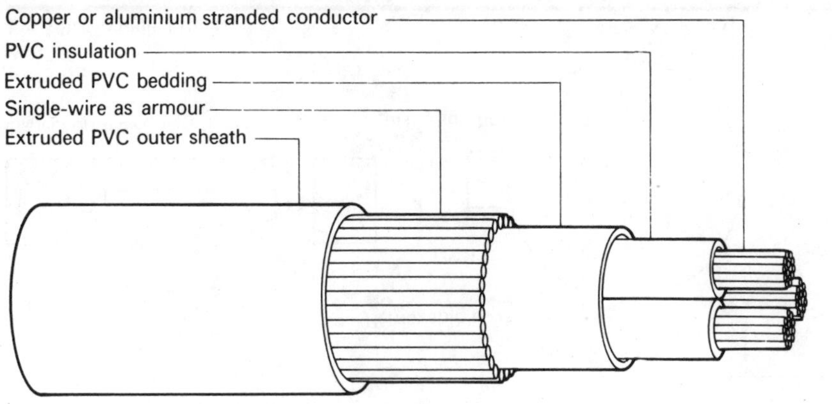
Copper or aluminium stranded conductor
PVC insulation
Extruded PVC bedding
Single-wire as armour
Extruded PVC outer sheath

Fig.3.13 Armoured three-phase, three-wire cable for laying underground

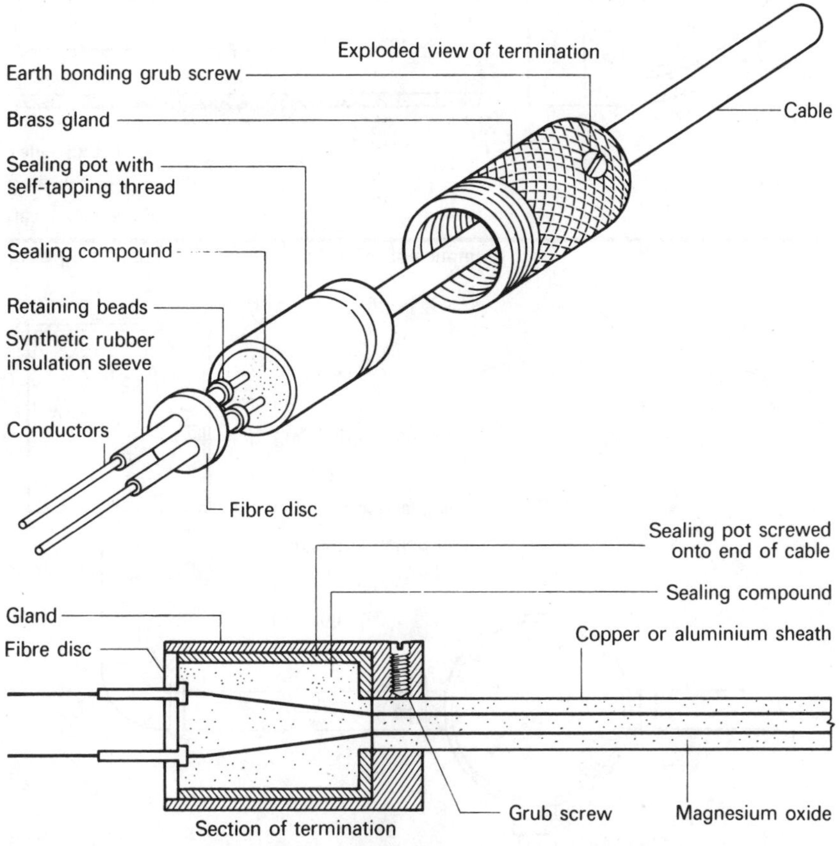
Exploded view of termination
Earth bonding grub screw
Cable
Brass gland
Sealing pot with
self-tapping thread
Sealing compound
Retaining beads
Synthetic rubber
insulation sleeve
Conductors
Fibre disc
Sealing pot screwed
onto end of cable
Sealing compound
Gland
Fibre disc
Copper or aluminium sheath
Grub screw
Magnesium oxide
Section of termination

Fig.3.14 Method of terminating mineral-insulated cable using earthing screw gland

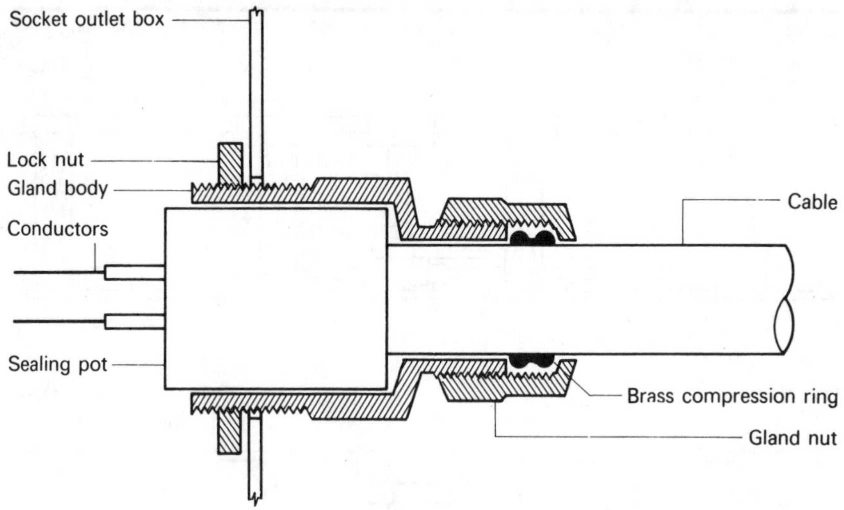
Socket outlet box
Lock nut
Gland body
Conductors
Cable
Sealing pot
Brass compression ring
Gland nut

Fig.3.15 Method of terminating mineral-insulated cable using universal ring-type gland

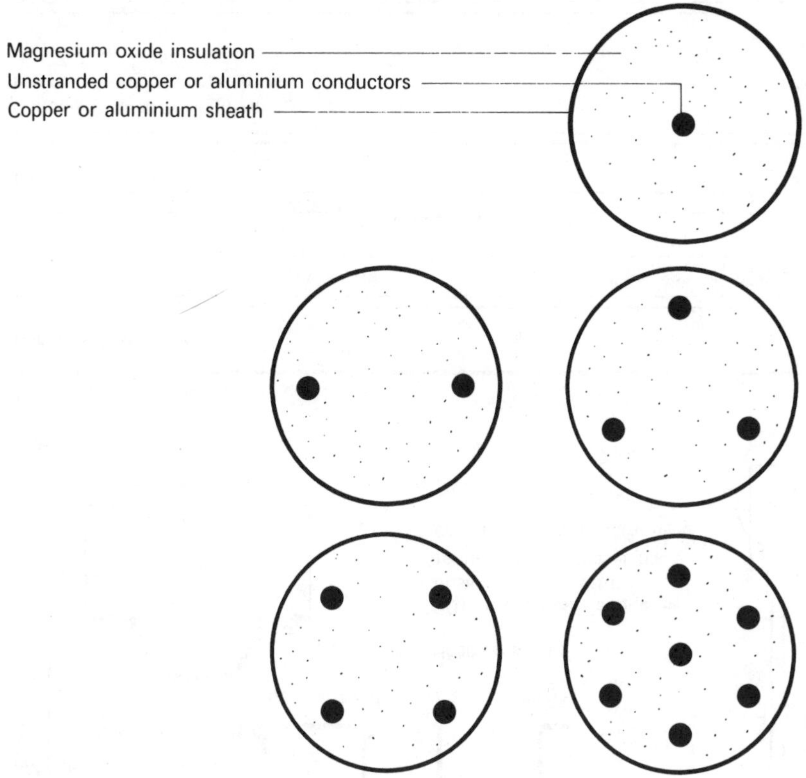
Magnesium oxide insulation
Unstranded copper or aluminium conductors
Copper or aluminium sheath

Fig.3.16 Core arrangements of mineral-insulated cables 660, 440, and 240 V

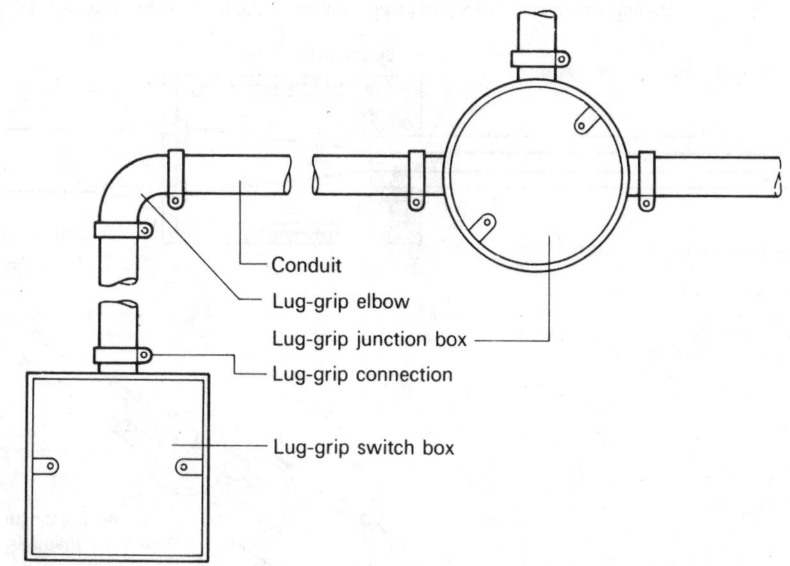
Conduit
Lug-grip elbow
Lug-grip junction box
Lug-grip connection
Lug-grip switch box

Fig.3.17 Lug-grip connection on light-gauge conduit

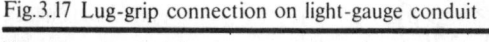

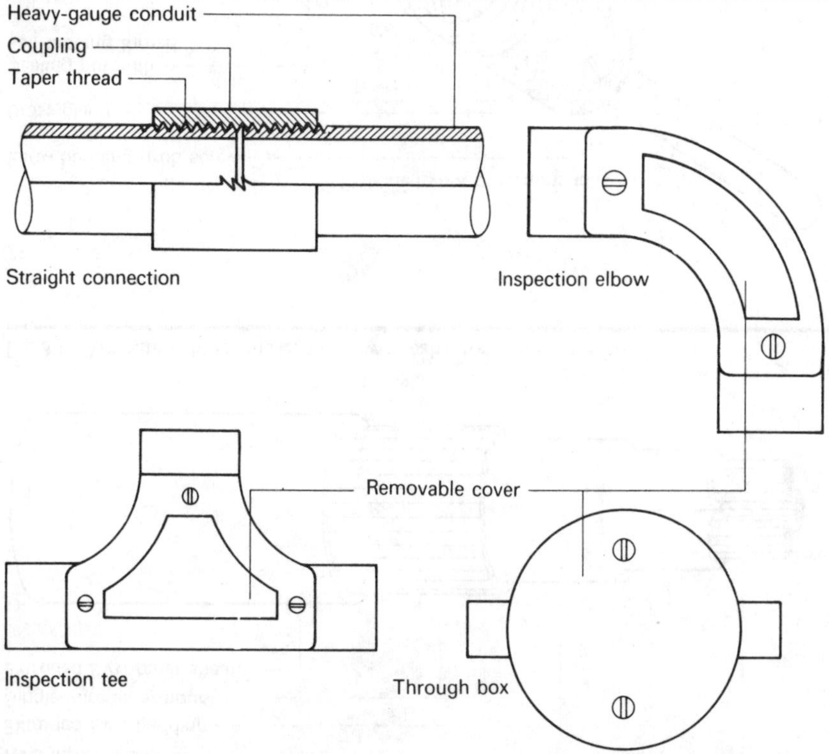
Heavy-gauge conduit
Coupling
Taper thread
Straight connection
Inspection elbow
Removable cover
Inspection tee
Through box

Fig.3.18 Conduit fittings

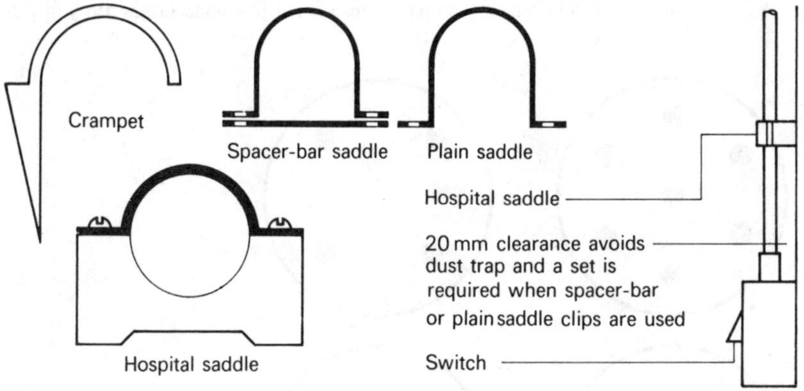
Crampet
Spacer-bar saddle
Plain saddle
Hospital saddle
20 mm clearance avoids dust trap and a set is required when spacer-bar or plain saddle clips are used
Hospital saddle
Switch

Fig.3.19 Methods of securing conduit

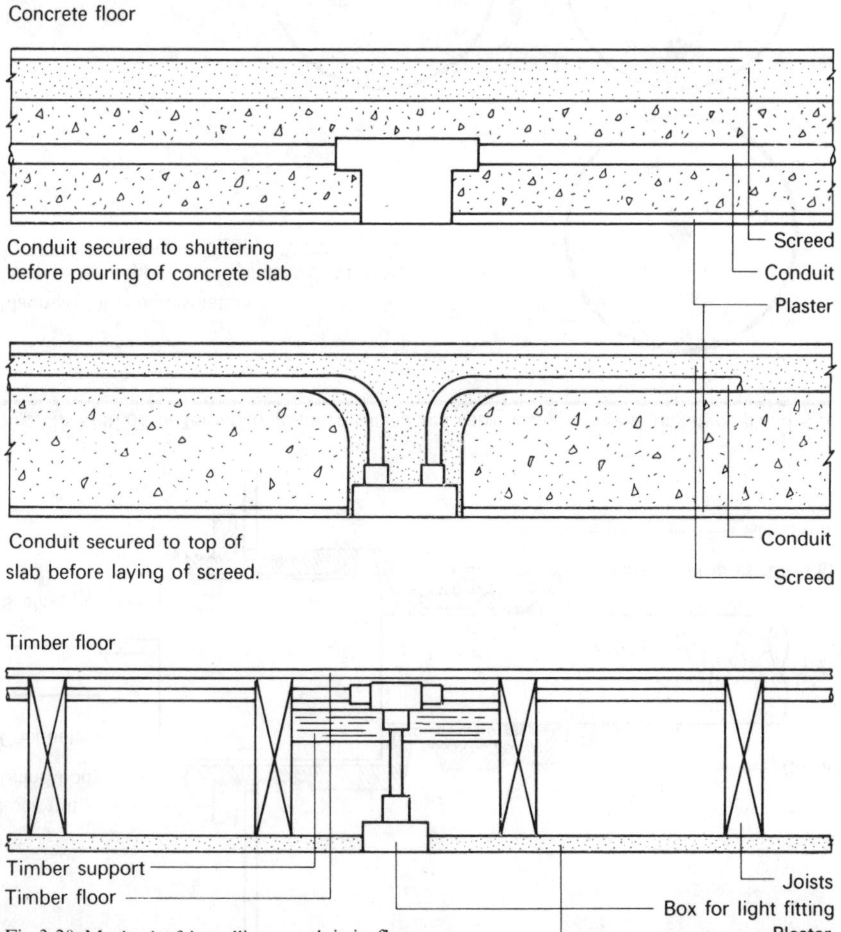
Concrete floor
Conduit secured to shuttering before pouring of concrete slab
Screed
Conduit
Plaster
Conduit secured to top of slab before laying of screed.
Conduit
Screed
Timber floor
Timber support
Timber floor
Joists
Box for light fitting
Plaster

Fig.3.20 Method of installing conduit in floors

Light-gauge steel conduit (Grade A)

This is commonly used for low-voltage (not exceeding 250 V) installations. The conduit is made from steel strip with the joint welded, brazed or butted, the latter being known as close-joint conduit. The conduit is protected from corrosion both inside and outside by black bituminous enamel paint. Because of the thin tube walls the conduit cannot be formed into bends in the same way as heavy-gauge conduit, and therefore special bends have to be used.

Fittings such as elbows and tees are fitted to this type of conduit by means of a lug-grip connection, which is tightened on to the conduit (see Fig. 3.17). All black enamelling must be removed from the end of the conduit before the fitting is secured, in order to ensure earth continuity.

Joints

Efficient joints in metal conduit systems are essential and the IEE Regulations require that the resistance through all the joints, from the main intake to the furthest point of the installation, must not exceed 1 Ω. The Regulations also state that the inner radius of a bend made in a heavy-gauge conduit shall not be less than two and a half times its external diameter.

Space factor

The Regulations also specify that after the cables have been drawn through the conduit, a space factor given in tables must not be exceeded. The space factor is the ratio, expressed as a percentage of the sum of the effective overall cross-sectional area of the cables, forming a bunch to the internal cross-sectional area of the tube in which they are installed. The effective overall cross-sectional area of a non-circular cable is taken as that of a circle of a diameter equal to the major axis of the cable.

To avoid the effects of induced currents in conduits, the cables of alternating current supplies must be bunched so that the neutral and live are all in the same conduit. This will ensure that the magnetic fields around the cables cancel out to zero.

Fittings

A large variety of fittings such as junction boxes, switch boxes, tees and bends, are available for use with both heavy- and light-gauge steel tubes (see Fig. 3.18).

Diameters

The conduit diameters most commonly used are 20 mm, 25 mm and 32 mm, and these can usually be accommodated in most screeds and slabs. For diameters above 32 mm it is generally neater and more economical to install cable trunking, and various conduits may be connected to this by means of smooth-bore brass bushes.

Drawing-in of cables

To prevent damage to the cable when drawing it through the conduit, the 'burr' inside the conduit which is left after cutting must be removed. It is essential that all conduit systems are completely installed before any cables are drawn in.

Methods of securing conduits

Metal conduits may be secured to the structure by saddles, which can be either hospital, spacer-bar or plain. Crampets and other types of saddles for supporting horizontal runs of conduit are also available. Figure 3.19 shows the method of securing conduit and Fig. 3.20 the method of installing conduit in both concrete and timber floors.

Copper conduit

Copper conduit is more resistant to corrosion than steel conduit and is therefore useful for outdoor use and for burying in concrete when water is liable to gain access to the conduit during installation. It has excellent electrical conductivity and provides good earthing, while its very high thermal conductivity tends to reduce condensation. The smooth bore of the copper tube permits the wires to be drawn through easily, both initially or after years of service.

Compression, capillary and bronze welded joints may be used, and inspection bends and junction boxes are cast in copper alloy, such as gunmetal. The tube can be bent by a machine or a spring.

Aluminium conduit

Aluminium conduit is a welded aluminium tube which is prepared for jointing by driving a pre-threaded steel tube into the end of the conduit. It can be installed with earthing requirements required by the 16th edition of the IEE Wiring Regulations.

The system of jointing enables the conduit to be used with standard fittings, and the lightness in weight of the tube offers a reduction in transport and installation costs. The steel inserts which have tapered shanks, are knurled so as to ensure mechanical and electrical connection when fitted into the tube. To prevent corrosion the steel inserts are sheradised. The conduit may be used for practically every type of installation, and in many respects it is equivalent to galvanised steel. As with copper, it has excellent electrical conductivity and provides good earthing, while its very high thermal conductivity tends to reduce condensation.

Aluminium conduit may be buried in the concrete or plaster, provided they remain dry after setting is complete. If the concrete or plaster is expected to remain damp for long periods, the conduit should be wrapped in adhesive PVC tape. The conduit may be easily bent by a machine and like copper has a smooth bore which enables wires to be drawn through easily, both initially or after years of service.

Plastic conduit

PVC is used which may be unplasticised for rigid conduit and plasticised for flexible tubes. The conduit may be obtained as heavy or light gauge and in round or oval sections. Heavy-gauge conduit may be jointed by screwed threads and light-gauge conduit by solvent welding. Injection-moulded PVC fittings are used and the range of fittings is wide enough for a conduit installation to be completely plastic and therefore all insulated. To ensure reliable and earthing continuity in an all-plastic installation, it is necessary to use a continuous earth wire.

PVC is generally unaffected by water, acids, oils and soils and can be safely buried in concrete, plaster or lime. It is light in weight, smooth in bore, quick to install and will not ignite. If the conduit is held in a flame for long periods it carbonises, but the carbon created is non-conductive. The coefficient of expansion is high (0.00035/°C), and if the conduit is installed in straight runs for lengths in excess of 6 m the use of expansion is recommended. Smaller diameter conduit may be bent cold by means of a bending spring and larger-diameter tube bent by heating and inserting a special rubber core. As plastic tends to 'spring back' it is necessary for the initial bend to be more acute than is finally required and the conduit secured as soon as possible afterwards. It is recommended that the conduit should be used in temperatures between 70 °C and −10 °C.

All-insulated sheathed cables (Fig. 3.21)

PVC sheathed cable is generally manufactured with one, two or three copper conductors insulated with PVC; the earth wire may be either insulated or un-insulated. The cables are flexible and are easily installed either on the surface or concealed within the structure. They can be secured by tinned brass buckle clips or polyvinyl clips. When the cables are run in the roof space or beneath a timber floor, they should be secured at the sides of the joists. When running cables across floor joists, they should be passed through holes preferably drilled in the neutral axis, but not less than 50 mm from the top of the joists (see Fig. 3.22). As with PVC conduit the cable is generally unaffected by water, acids, oils and most soils. If the cable is buried in the plaster a metal channel should be used to protect the cable and allow for rewiring (see Fig. 3.23).

Tough rubber sheathed cable (TRS)

The cable is generally manufactured with one, two or three copper conductors insulated with rubber and covered with a rubber sheath, the earth wire may be either insulated or uninsulated. They are installed in the same manner as the PVC sheathed cables. The cables do not resist direct sunlight, oil or chemical attack to the same extent as PVC sheathed cables and unlike the latter cables they cannot be obtained in a wide range of colours. The cables are not as popular as PVC sheathed cables, but are often used for extension leads.

Accessories

Accessories such as switches, ceiling roses and joint boxes used in conjunction with all-insulated sheathed cables are usually made from incombustible plastic and, like the cables, are all-insulated. The only metal parts are those required to carry the current and are not exposed; the cable sheath should extend well inside the fitting.

Identification

Colour identification of the cores of all-insulated sheathed cables is given in the Regulations of the IEE. The Regulations also specify the colours of the cores of the flexible cables (see Table 3.3).

Table 3.3 Colour identification of bare conductors and cable cores

Function	Colour of core of non-flexible*, all-insulated cables	Colour of core of flexible, all-insulated cables
Earthing	Green and yellow	Green and yellow
Live	Red (or yellow or blue†)	Brown
Neutral	Black	Blue

* The term 'non-flexible' is misleading, it relates to the main electrical installation, while the term 'flexible' relates to leads to appliances.
† As an alternative to red, if desired for large installations.

Prefabricated wiring systems

Essentially, the systems enable the electrical installation for a house to be pre-fabricated in a factory and are packed in kit form for delivery to the site. The systems reduce site labour, and since the kits are factory assembled to the highest standards the installation is automatically of a high standard.

Distribution systems

Lighting circuits

Every sub-circuit which originates from the lighting distribution fuse board is generally limited to a total load of 1000 W and requires 5 A fuses and switches. In large buildings 15 A fuses and wiring are sometimes used, due to the higher total load on the circuit. Wiring to lighting points should be carried out on what is known as the 'looping-in' method shown in Fig. 3.24.

One-way switch control

A single-pole switch is connected with wiring to control a lamp, as shown in Fig. 3.25. If required, several lamps may be controlled from one switch, as shown in the figure.

Two-way switch control

The two-way switch is, in principle, a single-pole changeover switch. When connected in pairs, the switches provide control of a lamp from two positions and may therefore be installed in bedrooms, landings and corridors. Figure 3.26 shows the generally accepted method of wiring two-way switches.

Intermediate switch control

An intermediate switch or switches used in conjunction with two, two-way switches, provide control of a lamp from three or more positions. Long corridors with several doors, long halls and multi-flight staircases, require intermediate switch control for reasons of both safety and convenience, so that every access point is provided with its own lighting control. Figures 3.27 and 3.28 show the wiring for intermediate switch control.

Earthing

The 16th edition of the IEE Regulations require that every switch (unless the switch is fixed inside an earthed metal box, in which the switch plate is in electrical contact with the box) shall be connected to an earth-continuity conductor. To comply with the Regulations, every ceiling rose must also be connected to an earth-continuity conductor.

Single-phase power circuits

The ring circuit

A ring circuit is a final sub-circuit consisting of two current-carrying conductors and an earth wire looped from one socket outlet to another, both ends of the circuit being connected to a 30 A miniature circuit-breaker or a fuse. If a metal conduit system is used the earth wire is unnecessary, providing that the conduit

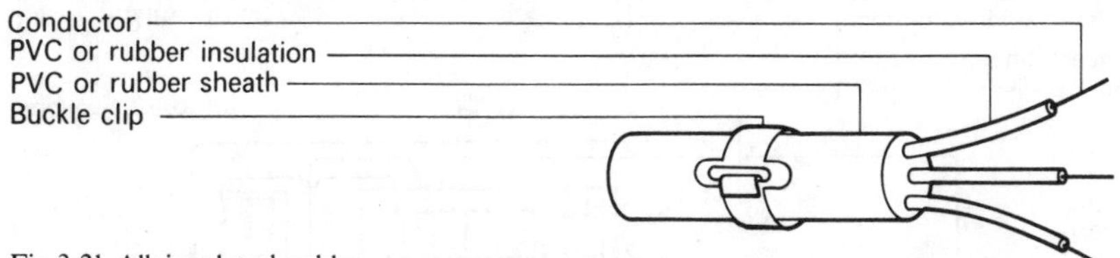
Conductor
PVC or rubber insulation
PVC or rubber sheath
Buckle clip

Fig.3.21 All-insulated cables

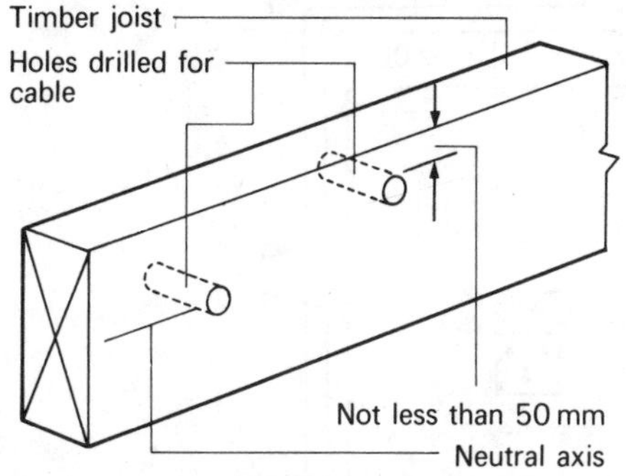
Timber joist
Holes drilled for cable
Not less than 50 mm
Neutral axis

Fig.3.22 Method of drilling timber joist

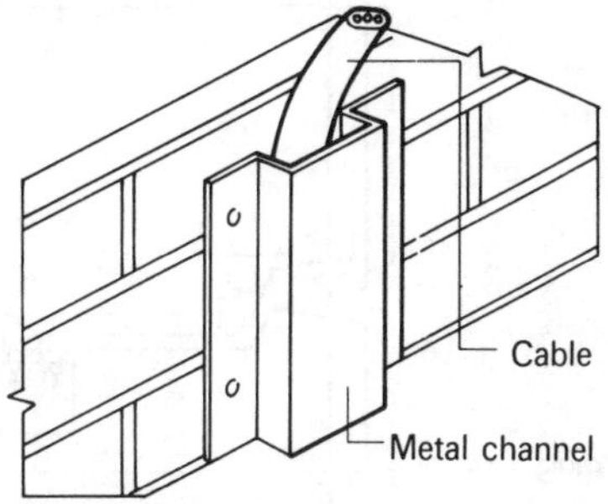
Cable
Metal channel

Fig.3.23 Protection of all-insulated sheathed cables

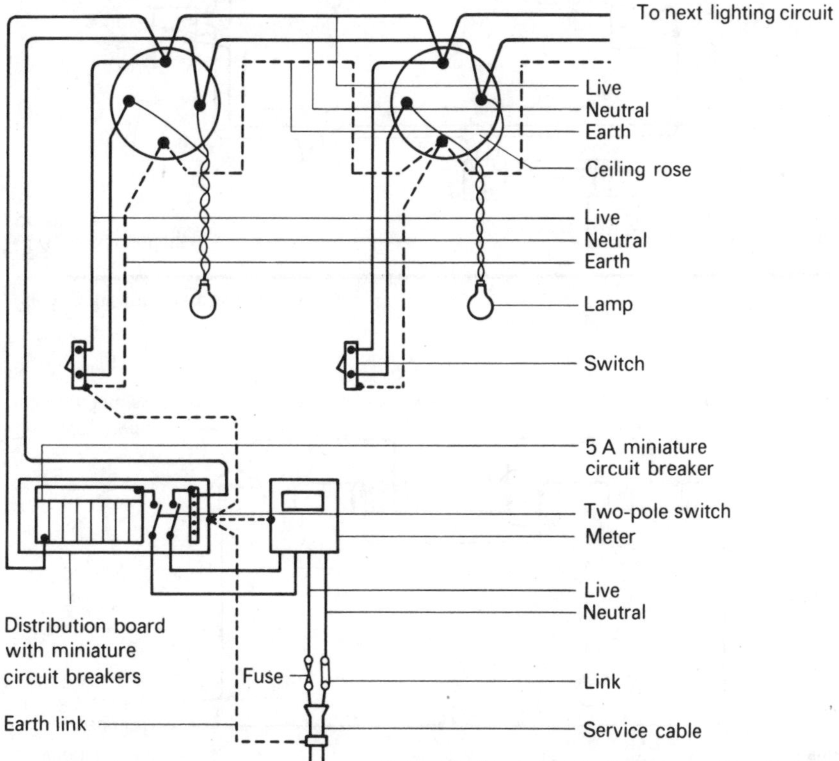
To next lighting circuit
Live
Neutral
Earth
Ceiling rose
Live
Neutral
Earth
Lamp
Switch
5 A miniature circuit breaker
Two-pole switch
Meter
Live
Neutral
Link
Service cable
Distribution board with miniature circuit breakers
Fuse
Earth link

Note: A ceiling rose shall not be installed in any circuit operating at a voltage normally exceeding 250 volts (IEE Regs)

Fig.3.24 Loop-in method of wiring

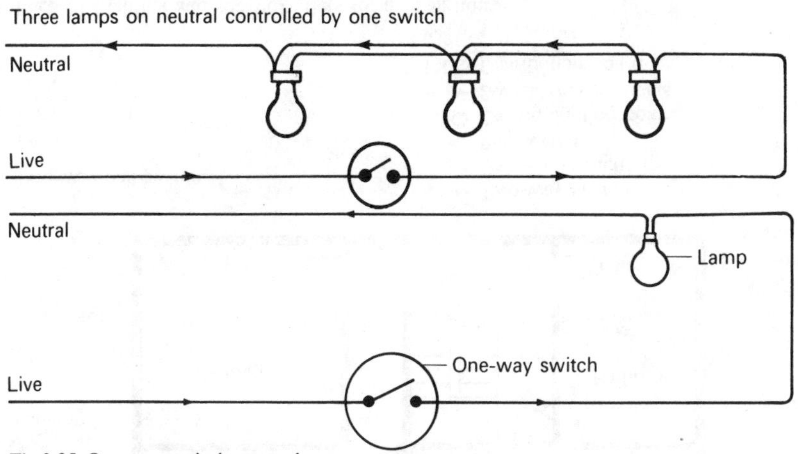
Three lamps on neutral controlled by one switch
Neutral
Live
Neutral
Lamp
Live
One-way switch

Fig.3.25 One-way switch control

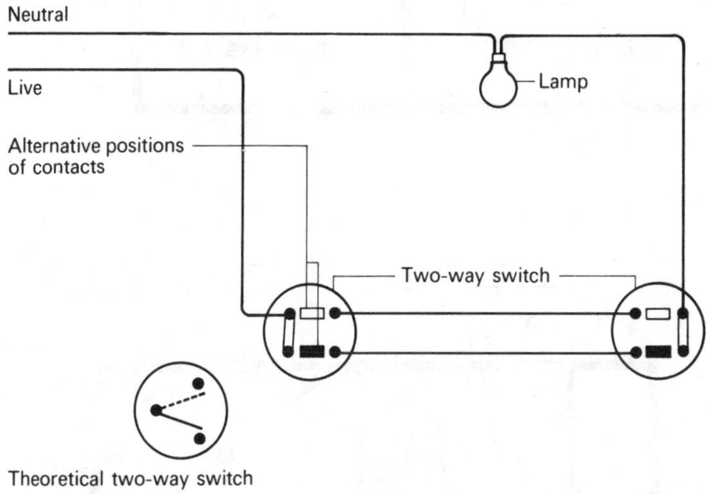
Neutral
Live
Lamp
Alternative positions of contacts
Two-way switch
Theoretical two-way switch

Fig.3.26 Two-way switch control

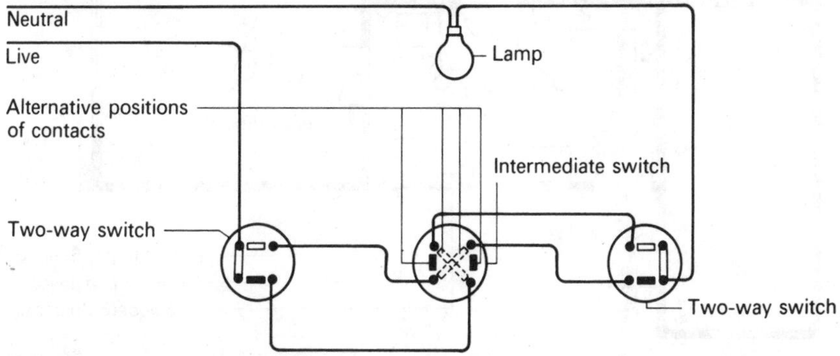
Neutral
Live
Lamp
Alternative positions of contacts
Intermediate switch
Two-way switch
Two-way switch

Fig.3.27 Intermediate switch control

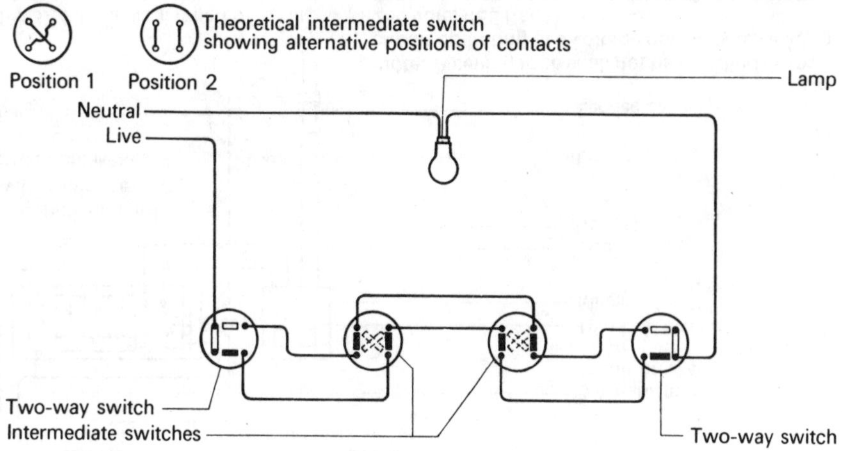
Theoretical intermediate switch showing alternative positions of contacts
Position 1
Position 2
Neutral
Live
Lamp
Two-way switch
Intermediate switches
Two-way switch

Fig. 3.28 Intermediate switch control

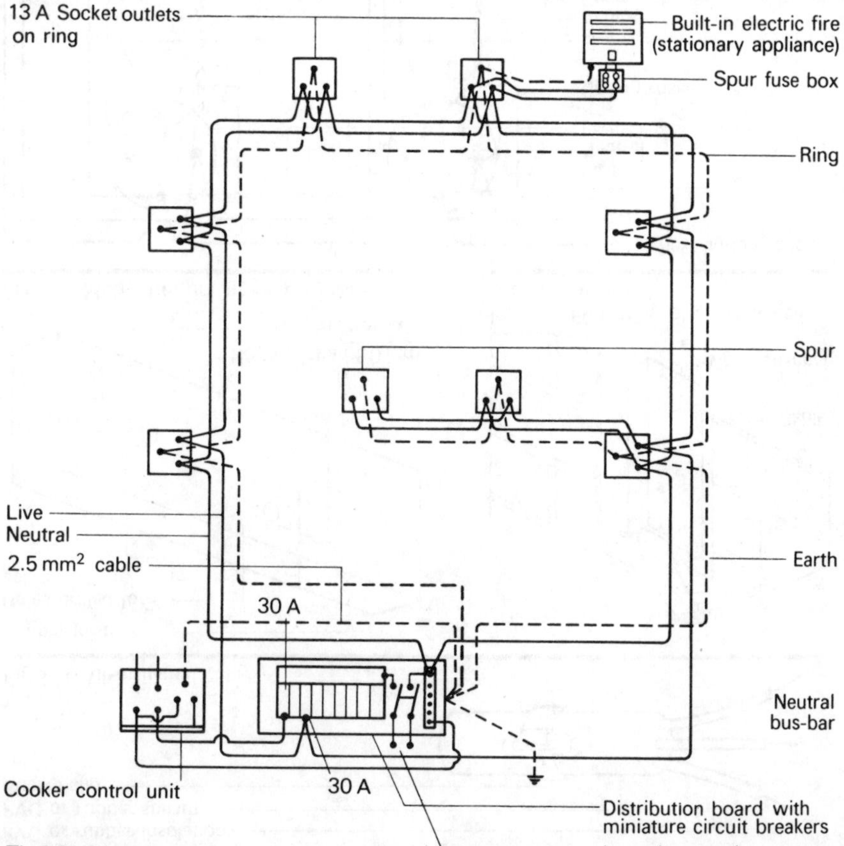
13 A Socket outlets on ring
Built-in electric fire (stationary appliance)
Spur fuse box
Ring
Spur
Live
Neutral
2.5 mm² cable
Earth
30 A
Neutral bus-bar
Cooker control unit
30 A
Distribution board with miniature circuit breakers
Incoming supply

Fig. 3.29 Ring circuit

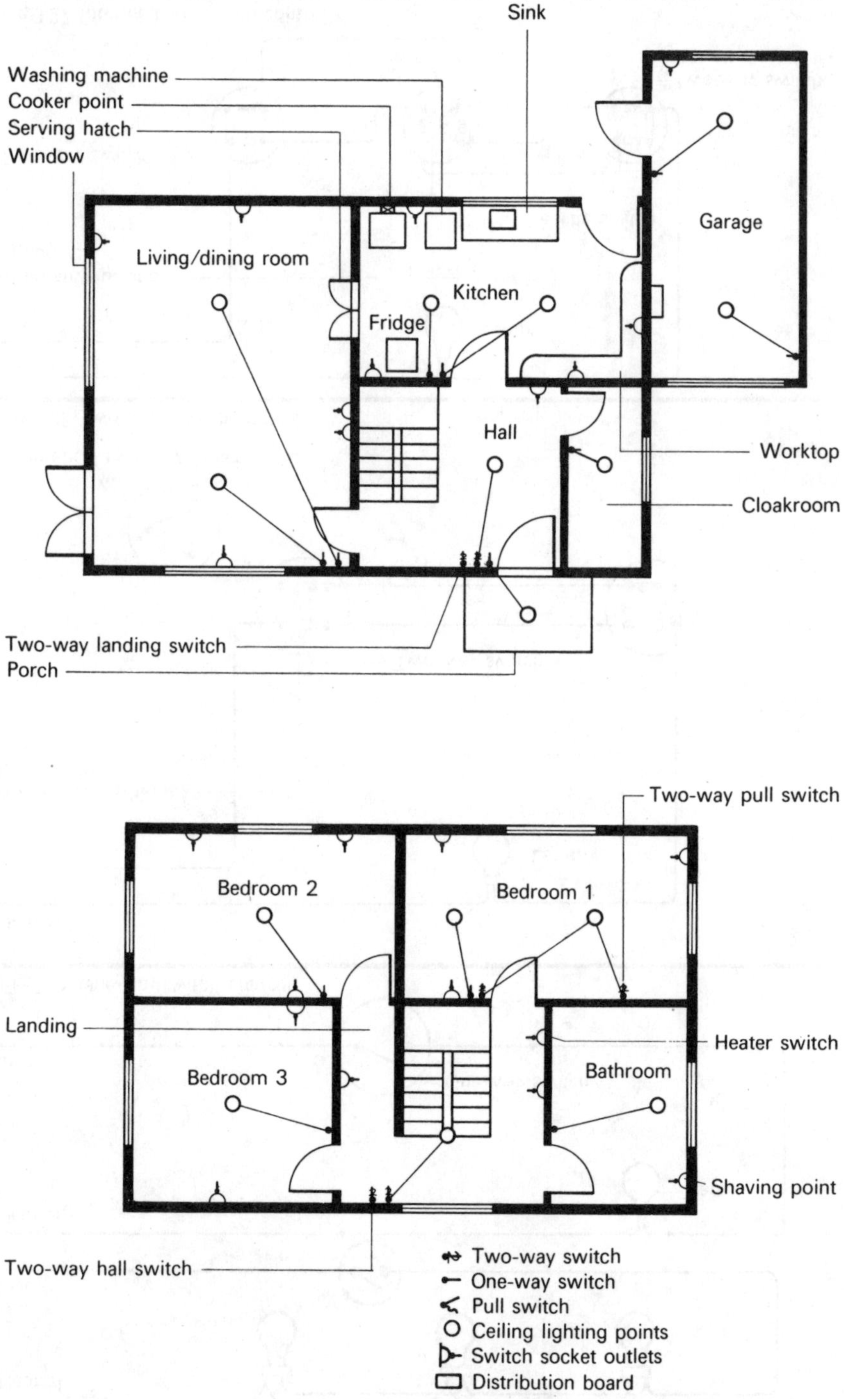
Sink
Washing machine
Cooker point
Serving hatch
Window
Living/dining room
Kitchen
Fridge
Garage
Hall
Worktop
Cloakroom
Two-way landing switch
Porch
Two-way pull switch
Bedroom 2
Bedroom 1
Landing
Bedroom 3
Bathroom
Heater switch
Shaving point
Two-way hall switch
Two-way switch
One-way switch
Pull switch
Ceiling lighting points
Switch socket outlets
Distribution board

Fig. 3.30 Lighting and socket outlets for a small house

joints are electrically sound and the conduit is earthed. It is permissible to cut the conductors where they loop into the terminals of the socket outlets, providing that the joints are electrically sound (Fig. 3.29 shows a wiring diagram of a ring circuit).

'Spurs'

A saving in cable used can sometimes be made by using 'spurs' instead of taking the ring to remote socket outlets. The total number of spurs must not exceed the total number of socket outlets and stationary appliances connected directly in the ring and there must not be more than two socket outlets per spur.

Fused plugs

The socket outlets on the ring are provided with fused plugs and the fusing can be either 13 A or 3 A to suit the apparatus connected to the plug, thus permitting the maximum use of the socket outlets and flexibility in the movement of the apparatus.

Table 3.4 Fuse ratings

13 A	**3 A**
Washing machine	Reading lamp
Kettle	Record player
Iron	Small mixer
Toaster	Coffee percolator
Vacuum cleaner	Blanket
Fire	Television
Freezer	
Large mixer	
Spin dryer	
Water heater	

Number of socket outlets

In domestic premises, a ring circuit may serve an unlimited number of socket outlets, provided that there is at least one ring circuit for each 100 m^2 of floor area. If there are two or more ring circuits the outlet points shall be reasonably distributed between the separate circuits.

For non-domestic buildings, the number of socket outlets shall be determined from the diversity factor, and the maximum loading of the apparatus shall not exceed that of the fuse or circuit-breaker protecting the circuit.

Switches

Although the Regulations do not require socket outlets to have a switch control, it is usually desirable for a switch to be fitted and the majority of 13 A socket outlets are now fitted with a single-pole switch.

Round-pin socket outlets

In older installations each socket outlet was served by an individual circuit from the distribution board. The plugs were round-pin for ratings of 2, 5 and 15 A and, since these were not standardised, a move from one house to another meant that the new owner often had to completely change the plugs for the appliances brought from the old house. There are still very many homes with the round-pin socket outlets, but the ring circuit with its standardised socket outlets is now used for new installations.

Lighting and socket outlets for houses

In order to avoid long runs of flexible cables and multi-point adaptors, the number of socket outlets must be adequate for the consumer's needs. The location of the items of furniture and electrical equipment should be considered when positioning socket outlets, lighting points and switches. Figure 3.30 shows the layout of socket outlets and lighting points for a small house. Table 3.5 gives the desirable number of socket outlets for a small house.

Table 3.5 Number of socket outlets (from Parker Morris report)

	Desirable number of socket outlets
Dining room	2
Living room	5
Double bedroom	3
Single bedroom	2
Kitchen (working area)	4
Hall	1
Landing	1
Garage	1

Table 3.6 gives the graphical symbols used for electrical installations.

Table 3.6 British Standard 1192: 1969. Graphical Symbols for Electrical Installations

Cooker control point	
Distribution board	
Meter	
Main control	
Power point	
Switch socket outlet	

Table 3.6 continued on p. 26

Table 3.6 British Standard 1192: 1969. Graphical Symbols for Electrical Installations— *continued*

One-way switch	
Two-way switch	
Discharge lamp	
Filament lamp	
Lighting column	
Pull or pendant switch	
Wall lamp	
Circuit-breaker	
Fuse	
Neutral link	

Bath and shower rooms

Special precautions are required in bathrooms and similar rooms to avoid danger from the electrical installations. Lighting fittings should enclose the lamp completely and there should be no exposed metalwork on the fittings. Switches inside bathrooms should be of the pull-cord type, mounted on the ceiling or high on the wall. Ordinary switches or other means of control (e.g. the immersion heater switch) should be situated so as to be normally inaccessible to a person using a fixed bath or shower. An electric shaver point must comply with the appropriate BS 3052 and be provided with an effective earthing terminal. In any room containing a fixed bath or shower, there must be no stationary appliance with elements that can be touched by a person in a fixed bath or shower, nor must there be any socket outlet for connecting a portable appliance. Double-pole switches should be used for stationary appliances (e.g. an electric towel rail) and the metalwork of the appliances should be effectively earthed.

Emergency supplies

In buildings such as theatres, cinemas and dance halls, where public assembly is involved, emergency lighting is required in case of failure of the normal lighting supplied from the electric mains. There should be battery-operated lights in the public spaces and exit corridors of sufficient power to enable people to see their way out, and there should be exit signs over each of the exit doors, also illuminated by lamps supplied from the batteries. The batteries are kept charged from the mains supply and there should be sufficient reserve power in the batteries to enable the emergency lighting to give light for 3 h after a mains failure.

There are three basic kinds of emergency lighting installations:

1. A system in which the battery-fed lights come on only after a mains failure.
2. The emergency lights are on wherever the building is occupied and are supplied from the transformer in the battery-charging plant. In the event of a mains failure, the lights automatically change over to the batteries.
3. The 'floating battery' system in which the lights are fed from the battery and remain on. The battery receives a charge from the mains slightly in excess of the load taken by the lights and this ensures that the battery is always fully charged. The size of the battery installation depends upon the load required by the emergency lamps and while small buildings may require a metal battery cabinet, large buildings may require a special battery room.

Lead/acid batteries are often used for emergency supplies, but nickel/alkali batteries have a longer life and do not give rise to corrosion problems as do lead/acid types. The delivery and storage of acid and distilled water must be remembered when designing the installation.

Diesel generators

For large buildings such as hospitals and departmental stores, stand-by diesel-electric generators have to be provided for emergency electrical supplies. The generator is automatically brought into operation on failure of the mains supply.

Chapter 4

Electrical installations in large buildings

Distribution principles

Due to the higher electrical load large buildings such as hospitals, factories and office blocks, will require a three-phase supply. The loading in some large buildings may be too high for the local low- or medium-voltage systems to provide and a private sub-station must therefore be installed, fed from the high-voltage cables from the Electricity Companies nearest switching station. Sub-stations are often required for factories and hospitals. Figure 4.1 shows the plan of a typical sub-station including the necessary electrical equipment.

The electrical installation in a large building is similar to a small building, but is divided into sections. There may be one main intake panel incorporating large fused switches or circuit-breakers, each of which control a feeder cable to subsidiary distribution panels in different parts of the building, or each separate building in a group. The subsidiary distribution panels are smaller versions of the main intake panel and they control distribution boards for each sub-section.

Distribution units

The intake and distribution units may be either made up on site from separate components, or standard factory-made cubicle switchboards may be used. The manufactured cubicle switchboards have the following advantages over control panels made up on site:

1. quicker to install;
2. neater in appearance;
3. usually take up less wall length;
4. a locked door prevents unauthorised access to equipment;
5. safer.

Current rating

The current rating of the cable and components must never be less than the protective device which controls it. A fused switch of 100 A rating can only serve a cable having a current rating of 100 A or more and the switchgear at the opposite end of the circuit must also be of 100 A rating up to the next smaller fuse or miniature circuit-breaker.

Rating discrimination

It is important that fuses or miniature circuit-breakers should provide discrimination, e.g. each subsidiary fuse or circuit-breaker should isolate a fault in its own section before the fuse or circuit-breaker one stage further back in the installation can operate. To ensure this, the ratings of two successive fuses or miniature circuit-breakers must differ by at least 30 per cent and preferably 50 per cent. To achieve this, it may sometimes be necessary to increase the cable and switchgear ratings so that the main fuse or miniature circuit-breaker cannot operate before the subsidiary one.

Voltage drop

In a small electrical installation the voltage drop in the circuit wiring is not usually significant, but in larger installations the voltage drop in the cables between the main intake and the subsidiary distribution panels may be very high. The 16th edition of the IEE Regulations require that the voltage drop between the main intake point and each supply point shall not exceed 2½ per cent of the nominal supply voltage.

Methods of distribution

There are three methods of distribution for large buildings: radial, ring main and rising main.

1. *Radial distribution*

The name comes from the fact that the services to subsidiary distribution panels radiate from the main intake panel. The main panel normally consists of a main switch connected to fused switches through a bus-bar chamber. Several separate feeder cables are run from the main intake panel to the subsidiary distribution panels, which may be situated in separate buildings, or at strategic points inside one building.

Figure 4.2 shows a radial distribution system for a four-storey block of flats. Figure 4.3 shows a detail of an overhead bus-bar distribution used in the factory and Fig. 4.4 shows a radial distribution system for a factory requiring three-phase and single-phase power supplies.

2. *Ring-main distribution*

In the case of a large development scheme having several buildings around the perimeter of the site, a ring-main circuit would be taken around the site with supplies taken into each building. Figure 4.5 shows the principle of the ring-main system and Fig. 4.6 a plan of a large site utilising a ring-main distribution system. For the same scheme, a radial distribution system would require a number of

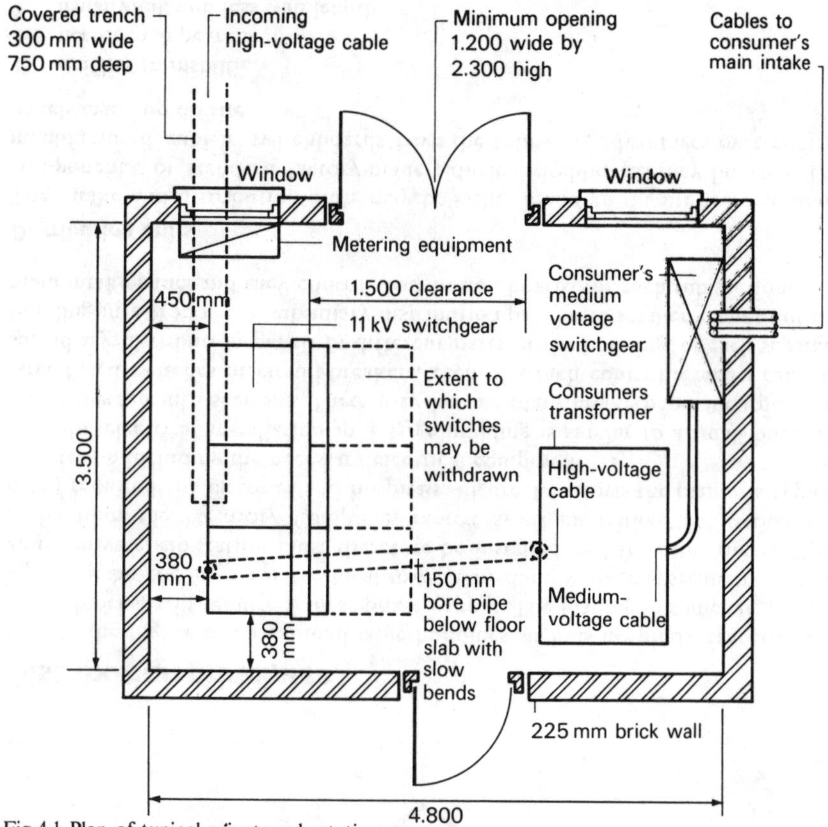
Covered trench
300 mm wide
750 mm deep
Incoming
high-voltage cable
Minimum opening
1.200 wide by
2.300 high
Cables to
consumer's
main intake
Window
Window
Metering equipment
1.500 clearance
450 mm
11 kV switchgear
Consumer's
medium
voltage
switchgear
Extent to
which
switches
may be
withdrawn
Consumers
transformer
High-voltage
cable
3.500
380
mm
150 mm
bore pipe
below floor
slab with
slow
bends
Medium-
voltage cable
380
mm
225 mm brick wall
4.800

Fig.4.1 Plan of typical private sub-station

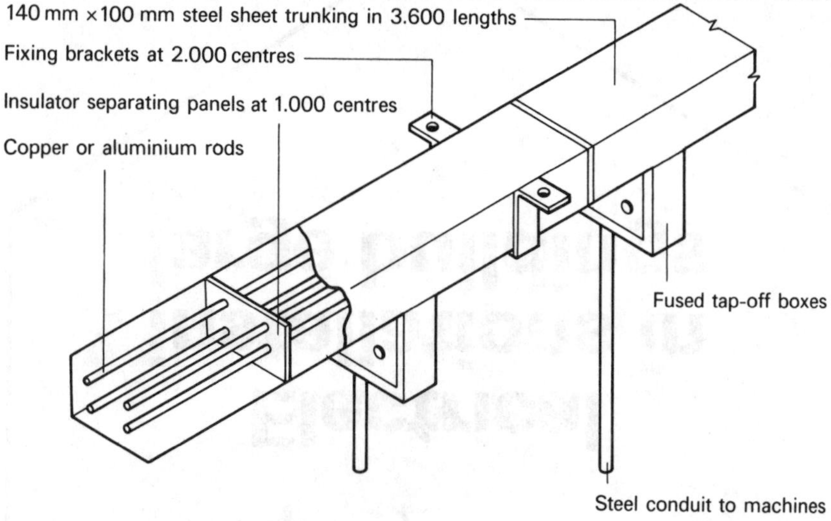
140 mm × 100 mm steel sheet trunking in 3.600 lengths
Fixing brackets at 2.000 centres
Insulator separating panels at 1.000 centres
Copper or aluminium rods
Fused tap-off boxes
Steel conduit to machines

Fig.4.4 Overhead bus bar distribution

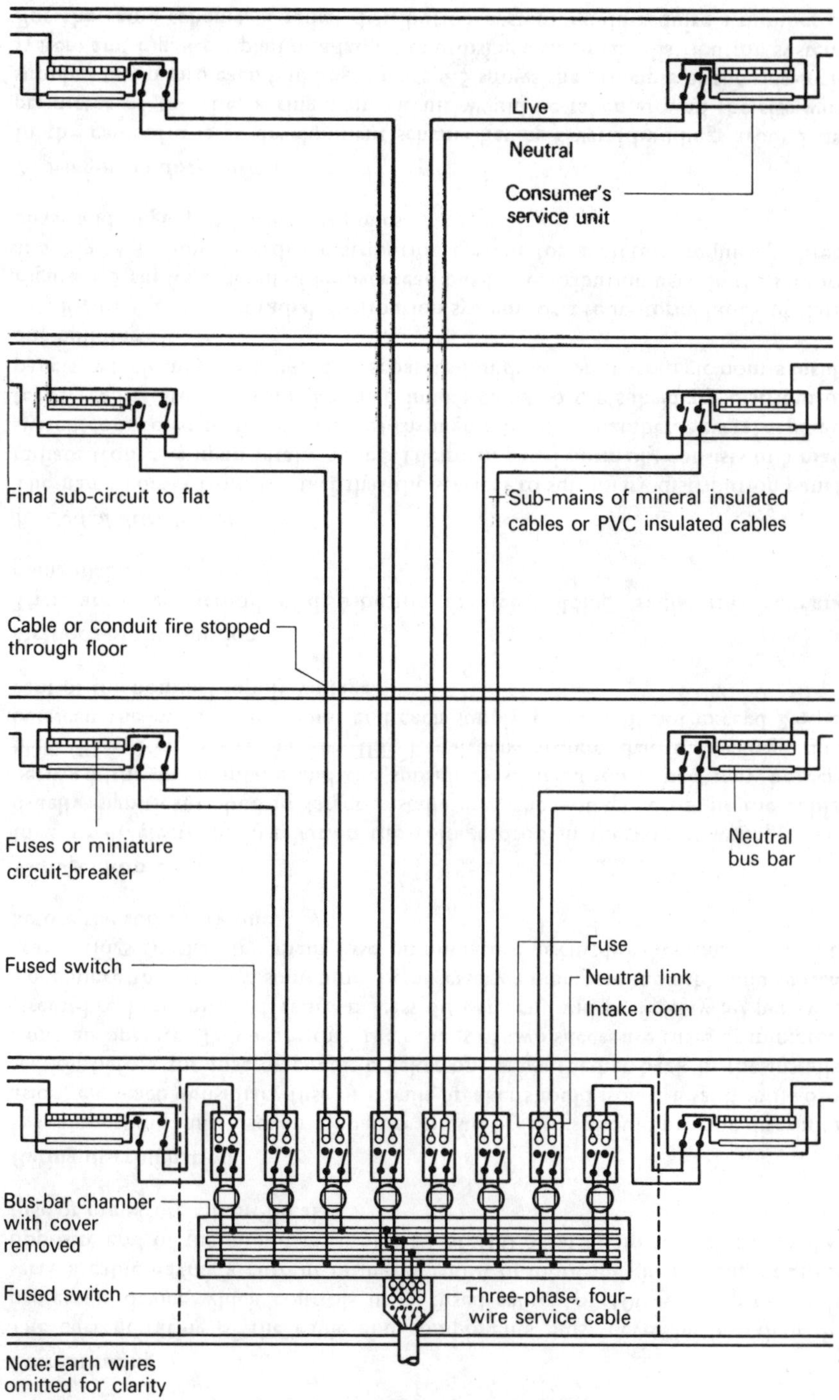
Live
Neutral
Consumer's
service unit
Final sub-circuit to flat
Sub-mains of mineral insulated
cables or PVC insulated cables
Cable or conduit fire stopped
through floor
Fuses or miniature
circuit-breaker
Neutral
bus bar
Fuse
Neutral link
Intake room
Fused switch
Bus-bar chamber
with cover
removed
Fused switch
Three-phase, four-
wire service cable
Note: Earth wires
omitted for clarity

Fig.4.2. Radial installation for a four-storey block of flats

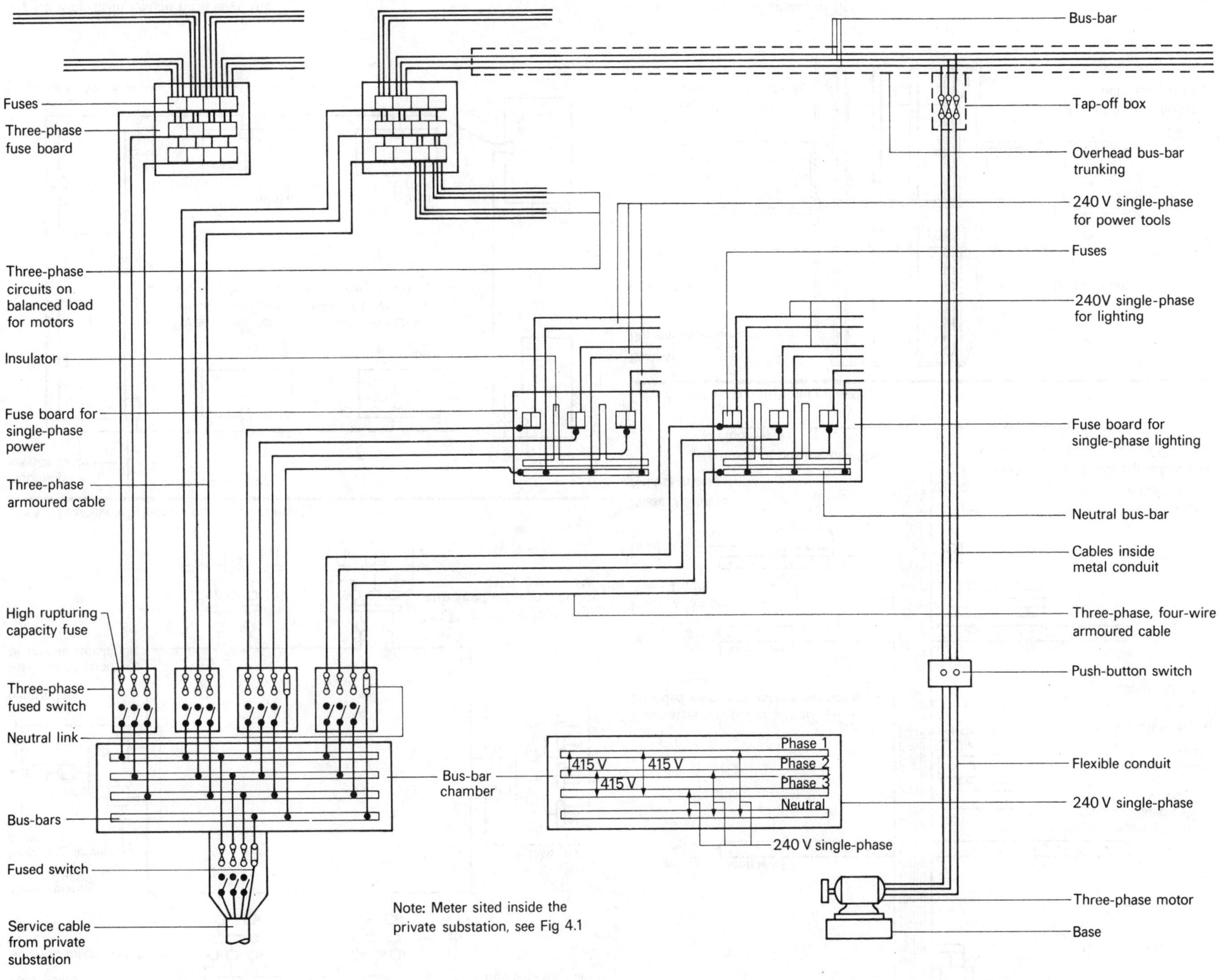

Bus-bar
Fuses
Tap-off box
Three-phase fuse board
Overhead bus-bar trunking
240 V single-phase for power tools
Fuses
Three-phase circuits on balanced load for motors
240V single-phase for lighting
Insulator
Fuse board for single-phase power
Fuse board for single-phase lighting
Three-phase armoured cable
Neutral bus-bar
Cables inside metal conduit
High rupturing capacity fuse
Three-phase, four-wire armoured cable
Push-button switch
Three-phase fused switch
Neutral link
Bus-bar chamber
Phase 1
Phase 2
Phase 3
Neutral
415 V
415 V
415 V
Flexible conduit
240 V single-phase
Bus-bars
240 V single-phase
Fused switch
Three-phase motor
Note: Meter sited inside the private substation, see Fig 4.1
Service cable from private substation
Base

Fig.4.3 Radial installation for a factory

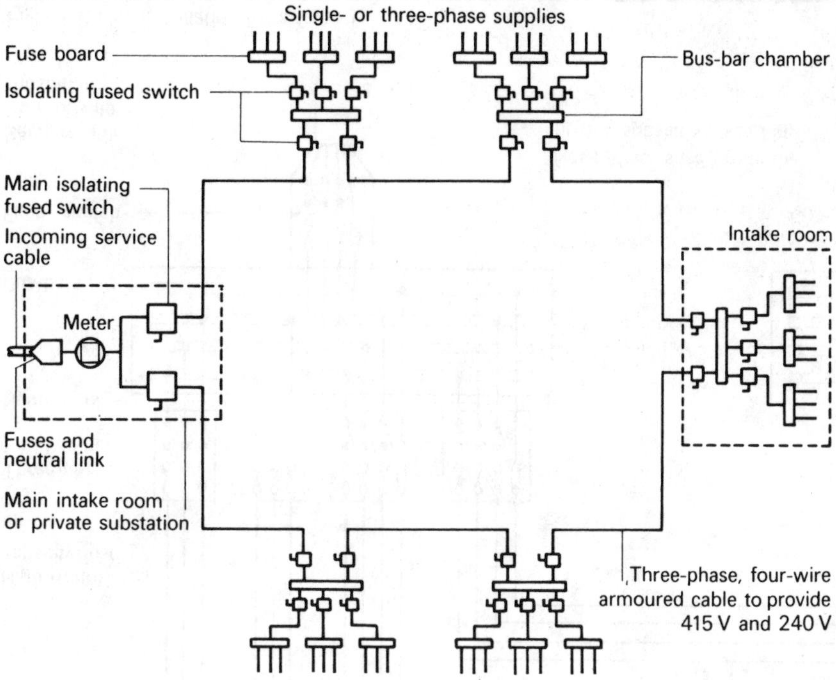
Single- or three-phase supplies
Fuse board
Isolating fused switch
Bus-bar chamber
Main isolating fused switch
Incoming service cable
Meter
Intake room
Fuses and neutral link
Main intake room or private substation
Three-phase, four-wire armoured cable to provide 415 V and 240 V

Fig.4.5 Principle of the ring main distribution

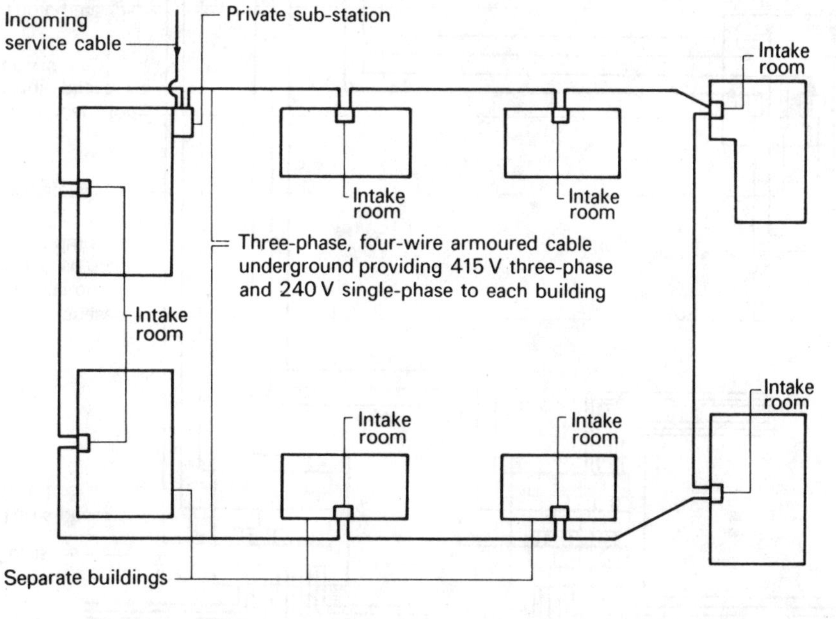
Incoming service cable
Private sub-station
Intake room
Intake room
Intake room
Intake room
Three-phase, four-wire armoured cable underground providing 415 V three-phase and 240 V single-phase to each building
Intake room
Intake room
Intake room
Separate buildings

Fig.4.6 Ring-main system for a large site

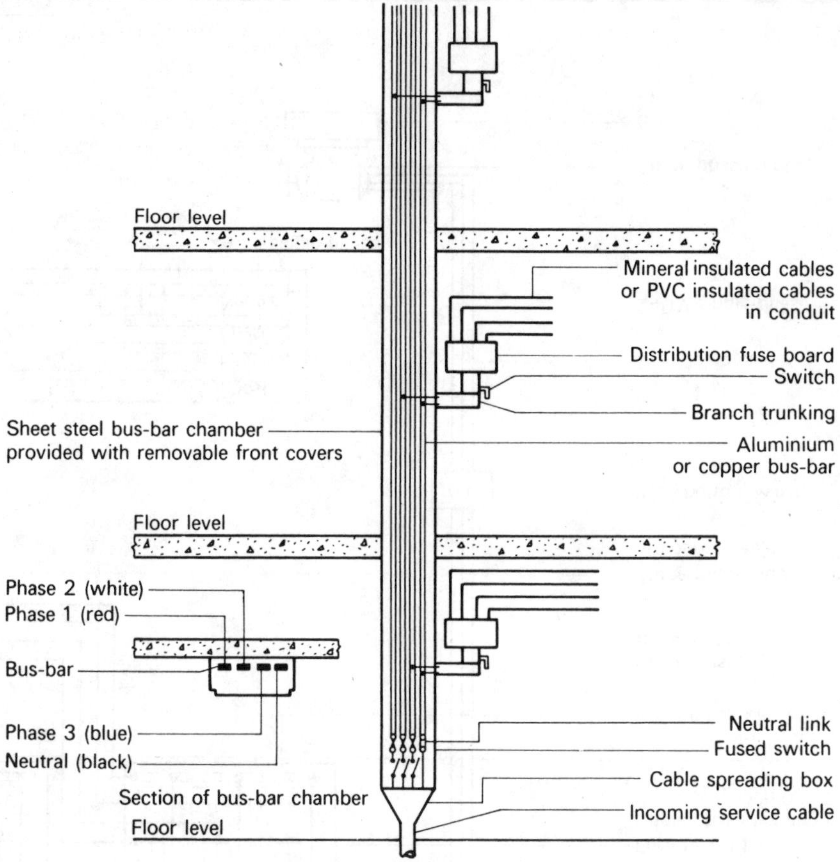
Floor level
Mineral insulated cables or PVC insulated cables in conduit
Distribution fuse board
Switch
Branch trunking
Sheet steel bus-bar chamber provided with removable front covers
Aluminium or copper bus-bar
Floor level
Phase 2 (white)
Phase 1 (red)
Bus-bar
Phase 3 (blue)
Neutral (black)
Section of bus-bar chamber
Neutral link
Fused switch
Cable spreading box
Incoming service cable
Floor level

Fig.4.7 Rising-main distribution system

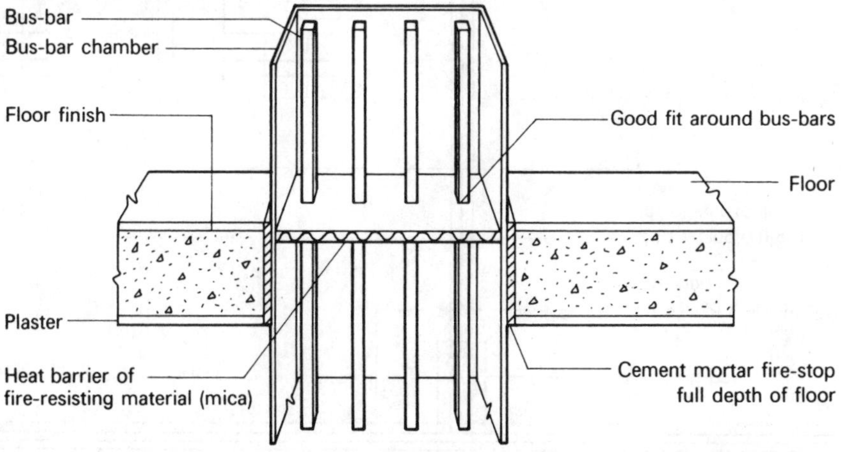
Bus-bar
Bus-bar chamber
Floor finish
Good fit around bus-bars
Floor
Plaster
Heat barrier of fire-resisting material (mica)
Cement mortar fire-stop full depth of floor

Fig.4.8 Prevention of spread of fire

feeder cables, each running separately from the main intake panel to the individual buildings.

The ring-main distribution system has the following advantages:

(*a*) Each building and individual sections of the ring may be isolated without switching off the entire installation.
(*b*) The current may flow in either direction which reduces the voltage drop.
(*c*) The ring may be sized to take account of the diversity factor for all the buildings, since a heavy load may be required for any one of the buildings, but it is unlikely that such a load will be required for all the buildings simultaneously.

3. *Rising main distribution*

For buildings above five storeys in height, it is normally preferable to pass conductors vertically through the building. The supply to each floor is connected to the rising main by means of tap-off subsidiary units.

Types of rising main: The rising main may be one of the following types:

(*a*) PVC or vulcanised rubber insulated cables mounted on porcelain cleats inside brick or concrete ducts, with hardwood or metal access doors on each floor.
(*b*) Paper, mineral, PVC, or vulcanised rubber insulated cables run in sheet steel vertical ducts.
(*c*) Uninsulated copper or aluminium bars run in steel sheet vertical ducts.

Supplies to floors

There are two main methods of supplying each floor:

(*a*) To pass conductors, whether cable or bars, into isolation or fused switches at each floor.
(*b*) To pass conductors into and out of a subsidiary distribution board at each floor.

With the first method the supply can be isolated at each floor, but not with the second method. Figure 4.7 shows a rising-main distribution system for a high-rise building, using a sheet metal duct and aluminium or copper bus-bars.

Precautions against spread of fire

The 16th edition of the IEE Regulations requires that where cables, conduits, ducts or trunking pass through walls, partitions or floors, the surrounding space left must be made good with cement mortar or similar fire-resisting material to the full thickness of the structure. In addition where cables, conduits or conductors are installed in channels, ducts, trunking or shafts which pass through floors or walls, suitable internal fire-resisting barriers shall be provided to prevent spread of fire. Figure 4.8 shows the method of preventing the spread of fire when a metal duct passes through a floor.

Installations in offices

High-rise office blocks require a rising-main distribution system with distribution fuse boards mounted alongside the riser to serve each floor. For an office used by one owner, a meter at the base of the riser would be required, while a block where sub-letting of floors is considered, a meter would be required for each floor or individual sections. Low-rise office blocks would require a radial distribution system similar to one used for low-rise blocks of flats.

Floor ducts

The main purpose of a floor duct system is to enable desks to be moved to any position in the office. The duct can be used to carry both low-voltage electrical supplies for machines and lighting and extra low-voltage supplies for private and public telephones. These low and extra low voltage circuits must be segregated inside the duct, and where an electrical service cable or conduit runs parallel with or crosses a telephone cable, a minimum clearance of 50 mm should be given between the two services. If a clear distance is impracticable, physical separation of at least 6 mm of insulating material will be required.

Types of floor ducts

There are three types of floor duct layouts:

1. *Grid:* This method provides adequate flexibility for telephone and electrical supplies and is used in open-plan offices (see Fig. 4.9). A suitable spacing of the ducts is 1.5–2.5 m, but other spacings may be used depending upon the degree of flexibility required.

2. *Branching:* This method uses a central feeder duct with branches to each window bay. The branches may either terminate just short of the wall, or extend to wall outlets. Figure 4.10 shows a branching duct layout with wall outlets. The layout provides reasonable flexibility for open-plan offices, but is also used for partitioned offices with the central feeder duct in the corridor.

3. *Perimeter:* This is the cheapest method but does not provide the flexibility obtained by the grid and branching layouts. A main feeder duct is located about 450 mm from the outside wall with short branches taken from junction boxes to wall outlets for telephones, or both telephones and electrical supply. Figure 4.11 shows a perimeter layout suitable for a partitioned office.

Materials for ducts

Steel and pitch fibre ducts are in general use. Figure 4.12 shows a steel sheet duct having twin compartments, one for telephones and the other for electric power. The overall duct dimensions are 100 mm x 25 mm. Figure 4.13 shows a sheet steel junction box, wall bend and wall outlet. The junction box contains partitions which effectively separate the services passing through the box.

Floor outlets

These can be pre-fixed at suitable spacings, or mounted at any point on the duct and connected with the necessary cables, by drilling holes through the floor finish and drawing cables through from the nearest junction box.

Skirting ducts

Hollow skirtings may be used to distribute and conceal the cables. They may be

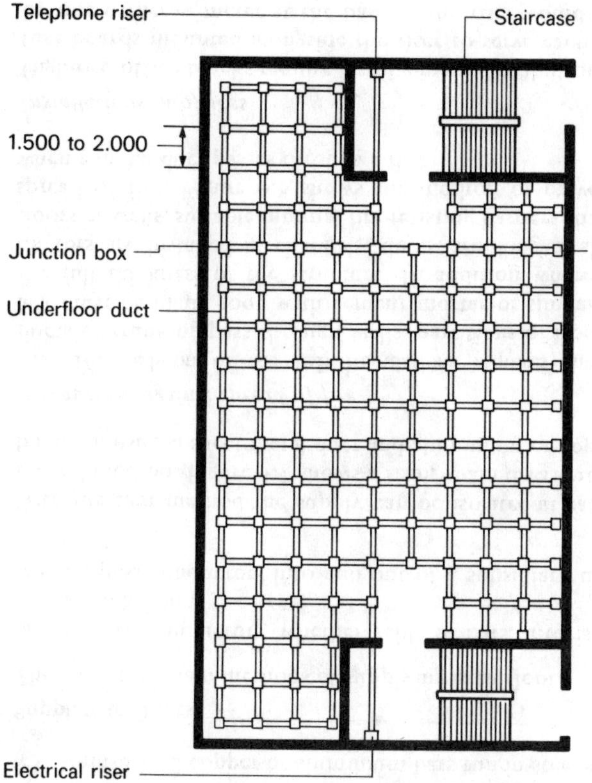
Telephone riser
Staircase
1.500 to 2.000
Junction box
Duct end
Underfloor duct
Electrical riser

Fig.4.9 Grid layout for open-plan office

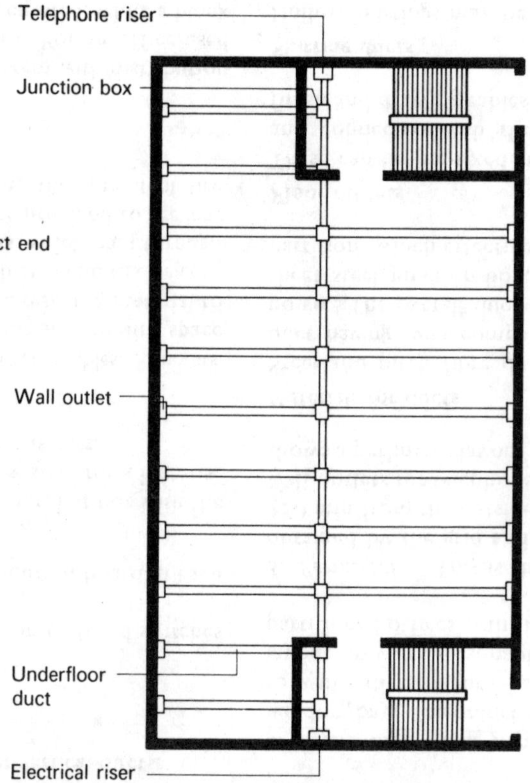
Telephone riser
Junction box
Wall outlet
Underfloor duct
Electrical riser

Fig.4.10 Branching layout for partitioned or open-plan office

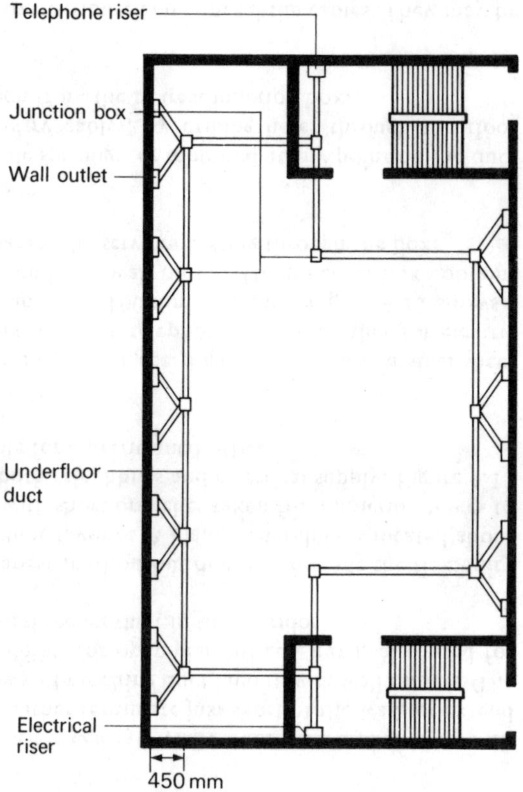
Telephone riser
Junction box
Wall outlet
Underfloor duct
Electrical riser
450 mm

Fig.4.11 Perimeter layout for partitioned office

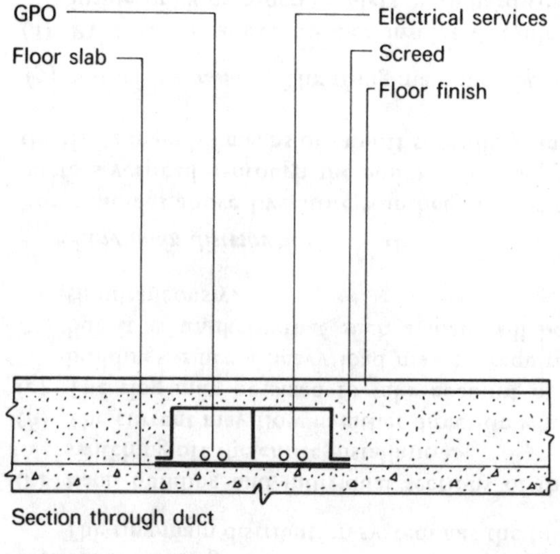
GPO
Floor slab
Electrical services
Screed
Floor finish
Section through duct

Fig.4.12 Floor duct with twin compartments

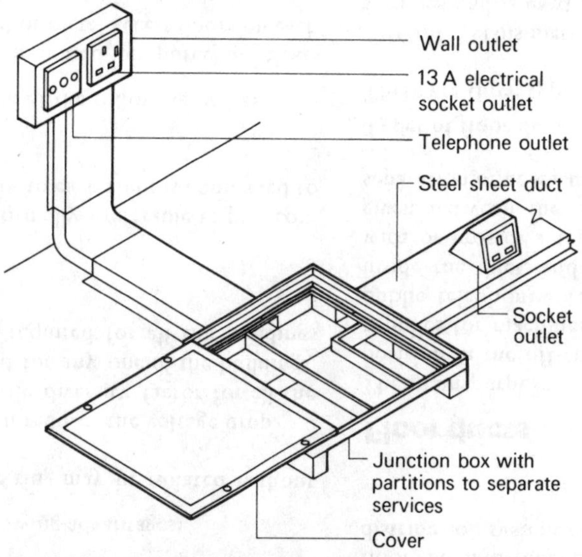
Wall outlet
13 A electrical socket outlet
Telephone outlet
Steel sheet duct
Socket outlet
Junction box with partitions to separate services
Cover

Fig.4.13 View of junction box and wall outlet

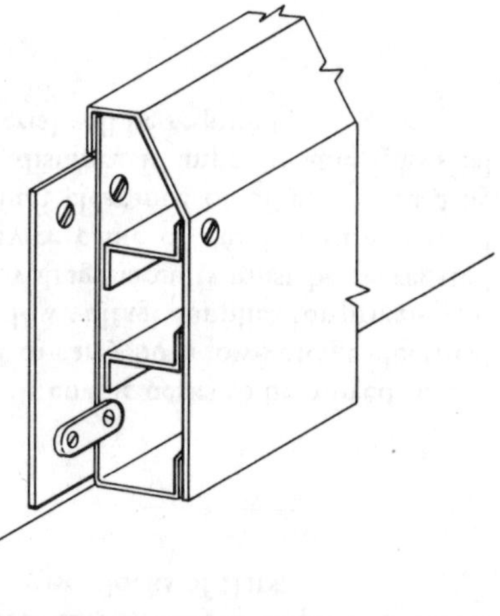

Fig.4.14 Metal skirting duct

used in conjunction with an underfloor duct system, so that cables can be taken from the underfloor duct, via an outlet at the back of the skirting and run around the room with the skirting to the required position. The duct may be divided by suitable partitions to carry electrical power and telephone cables (see Fig. 4.14). Extended use of the duct is rendered difficult by piers and columns projecting from the walls, which necessitate many bends and removable panels to negotiate the obstructions.

Materials

Skirting ducts are available in sheet steel, fibre or fabricated in timber to the architect's design. Metal ducts must be effectively earthed.

Chapter 5

Site electricity, space heating, telephones

Site electricity

The use of electricity on building sites increases productivity for the following reasons:

1. Reduction of manual handling of materials.
2. Reduction of manual effort by the use of portable power tools.
3. The improvement of working conditions and safety by the use of interior and exterior electric lighting.
4. The use of electric heating which enables work to proceed at low temperatures.

Type of supply

The most convenient and usually the most economical supply is from the Electricity Company, but private generators may be used and specially soundproofed types to reduce noise from building sites are available. Private generating plant must comply with British Code of Practice 323.

Notices

Electricity Companies can usually provide a supply wherever it is needed, but this cannot always be arranged at short notice. It is therefore necessary to give notice in writing to the Electricity Company as soon as possible, with the following particulars:

1. Address of site.
2. Description of building to be constructed.

3. Total floor area.
4. Site location plan.
5. Name, address and telephone number of the developer and building.
6. Estimated commencement and completion dates.
7. Estimated maximum demand for a permanent supply.
8. Estimated maximum loading for site electricity supply (see Table 5.1).

Table 5.1 gives the power required for site electrical equipment (BRE Digests, Nos. 87 and 88).

Table 5.1 Electrically operated plant

Plant	Power (kW)
Tower crane hoist motor	2.25–60
Tower crane slew motor	1.00–11.00
Tower crane travel motor	0.75–7.50
Tower crane crab or derricking motor	0.75–7.50
Weigh-batching unit	6.00–15.00
Concrete mixer	1.00–15.00
Goods hoist	1.50–33.50
Passenger/goods hoist	5.50–33.50
Compressor	5.50–75.00
Conveyor	2.25–7.50
Pump	0.25–30.00
Timber sawbench	0.75–7.50
Concrete saw	0.75–2.25
Floor grinder and polisher	0.375–2.25

Voltages

The use of electricity on building sites requires greater attention to electrical safety than most other uses, and low voltages should be used wherever possible. Building operatives are nearly always in contact with earth in damp situations and are therefore vulnerable to electric shock: lower voltages will reduce the shock.

Table 5.2 gives the usual voltages for various uses on building sites.

Table 5.2 Voltages and uses

Voltage at 50 Hz	Phase	Source of supply	Special provision	Uses
415	Three	Electricity Board	–	Cranes, hoists, concrete mixers, compressors requiring above 3.73 kW
240	Single	Electricity Board	–	Site offices, external flood-lighting
110	Three	415 V transformer	Secondary winding phase to earth 64 V	Transportable equipment, vibrators, site lighting pumps and hand tools requiring up to 3.73 kW
110	Single	240 V transformer	Secondary winding outers to earth 55 V	All portable and transportable tools up to 1.8 kW and site lighting
50	Single	Transformer	Secondary winding outers to earth 25 V	Dangerous situations, e.g. tunnels, tanks inside boilers
25	Single	Transformer	Secondary winding outers to earth 12.5 V	As above

Note: Self-contained battery-operated cap or hand lamps may also be used in dangerous situations.

Hand lamps

Inspection lamps are often subject to rough usage and should therefore be of robust construction, with a body of strong insulating material and conform to BS 1980. A guard over the lamp should always be provided and the lampholder must not be in metallic contact with the guard. In normal situations the lamp should be supplied at 110 V alternating current (55 V to earth). In dangerous situations lamps should be operated at either 25 V or 12.5 V or battery-operated. The use of battery-operated lamps prevent entanglement with the cable, which is a great danger in restricted areas.

Portable tools

Portable tools should be operated at 110 V with single-phase alternating current and conform to BS 2769. All-insulated or double-insulated portable electric tools should not be regarded as an alternative to the safe low-voltage system. The tools themselves are safe, but the danger of shock from a faulty cable remains. Each tool or portable equipment should be examined by the operator before use and inspected by a competent electrician at intervals not exceeding 7 days, and also each time it is returned to the store.

Electricity intake

The electricity service cable should be terminated in a watertight building which for a small site is usually the contractor's hut, for larger sites a separate hut is erected. It is essential that at the intake position the contractor provides main switchgear which should incorporate:

1. Means of interrupting the supply in the event of excess current.

2. Means of interrupting the supply in the event of dangerous earth leakage.
3. Means of isolation.

At the agreed intake position the Electricity Company will install a meter, but all the electrical supply equipment from the outgoing side of the meter is the responsibility of the building contractor.

Excess current

A fuse is the most common method of interrupting the supply in the event of excess current, but miniature circuit-breakers are gaining popularity and prevent the risk of abuse. If a fault does occur and this is rectified, it is necessary only to reset the breaker, whereas fuses have to be replaced.

Earthing

The metal sheath of underground service cables may be used for earthing purposes, but where the service is provided by overhead lines earthing will have to be made by means of an earth electrode.

Should the nature of the soil be such as to make a good earth connection impossible, alternative means must be provided to interrupt the supply in the event of dangerous earth leakage. This can be achieved by a current-operated earth leakage circuit-breaker with a trip coil which will break the circuit when the voltage across the coil exceeds 40 V. The circuit-breaker will also act as a means of isolation, but when a current-operated circuit-breaker is not used, a fused switch would provide means of isolation. In the interest of safety the circuit-breaker or switch should be clearly marked and accessible at all times.

Cables

All cables should be capable of carrying the required load and maintained to prevent danger. PVC or vulcanised rubber insulated and sheathed cables are used and should be fixed at suitable intervals, cables should be kept off the floors unless they are adequately protected.

Motors

Every electric motor should be provided with means of starting and stopping – the latter so placed as to be easily operated by the person in charge of the motor. It is also essential for every electric motor having a rating exceeding 370 W to be provided with control apparatus to protect the motor or cables against the effects of excess current. Electric motors should be provided with means of preventing restarting after stopping due to a drop in voltage or failure of supply, where unexpected restarting of the motor might cause danger, e.g. timber sawbench.

Overhead cables

Danger sometimes arises from workmen approaching overhead power lines when operating cranes or erecting steelwork. Moving of the power lines by the Electricity Board to a safe distance before operations begin is the best precaution, but if this is not possible a system of barrier fences and 'goalposts' should be erected to prevent cranes and other machines from fouling the power lines (see Fig. 5.1).

Underground cables

Most underground cables are owned by either the Regional Electricity Company, or by the private generating businesses. Before any construction begins, the Regional Companies engineer and the regional office of the private generating businesses should be consulted in order to locate the cables. Other consumers' cables may be buried and for this reason care should be taken during excavation. During the period of the excavation, all the underground cables should be made dead.

Distribution system

Long trailing cables should be avoided by planning the installation so as to bring socket outlets near the working areas. A distribution system that gives the maximum flexibility has been developed by the Building Research Establishment (see Fig. 5.2). It consists of three packaged units:

1. Transformer unit.
2. Distribution unit.
3. Outlet unit.

The system provides a 110 V supply for site work with interconnecting cables. Distribution units are optional and outlet units can be connected directly to the transformer units.

Transformer unit

The transformer unit consists of a ventilated sheet steel cubicle fitted with a lockable door. Eleven double-pole circuit-breakers are fitted inside to protect outgoing circuits to distribution units, outlet units and lighting circuits. There is also a 60 A isolating circuit-breaker, and two 15 A and two 30 A socket outlets are fitted on the outside of the case. The cubicle is provided with handles and lifting rings, and houses a 240/110 V, single-phase, 50 Hz, double-wound transformer with earth screen. The largest unit is 15 kVA, but some smaller units may be fitted if required.

Distribution unit

The distribution unit consists of a sheet metal box mounted on a tubular skid stand and provided with a carrying handle. It is fitted with two 30 A socket outlets and two 15 A socket outlets, each protected by double-pole moulded circuit-breakers. The provision of the 15 A socket outlet is intended for the connection of local general lighting circuits.

Outlet unit

The outlet unit consists of a sheet metal box with a lockable front cover, mounted on a tubular skid stand and with a carrying handle. The unit is fitted with six 5 A socket outlets, two 15 A socket outlets and one 30 A socket outlet, each protected by moulded circuit-breakers. It is connected with 22.5 m of three-core cable with a 30 A plug for connection to one of the distribution units.

Socket outlets

Socket outlets on all three units are to BS 196: 1961 and are fitted with hose-proof spring-back covers for protecting contacts when not in use. The sockets will only accept plugs for 110 V, single-phase, 50 Hz supply and all plugs and

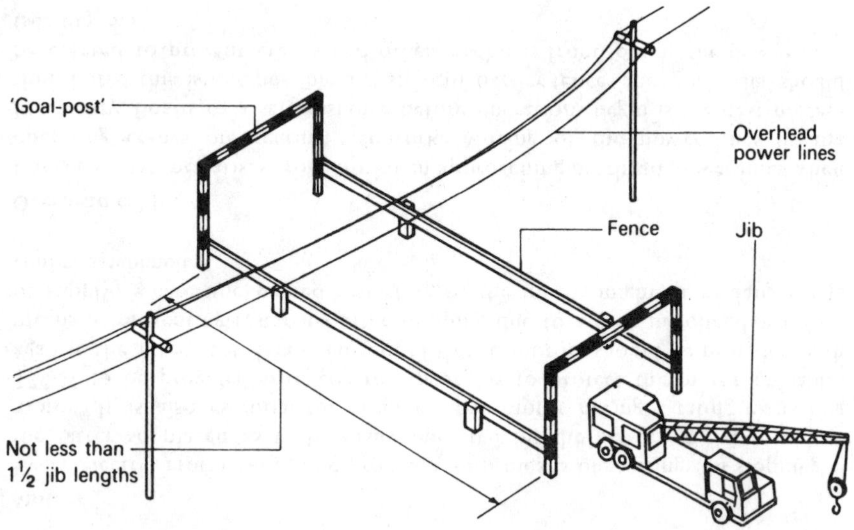
'Goal-post'
Overhead power lines
Fence
Jib
Not less than 1½ jib lengths

Fig.5.1 'Goalposts' to give protection to overhead power lines

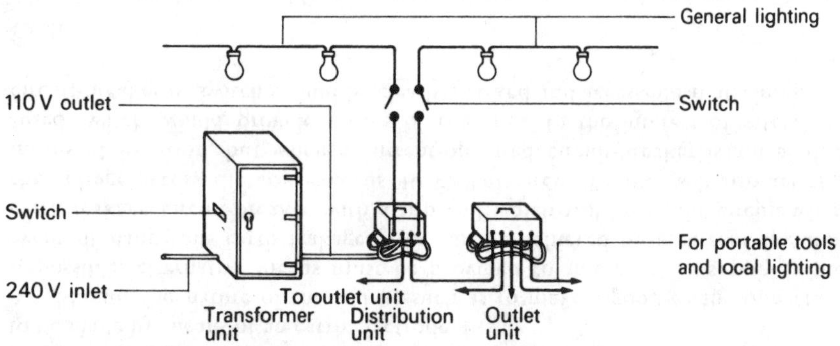
General lighting
110 V outlet
Switch
Switch
For portable tools and local lighting
240 V inlet
To outlet unit
Transformer unit
Distribution unit
Outlet unit

Fig.5.2 Low-voltage distribution

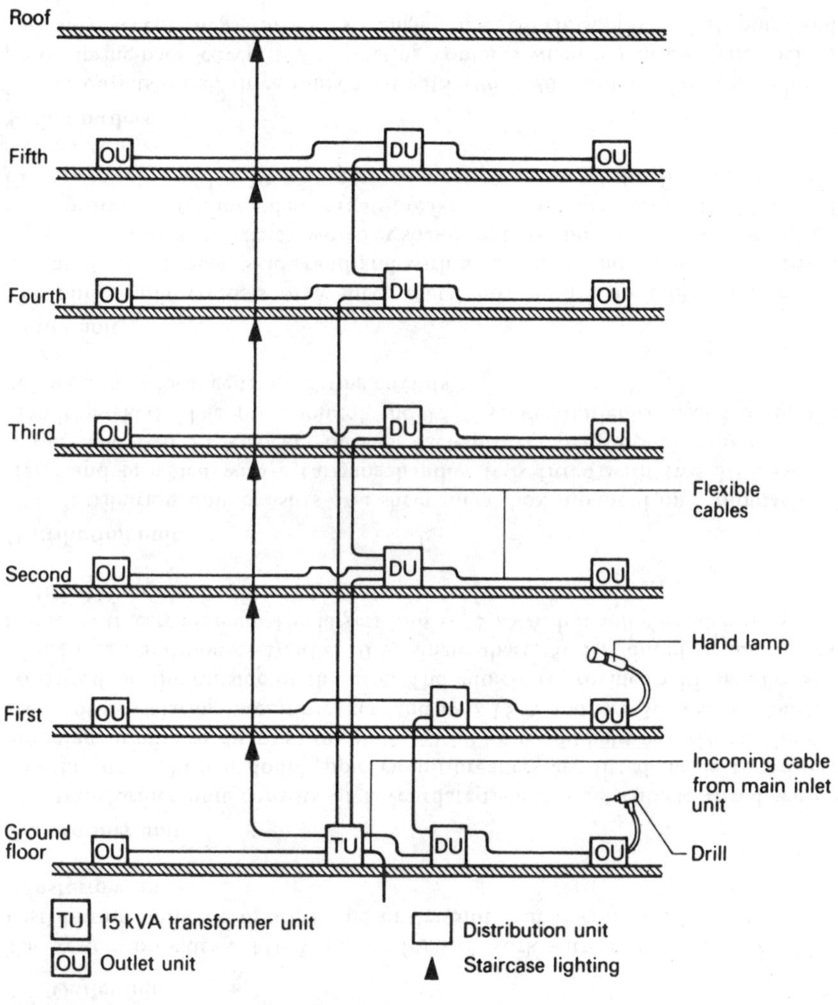
Roof
Fifth
Fourth
Third
Second
First
Ground floor
OU
DU
TU
Flexible cables
Hand lamp
Incoming cable from main inlet unit
Drill
TU 15 kVA transformer unit
OU Outlet unit
Distribution unit
Staircase lighting

Fig.5.3 System for a five-storey building

sockets are engraved for the working voltages and current. Figure 5.3 shows the position of these units for a five-storey building.

Regulations

The 16th edition of the IEE Regulations require that an electrical installation on a construction site, and a temporary installation other than an installation in a private dwelling, shall be in the charge of a competent person who shall accept full responsibility for the safety of the installation and for any alteration or extension. His name and designation shall be prominently displayed close to the main switch or circuit-breaker.

The installation must be tested at intervals of 3 months, or at shorter periods depending upon the particular nature of the installation. The Health and Safety Regulations require that posters giving instructions on the treatment of persons suffering electric shock shall be prominently displayed on every building site where electricity is used, either for light or power.

Summary of recommendations

1. Give adequate notice and information to the Electric Company.
2. Consult the Regional Engineer and the National Grid Company (N.G.C.) to establish the location of underground cables.
3. Use portable tools operated at 110 V.
4. In dangerous situations use battery-operated lamps.
5. Avoid trailing cables along the floors.
6. Use only waterproof equipment which comply with British Standards.
7. Provide protection from overhead power lines.
8. Have the installation regularly inspected and tested by a competent person.

9. Provide protection from earth leakage and excess current.
10. Display first-aid instructions in prominent positions.

Electric space heating

Electric space heating can be broadly classified into direct and indirect systems. Direct systems use appliances such as fires and convector heaters, using electricity at the standard rate; they have the advantage of flexibility, immediate response to heating, and utilise 100 per cent of their input of power.

Indirect or storage systems are operated on 'off-peak' supplies and are therefore cheaper to operate than direct systems but require some time, due to thermal lag, before they can be used for heating. Indirect systems include block-type storage heaters and underfloor heating. The direct and indirect systems may be used independently, or may be complementary to one another to meet any particular heating requirement.

Advantages of electric heating

1. Absence of products of combustion or fumes near the building.
2. Clean and silent.
3. Economical in builders' work, as no fuel storage, boiler room and chimney are necessary.
4. Ease of maintenance and automatic control.
5. Portable appliances may be used.
6. No risk of freezing or leaks.
7. Pumping not required and heating may be achieved at any point regardless of height.

Direct systems

A wide range of direct-heating electrical appliances are available such as fires, convectors, tubular heaters, oil-filled radiators, fan heaters, and high-temperature panels. The appliances can be fixed or portable and some are available in both forms. Direct electric heating are normally more expensive to run than indirect storage heating, but is easier to control. Electric fires, tubular heaters, convectors and overhead radiant heaters, are commonly used in offices, factories, shops and residential buildings. They can also be used for 'topping-up' where storage systems provide background heating. Figures 5.4–5.9 show various types of direct-heating appliances.

Ceiling heating

Its low thermal capacity makes an electrically heated ceiling very flexible and convenient, as it responds quickly to thermostatic and time control. Because the ceiling provides radiant heating, a room is comfortably warm when the air temperature is only 17 °C to 18 °C. When the heating is by convection, an air temperature of 20 °C to 21 °C is required to provide the same level of thermal comfort. The lower air temperature of radiant heating systems means that heat losses from air escaping through open doors and windows is reduced. The occupants of the room also feel fresher.

Ceiling heating is suitable for most well-insulated buildings, including shops, hotels, flats, houses, churches and hospitals. The system consists of a special flexible sheet material, made by coating glasscloth with a conducting silicone elastomer. Each ceiling heating element is fitted with copper strip electrodes along each side and enclosed in an electrically insulating sheath. The heating elements are installed immediately above the ceiling face, usually by stapling them to joists or battens. A layer of mineral wool or glassfibre is laid above the elements to provide insulation (see Figs. 5.10 and 5.11). Connections to the mains electricity supply – and between individual elements – is made from one end of the elements only so that only one supply point is needed in each room. A thermostat is incorporated in the circuit. The power loading depends upon the height of the ceiling and the dimensions of the room, but it is usually between 200 W and 250 W/m^2. The ceiling temperature is between 32 °C and 38 °C, and in order to avoid too high a temperature at head level, the ceiling height should not be less than 2.3 m.

Indirect storage systems

In order to encourage the use of electricity when power stations are operating on low loads, the Electricity Companies offer low-rate tariffs for power consumed during these off-peak times, usually between 23.00 h and 07.00 h. The power consumed is recorded by a white meter and a time switch controls the time when the current is switched ON and OFF. The heat given out by the electrical elements is stored during the time the current is switched ON and this heat is released when the current is switched OFF. The main types of indirect storage appliances are as follows:

1. *Block storage heaters:* (*a*) The heat is stored in refectory blocks contained in an insulated metal casing and given out by radiation and natural convection (see Fig. 5.12).

2. *Block storage heaters:* (*b*) The heat is stored in refectory blocks contained in an insulated metal casing and given out again by radiation and forced air flow produced by a fan (see Fig. 5.13).

3. Warm-air system: A block storage heater with a fan is used and the heat forced by the fan through ductwork to the various rooms (see Fig. 5.14).

4. *Hot-water thermal storage system:* 'Off-peak' electrical energy is converted into heat energy and used to heat water, which is transferred to well-insulated horizontal or vertical thermal storage cylinders (see Fig. 5.15).

The heat generator can be an electrode boiler shown in Fig. 5.16, or an electric immersion heater inside the cylinder. Operation is as follows:

(*a*) The electrode boiler or immersion heater heats the water in the cylinder.
(*b*) This heat takes place during the 'off-peak' period, and in order to reduce the size of the storage cylinder the water temperature is raised as high as possible without generating steam. The water temperature may be as high as 185 °C, depending upon the pressure of water acting upon the cylinder. The water pressure may be increased by the use of a gas cushion inside a cylinder connected to the return pipe.
(*c*) On completion of the heating period, the current to the heater is cut off automatically by a switch and there is sufficient heat energy in the storage cylinder to maintain the building's space heating and hot-water requirements until the next period.

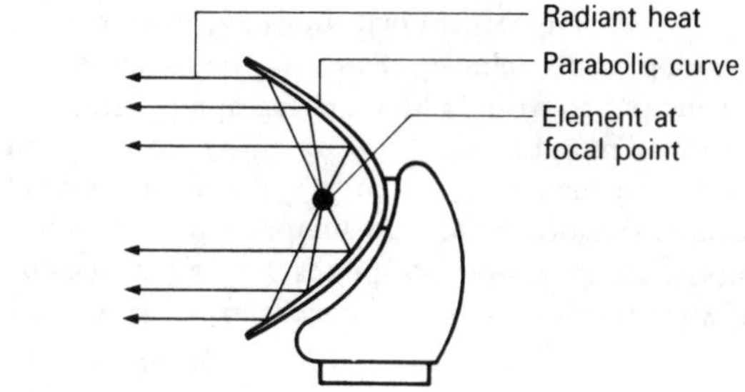
Radiant heat
Parabolic curve
Element at
focal point

Fig.5.4 Electric fire with parabolic reflector

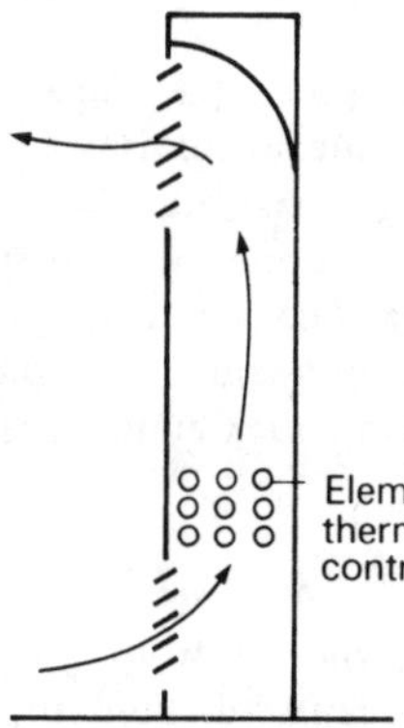
Elements on
thermostat
control

Fig.5.5 Electric natural convector heater

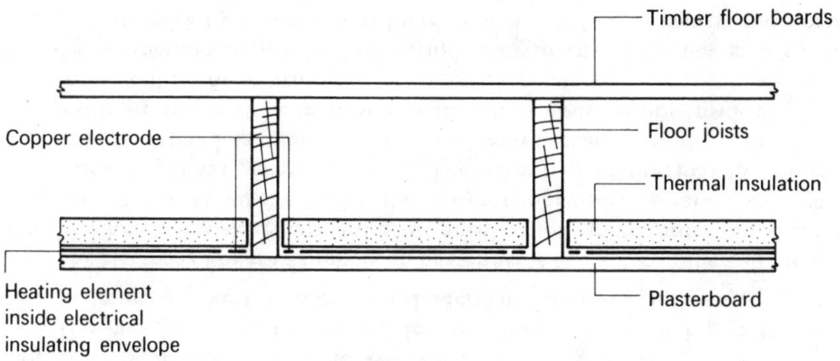
Timber floor boards
Copper electrode
Floor joists
Thermal insulation
Heating element
inside electrical
insulating envelope
Plasterboard

Fig.5.10 Ceiling heating in timber floor

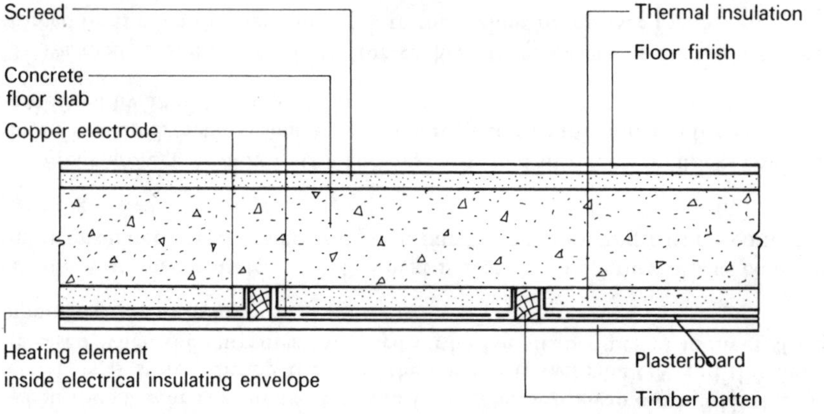
Screed
Thermal insulation
Floor finish
Concrete
floor slab
Copper electrode
Heating element
inside electrical insulating envelope
Plasterboard
Timber batten

Fig.5.11 Ceiling heating in concrete floor

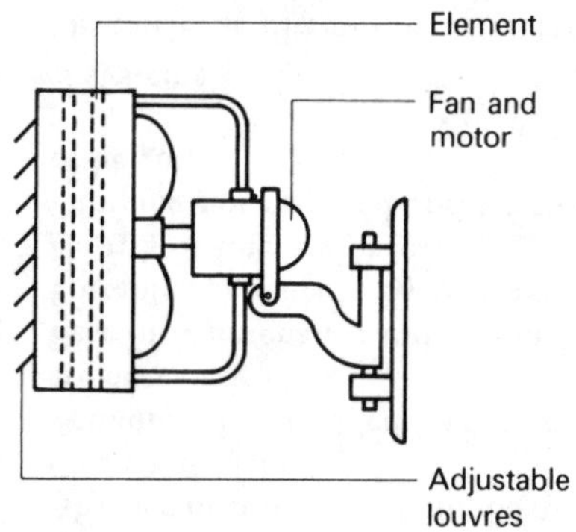
Element
Fan and
motor
Adjustable
louvres

Fig.5.6 Electric unit heater

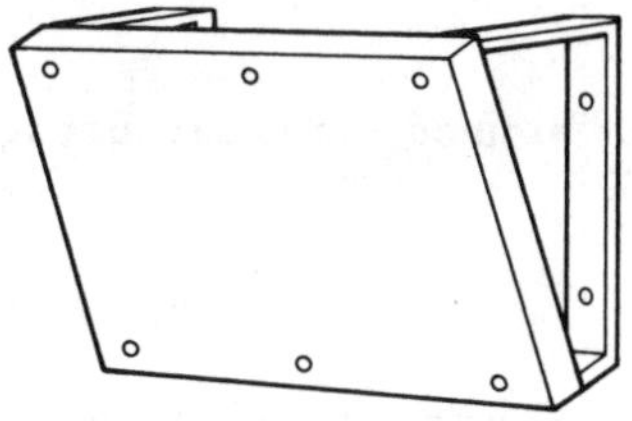

Fig.5.7 High-temperature electric panel heater

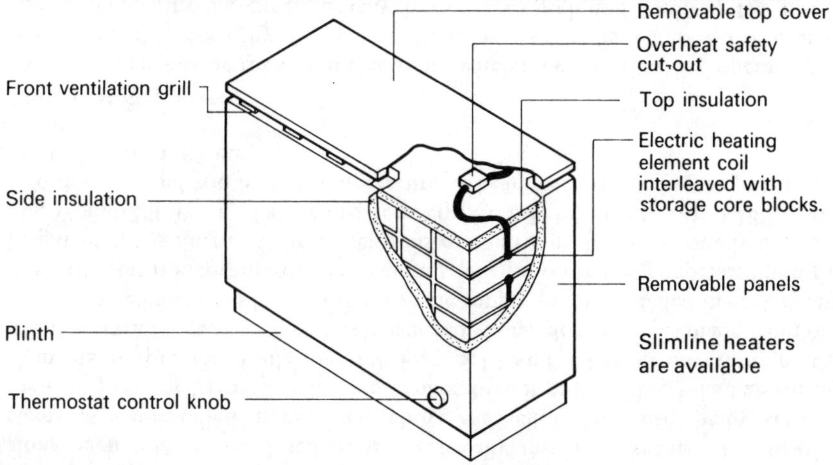
Removable top cover
Overheat safety
cut-out
Front ventilation grill
Top insulation
Electric heating
element coil
interleaved with
storage core blocks.
Side insulation
Removable panels
Plinth
Slimline heaters
are available
Thermostat control knob

Fig. 5.12 Block storage heater (without fan)

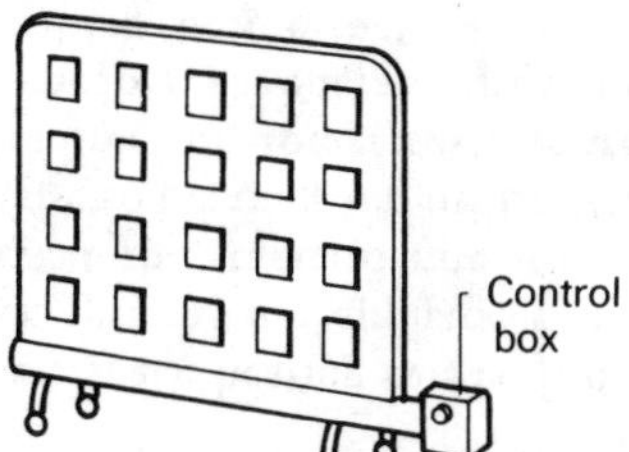
Control
box

Fig.5.8 Oil-filled portable radiator

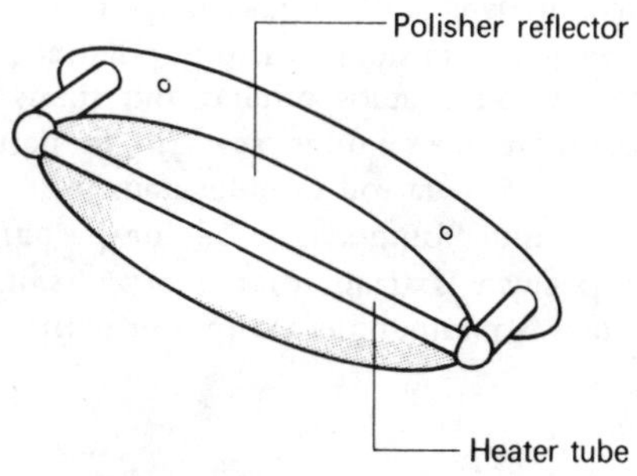
Polisher reflector
Heater tube

Fig.5.9 Wall-mounted infrared heater

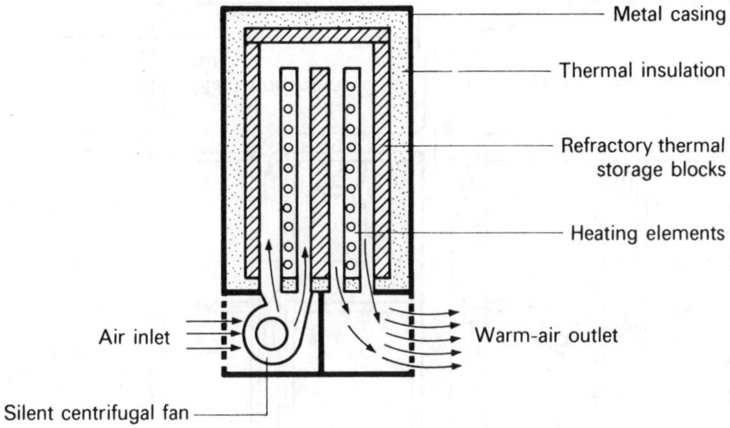

Metal casing
Thermal insulation
Refractory thermal storage blocks
Heating elements
Air inlet
Warm-air outlet
Silent centrifugal fan

Fig .5:13 Block storage heater (with fan)

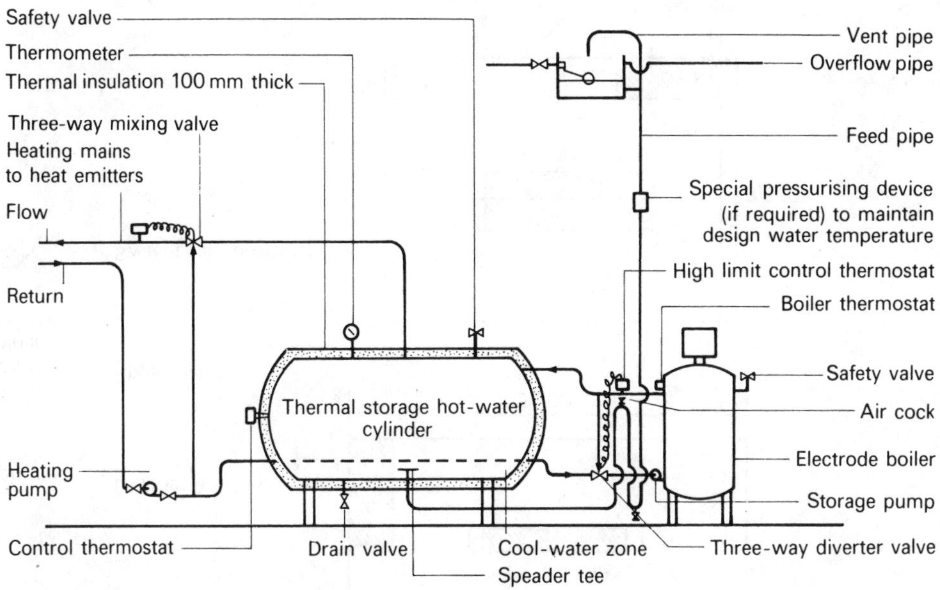

Safety valve
Thermometer
Thermal insulation 100 mm thick
Three-way mixing valve
Heating mains to heat emitters
Flow
Return
Heating pump
Control thermostat
Drain valve
Thermal storage hot-water cylinder
Cool-water zone
Speader tee
Vent pipe
Overflow pipe
Feed pipe
Special pressurising device (if required) to maintain design water temperature
High limit control thermostat
Boiler thermostat
Safety valve
Air cock
Electrode boiler
Storage pump
Three-way diverter valve

Fig.5.15 Hot-water thermal storage system

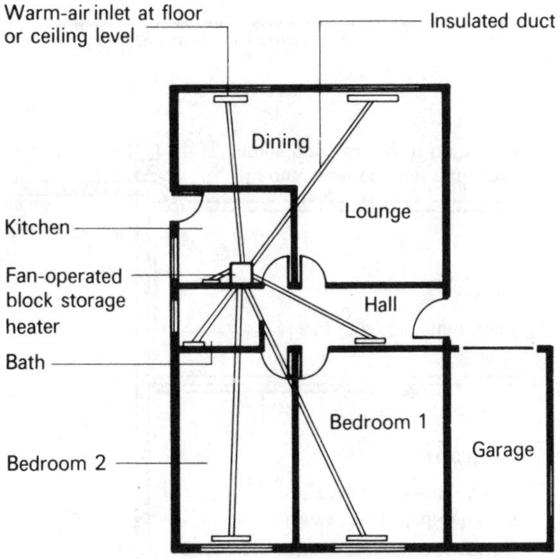

Warm-air inlet at floor or ceiling level
Insulated duct
Dining
Lounge
Kitchen
Fan-operated block storage heater
Hall
Bath
Bedroom 1
Garage
Bedroom 2

Fig.5.14 Electric warm-air heating system (with radial ducts)

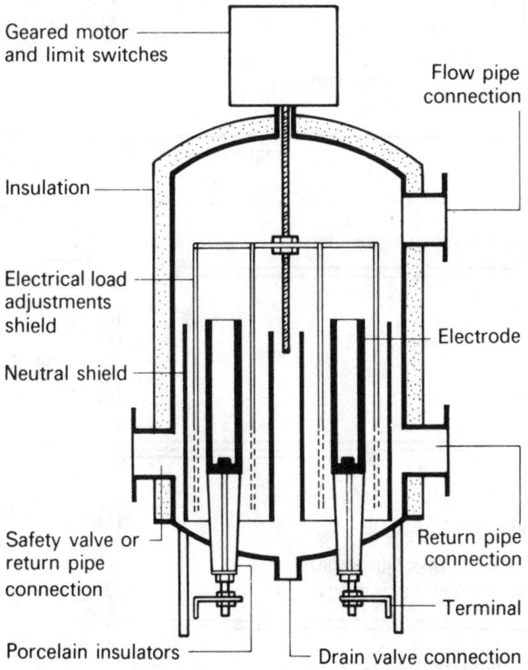

Geared motor and limit switches
Flow pipe connection
Insulation
Electrical load adjustments shield
Electrode
Neutral shield
Safety valve or return pipe connection
Return pipe connection
Terminal
Porcelain insulators
Drain valve connection

Fig.5.16 Electrode boiler

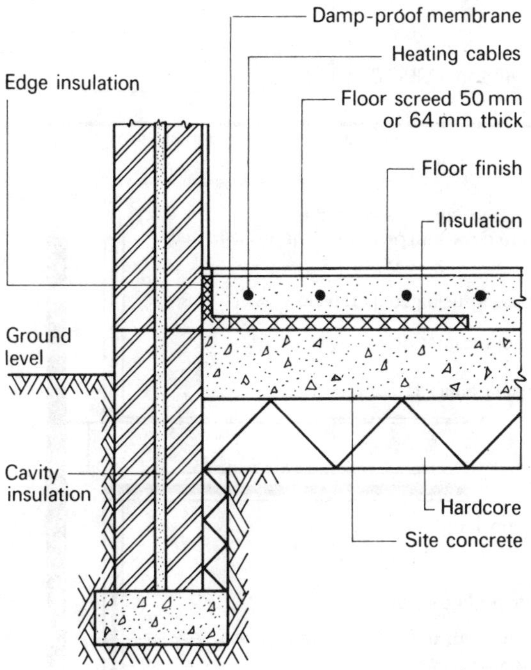

Damp-proof membrane
Heating cables
Edge insulation
Floor screed 50 mm or 64 mm thick
Floor finish
Insulation
Ground level
Cavity insulation
Hardcore
Site concrete

Fig.5.17 Electric underfloor heating

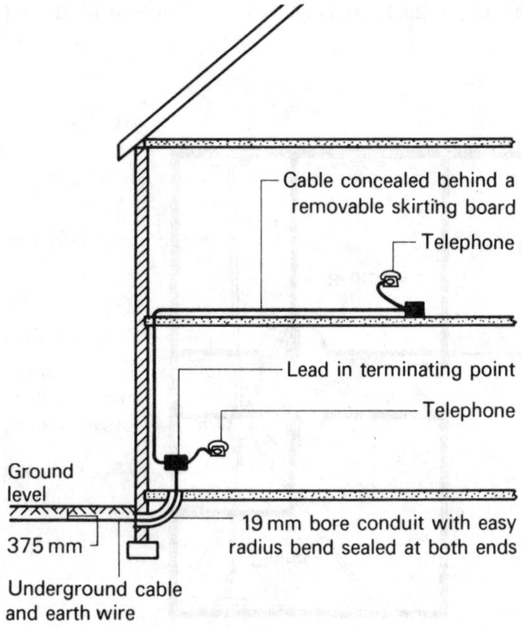
Cable concealed behind a removable skirting board
Telephone
Lead in terminating point
Telephone
Ground level
375 mm
19 mm bore conduit with easy radius bend sealed at both ends
Underground cable and earth wire

Fig.5.18 Underground telephone cable

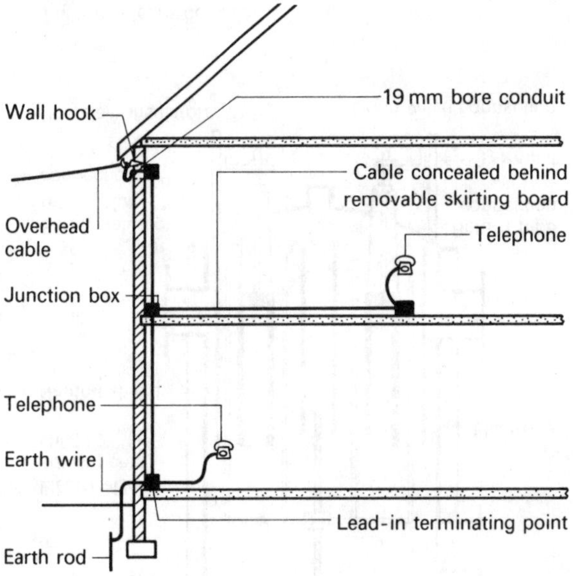
Wall hook
19 mm bore conduit
Cable concealed behind removable skirting board
Overhead cable
Telephone
Junction box
Telephone
Earth wire
Lead-in terminating point
Earth rod

Fig.5.19 Overhead telephone cable

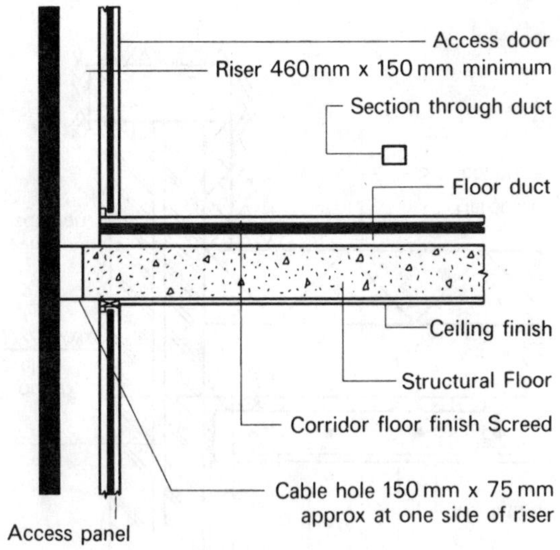
Access door
Riser 460 mm x 150 mm minimum
Section through duct
Floor duct
Ceiling finish
Structural Floor
Corridor floor finish Screed
Cable hole 150 mm x 75 mm approx at one side of riser
Access panel

Fig.5.21 Cross-section through rising duct

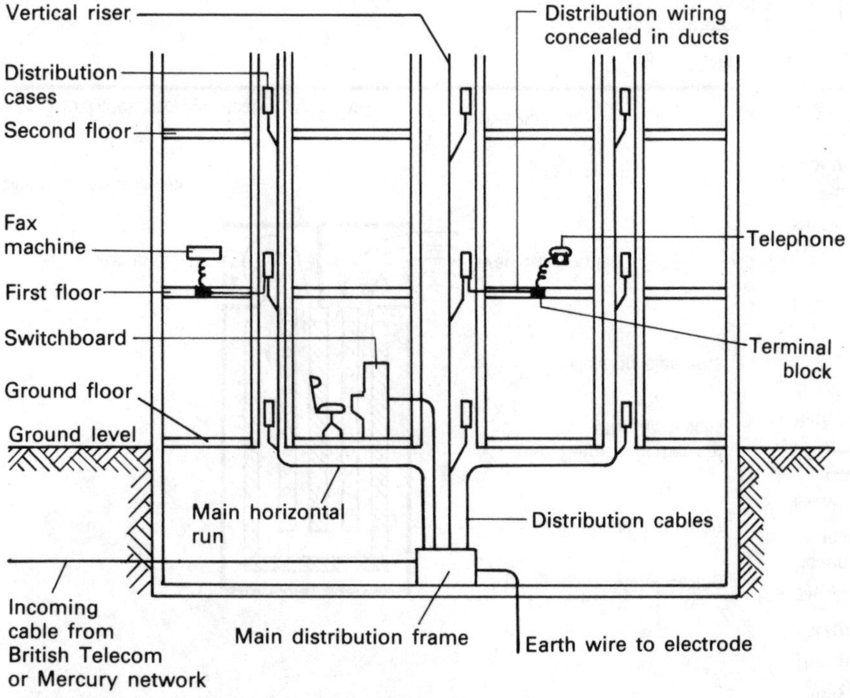
Vertical riser
Distribution wiring concealed in ducts
Distribution cases
Second floor
Fax machine
Telephone
First floor
Switchboard
Terminal block
Ground floor
Ground level
Main horizontal run
Distribution cables
Incoming cable from British Telecom or Mercury network
Main distribution frame
Earth wire to electrode

Fig.5.20 Use of rising ducts in a large building

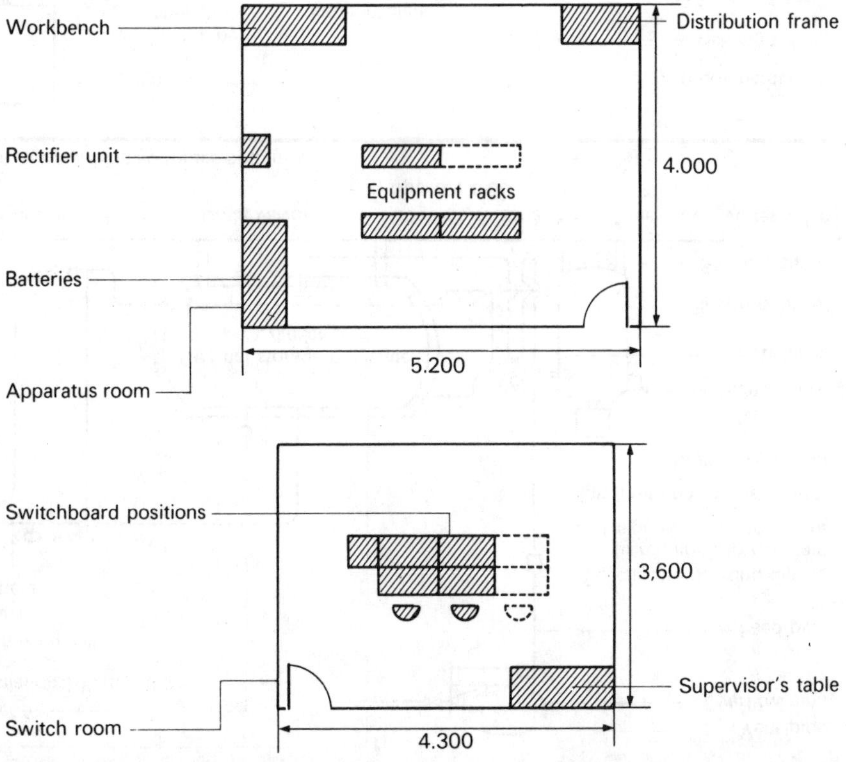
Workbench
Distribution frame
Rectifier unit
Equipment racks
4.000
Batteries
5.200
Apparatus room
Switchboard positions
3,600
Supervisor's table
Switch room
4.300

Fig.5.22 G Floor plans PABX with 100 — 200 lines

5. *Floor-warming system:* Electric heating cables are either directly embedded in the floor screed, or passed through metal conduit which is also embedded in the screed. Most floor-warming systems, employ the direct embedded-type element (see Fig. 5.17). The concrete floor provides the necessary heat storage, but it is essential that insulation is incorporated in the floor slab to prevent loss of heat. Installations are designed to provide an average floor surface temperature of 24 °C and the loading per square metre is between 100 and 150 W. The heating cables should cover the whole floor at spacings not more than 100 m or less than the screed depth, so as to avoid temperature variations on the floor surface.

Telephone installations

The entry of telephone cables into a building should be agreed with the British Telecom engineers at the design stage of the building. For large buildings, a discussion with the Telecom engineers in respect of their requirements for the location of switch boards and other apparatus will ensure that the telephone services provided for the building will be sufficient to meet immediate and future needs.

Note: Water-sprinkler systems should be excluded from rooms containing the main telephone equipment. However, dry powder or carbon dioxide fire-control systems may be used for these rooms.

Telephones for housing estates

The Telephone Manager will require the following information as soon as possible:

1. Name and address of developer.
2. Proposed numbers and types of dwellings.
3. Layout of site.
4. Estimated starting time for site works.
5. Proposed rate of progress of the buildings.
6. The sequence of development.

From this information, the Post Office engineers will prepare plans for the installation of telephone cables. The Telephone Manager will also appoint a liaison officer who will ensure that the telephone installation work goes as smoothly as possible.

Type of telephone arrangement

The simplest arrangement for a public telephone is one instrument connected to an exchange line. If two telephones are required for the dwelling, an extension can be added which is the exact duplicate of the main instrument so that both bells will ring for an incoming call and may be answered from either telephone.

Note: Telephones cannot be provided from a common main, like electricity and water, a pair of wires all the way to the telephone exchange is needed for each telephone.

Underground cables

The main telephone cables are laid in conduits with surface joint boxes. In minor roads, the cables are usually buried directly in the ground under the footpath or verge. The necessary cross-connection between underground cables are provided at cabinets above ground. The cable for the house is brought inside, as close as possible to the position of the telephone at ground-floor level (see Fig. 5.18).

Earthing

Many telephone installations in residential areas, require an earth connection and it is just as important to conceal the earth wiring as it is for other wiring. A conduit may be used if convenient, but it is not essential and the earth wire may be buried in a solid floor or plaster or laid in a cavity wall. It should be run from the lead-in terminating point to the earth connection.

The Telephone Manager will supply a copper earth wire having an area of not less than 1.5 mm^2 and advise on the type and location of the earth connection.

Location of telephone cable

The cable may be laid in the water-pipe trench or in a 'common trench' used for the water, gas and electricity mains. The British Telecom will look into the possibilities of using a shared or common trench with other public utilities.

Overhead cable (see Fig. 5.19)

External distribution for dwellings by overhead cable has the following advantages over underground distribution:

1. An overhead cable is smaller and cheaper than an underground cable.
2. Overhead distribution involves a cable along one side of the road only.
3. It is more convenient to provide additional telephones to existing dwellings, whereas underground cables must be provided for every dwelling when the property is being built. However, underground cables are out of sight and do not spoil the outlook.

Note: Telephone service to large blocks of flats is provided by underground cables because it is more economical.

Telephone installations for large buildings

Distribution cables of suitable sizes are taken from a main distribution frame in the basement to the various floors, and to conceal these cables it is necessary to provide one or more vertical ducts which rise from the basement to the top of the building. To obtain the maximum benefit from the distribution system, each vertical duct should be placed centrally with respect to the distribution area which it is to serve. There is a limit to the number of cables which can be taken from any one horizontal duct from a riser to the telephone instruments. To keep the number of cables within reasonable limits, the number of risers and their siting should be such that no telephone outlet needs more than 328 m of cable to connect it to a riser.

Size of shafts

The minimum internal dimensions of the vertical ducts or risers should be 150 mm. For larger buildings the dimensions will have to be increased to

600 mm x 225 mm, because of the larger cables. An aperture of 150 mm x 75 mm is required to take the cables through a horizontal duct in the structural floor.

Sharing of risers with other services

The telephone cables may be installed in a riser used for other services, providing that the cables are segregated from the electrical mains. Partitions giving full separation are preferred to separating by distance only.

Access to riser

Access to each riser will be necessary on each floor and should be available from a corridor or other common space. Hinged doors are preferable to screw-on panels, especially where several services are using the same riser. Figure 5.20 shows the layout of telephone cables and equipment for a large building, and Fig. 5.21 shows a cross-section through vertical and horizontal ducts required for the building.

Facilities required for private telephone switchboards

The type and size of switchboard and apparatus will depend upon the requirements of the building. Telephones operate on 50 V and are usually operated by batteries which are automatically recharged through the mains supply by an automatic recharger; a rectifier will therefore be needed to change the alternating current to direct current for this purpose. An apparatus room will be required for this and other equipment, and this should be sited as near as practicable to the switchroom. It should be reasonably dust-proof and should have normal heating and ventilating.

The building may require a Private Automatic Branch Exchange (PABX) or a Private Manual Branch Exchange (PMBX), depending upon the size of building and costs. There may be one or more simple switchboards in the case of a small or multi-occupier building, with the associated apparatus in one room. For larger buildings, a large and complex manual or automatic telephone exchange with separate rooms for the switchboards and apparatus will be required (see Fig. 5.22).

Internal telephones

There are two basic types of internal telephones, the push-button type and the dial type. The push-button type involves running a rather large cable to all telephones, but it does not require an exchange. Communication between telephone instruments is effected by lifting the receiver and pushing a button of which there will be one on each instrument, corresponding to each of the other instruments. Where there are more than six telephones the system can be very clumsy and does not give secrecy because conversations can sometimes be overheard from other telephones. Dial telephones employ a Private Automatic Exchange (PAX) system, separate from the Post Office system. The instruments are similar to those used by the Post Office, but are usually less robust in construction.

Private circuit

If a business has two or more locations there will be a lot of conversation, fax and data information transmitted between them. British Telecom can install a private circuit or network to transmit this information which will save time and costs. A fixed annual rent is charged for this service.

Chapter 6

Gas supply and distribution

Gas supply

Natural gas

Almost all the gas used in Great Britain is natural gas from beneath the North Sea. Liquid natural gas at a temperature of −160 °C is imported in special ships from Algeria and is stored in frozen earth underground tanks on Canvey Island and Ambergate in Derbyshire (see Fig. 6.1). These frozen earth tanks are kept hard by the liquid gas at −160 °C contained in them. The tanks are approximately 40 m in diameter and 40 m deep and hold about 21 000 t (tonnes) of liquid natural gas which is 600 times as much gas as they would hold if the gas was not liquefied. The liquid is converted into a gas and transmitted through a pipeline to Leeds, with branches to provide bulk supplies to eight of the twelve Area Gas Boards. This Algerian gas is used with natural gas from beneath the North Sea and it represents about 10 per cent of the gas supply for the UK.

Formation

Plankton and other minute organisms that flourished in the seas, and the trees and plants which grew in the earth's swamps millions of years ago, died and sank to the sea-bed. They were buried and sank deeper below the surface of the sea-bed, became compacted and subjected to increased pressure and temperature. The minute organisms, trees and plants were converted into coal, petroleum and natural gas. Far beneath the bed of the North Sea, the gas is trapped below an impermeable layer of rock and salt. It is tapped, brought ashore by a pipeline and fed into the pipe mains supplying the eight Area Gas Companies in England,

Scotland and Wales. Figure 6.2 shows the rock strata below the North Sea with dome traps for natural gas.

Composition and characteristics

Table 6.1 gives the compositions of natural gas and coal gas and Table 6.2 a comparison of their characteristics.

Table 6.1 Approximate composition of fuel gases

Constituents	Natural gas (%)	Coal gas (%)
Hydrogen	–	45
Methane	94.7	24
Carbon monoxide	–	14
Carbon dioxide	0.05	–
Butane	0.04	–
Propylene	–	3
Ethane	3.00	–
Propane	0.51	–
Hydrocarbons	0.27	–
Oxygen	–	2
Helium	0.03	–
Nitrogen	1.30	9

Table 6.2 Characteristics of natural and coal gas

Natural gas	Coal gas
Gross calorific value 37 MJ/m^3	Gross calorific value 19 MJ/m^3
Relative density (air = 1.00) 0.55	Relative density (air = 1.00) 0.475
Density (kg/m^3) = 0.7 (approximately)	Density (kg/m^3) = 0.6 (approximately)
Contains no sulphur and therefore less corrosion of appliances	May contain minute quantities of sulphur
Contains no moisture and therefore condensation boxes may be omitted	Contains moisture and precautions must be taken to prevent accumulation of water in the gas supply pipe
Contains no carbon monoxide and therefore non-poisonous	Contains about 14 per cent of carbon monoxide and is therefore poisonous
Low flame velocity of 350 mm/s	High flame velocity of 1000 mm/s
May contain dust and filters may be be required	Contains virtually no dust
Requires approximately 9.5 m^3 of air per 1 m^3 of gas for complete combustion	Requires approximately 4.5 m^3 of air per 1 m^3 of gas for complete combustion

Distribution of gas

British Gas has a distribution system linking the consumers' service pipework with the gas manufacturing plant, or the natural gas. The National Transmission System links up the Area Boards' system and also the shore terminals of the natural gas fields (see Fig. 6.3).

Gas burners

When a light is applied to a mixture of gas and air, the flame formed at the point of ignition spreads through the mass of gas and air at a definite maximum speed known as the 'burning velocity' of the gas.

Natural gas contains a high percentage of methane and none of the hydrogen present in coal gas, so it ignites and burns differently. Methane has a slower burning velocity than hydrogen, and since natural gas requires twice the amount of air for complete combustion a special aerated type of burner is required or otherwise the flames would lift off the burner and emit smoke. A natural-gas burner is provided with a small retaining flame from separate parts of the burner. Such a flame – which in itself is quite stable – is called a 'retention flame' and it completely stabilises the main flame, preventing any tendency to lift off the burner.

Figure 6.4 shows a burner without a retention flame and with the gas pressure too low. The velocity of the gas passing through the injector is also too low, thus allowing only half the amount of air required for correct combustion to be mixed with the gas, resulting in a smoky flame.

Figure 6.5 shows a burner without a retention flame, but having the correct gas pressure; the flame is clean, but lifts off the burner and is therefore unstable. Figure 6.6 shows a burner incorporating a retention flame and also the correct gas pressure, thus resulting in a stable, clean flame.

Note: The tendency of natural gas flames to lift off the burner may also be prevented by increasing the number of burner ports and therefore the total flame port area. This slows down the speed of the gas/air mixture and stabilises the flames.

Gas services in buildings

Definitions

Condensate receiver: A receptacle fitted at the lowest part of the gas-line pipe to collect any liquid or deposit which might otherwise form in the pipe.

Control cock: A cock or valve fitted at the inlet of the meter to control the supply of gas from the Board's main to the consumer.

Governor: A regulating device for reducing the high transmission pressures to normal requirements.

Installation pipe: Every pipe between any meter and the point at which the appliance is connected.

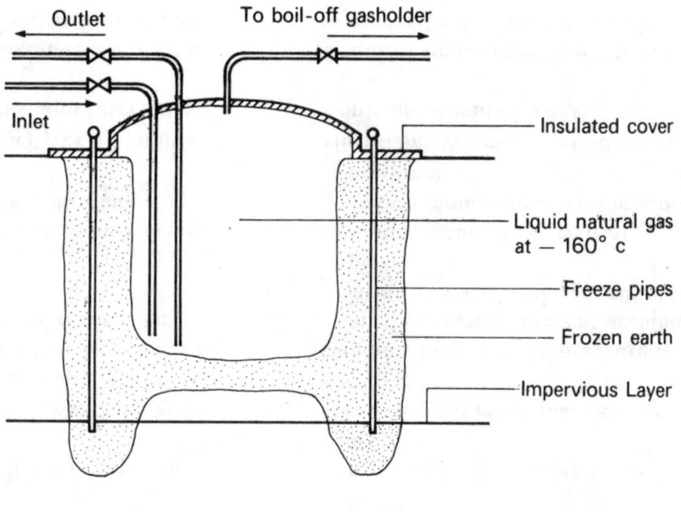
Outlet
To boil-off gasholder
Inlet
Insulated cover
Liquid natural gas
at — 160° c
Freeze pipes
Frozen earth
Impervious Layer

Fig.6.1. Frozen ground liquid gas storage tank

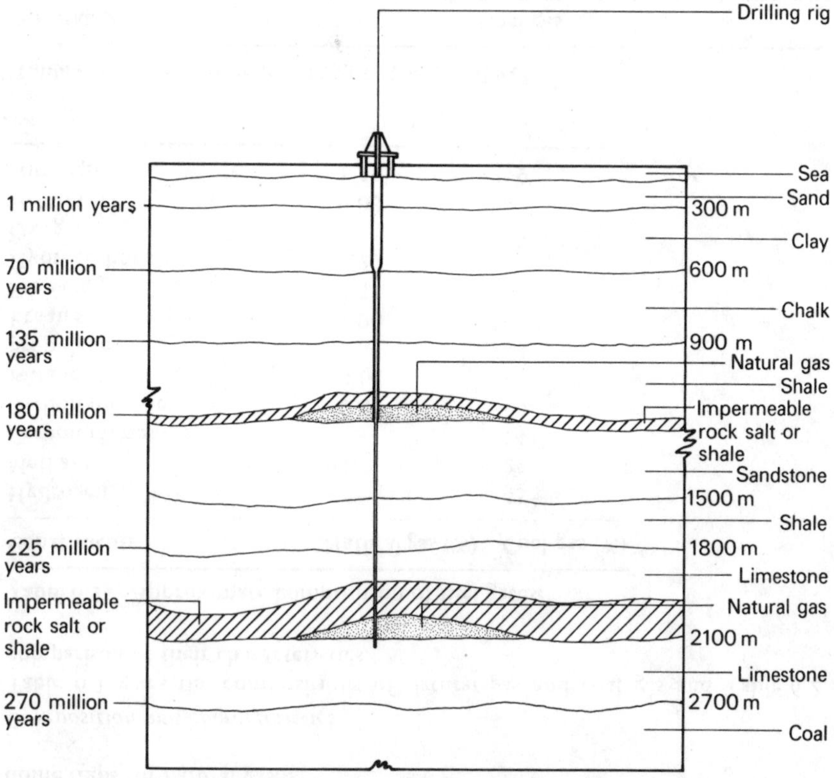
Drilling rig
Sea
Sand
1 million years
300 m
Clay
70 million years
600 m
Chalk
135 million years
900 m
Natural gas
Shale
180 million years
Impermeable rock salt or shale
Sandstone
1500 m
Shale
225 million years
1800 m
Limestone
Impermeable rock salt or shale
Natural gas
2100 m
Limestone
270 million years
2700 m
Coal

Fig.6.2 Rock strata below the North Sea with salt dome traps for gas

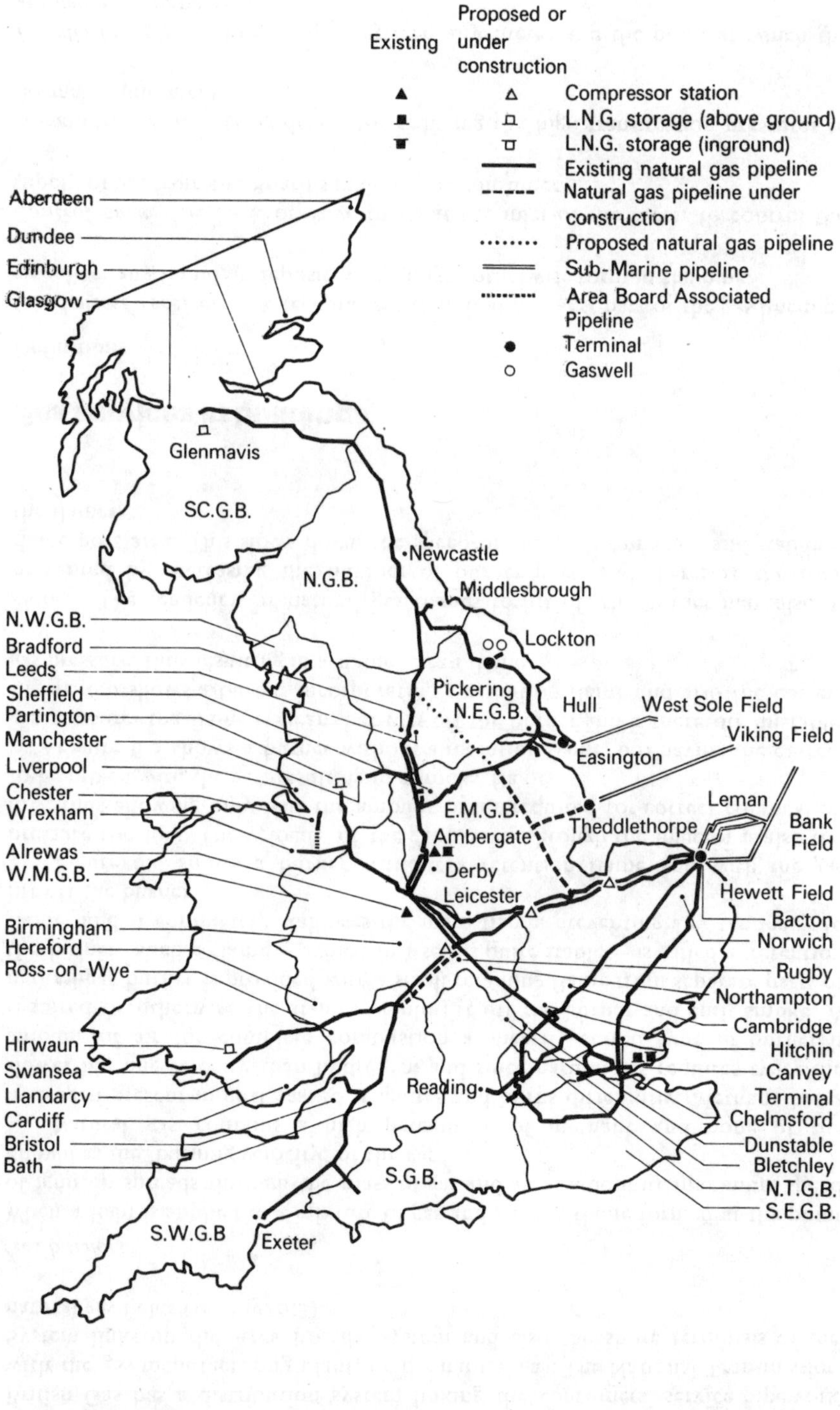
Existing
Proposed or under construction
Compressor station
L.N.G. storage (above ground)
L.N.G. storage (inground)
Existing natural gas pipeline
Natural gas pipeline under construction
Proposed natural gas pipeline
Sub-Marine pipeline
Area Board Associated Pipeline
Terminal
Gaswell
Aberdeen
Dundee
Edinburgh
Glasgow
Glenmavis
SC.G.B.
Newcastle
N.G.B.
Middlesbrough
N.W.G.B.
Bradford
Leeds
Sheffield
Partington
Manchester
Liverpool
Chester
Wrexham
Alrewas
W.M.G.B.
Lockton
Pickering
N.E.G.B.
Hull
West Sole Field
Viking Field
Easington
E.M.G.B.
Ambergate
Theddlethorpe
Leman
Bank Field
Derby
Leicester
Hewett Field
Bacton
Norwich
Birmingham
Hereford
Ross-on-Wye
Rugby
Northampton
Cambridge
Hitchin
Hirwaun
Swansea
Llandarcy
Cardiff
Bristol
Bath
Reading
Canvey
Terminal
Chelmsford
Dunstable
Bletchley
N.T.G.B.
S.E.G.B.
S.G.B.
S.W.G.B
Exeter

Fig.6.3 The National Transmission System

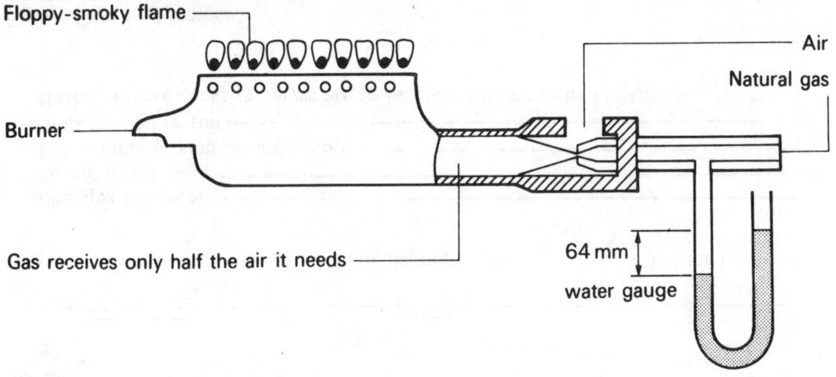

Floppy-smoky flame
Air
Natural gas
Burner
Gas receives only half the air it needs
64 mm
water gauge

Fig.6.4 Gas pressure too low

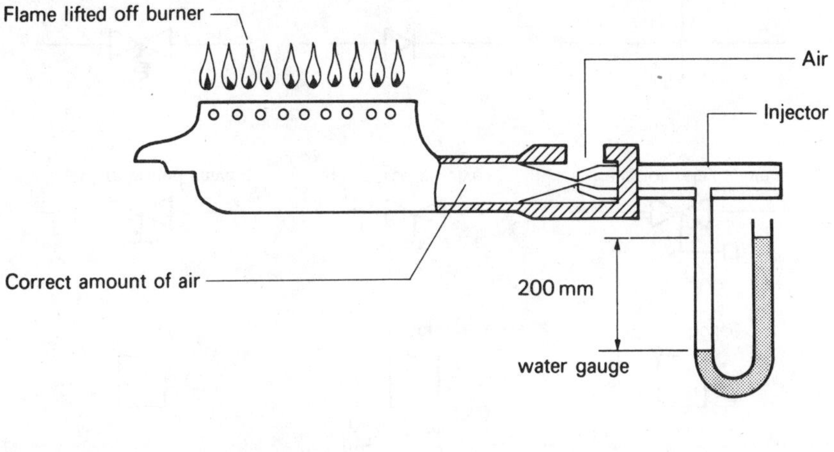

Flame lifted off burner
Air
Injector
Correct amount of air
200 mm
water gauge

Fig.6.5 Gas pressure correct but no retention flame

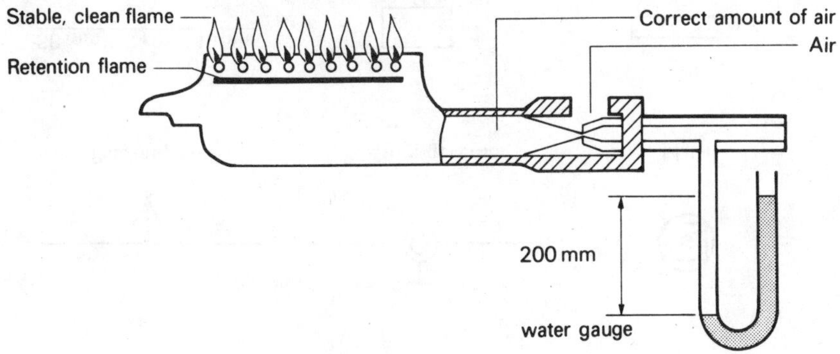

Stable, clean flame
Correct amount of air
Air
Retention flame
200 mm
water gauge

Fig.6.6 Gas pressure correct and retention flame incorporated in the burner

Meter compartment: An enclosure in which one or more meters are accommodated. The compartment should be well ventilated and if situated on a public stairway or corridor, be constructed of fire-resisting materials.

Meter bypass: An arrangement of pipes, and cocks or valves whereby the gas may be led round instead of through the meter. Used in schools, factories and hospitals to ensure a continuous supply of gas should it be necessary to disconnect the meter.

Ordinary meter: A meter which provides a supply of gas for which the amount registered is periodically charged to the customer on a monthly or quarterly basis.

Prepayment meter: A meter fitted with a mechanism which, on insertion of a coin, permits the passage of the prepaid quantity of gas.

Pressure gauge: An instrument used for finding the pressure of gas – usually a U tube containing water.

Pressure point: A small plug-type fitting on a gas pipe to which a pressure gauge may be attached for taking pressure readings.

Primary meter: A meter connected to the service pipe, the index reading of which constitutes the basis of the charge by the Board for gas used on the premises, and where there may or may not be a secondary or subsidiary meter.

Secondary meter: A subsidiary meter for measuring gas used on separate parts of premises or on separate appliances, the whole of which has passed through the primary meter.

Secondary supply pipe: The pipe connecting the inlet of a secondary meter to the outlet pipe from a primary meter.

Service pipe: The pipe between the gas main and the customer's control cock (normally at the meter). In large buildings with more than one tenancy, there may be several service pipes entering at different points in the building.

Service valve: A valve inserted in the service pipe close to the site boundary. The valve allows the supply of gas to the building to be turned off in an emergency. A key to operate the valve must be kept available.

Design considerations

Consultation with the Gas Board in the early stages of design and planning of a building is essential to ensure that adequate provision is made for the gas services. All necessary information regarding the position of the meter, controls and installation points to serve the appliances should be made as early as possible by means of drawings and specifications. Figure 6.7 shows the graphical symbols used for gas installation drawings.

The drawings should show the following:

1. The position of ducts, chases and channels required when the pipes are to be concealed.

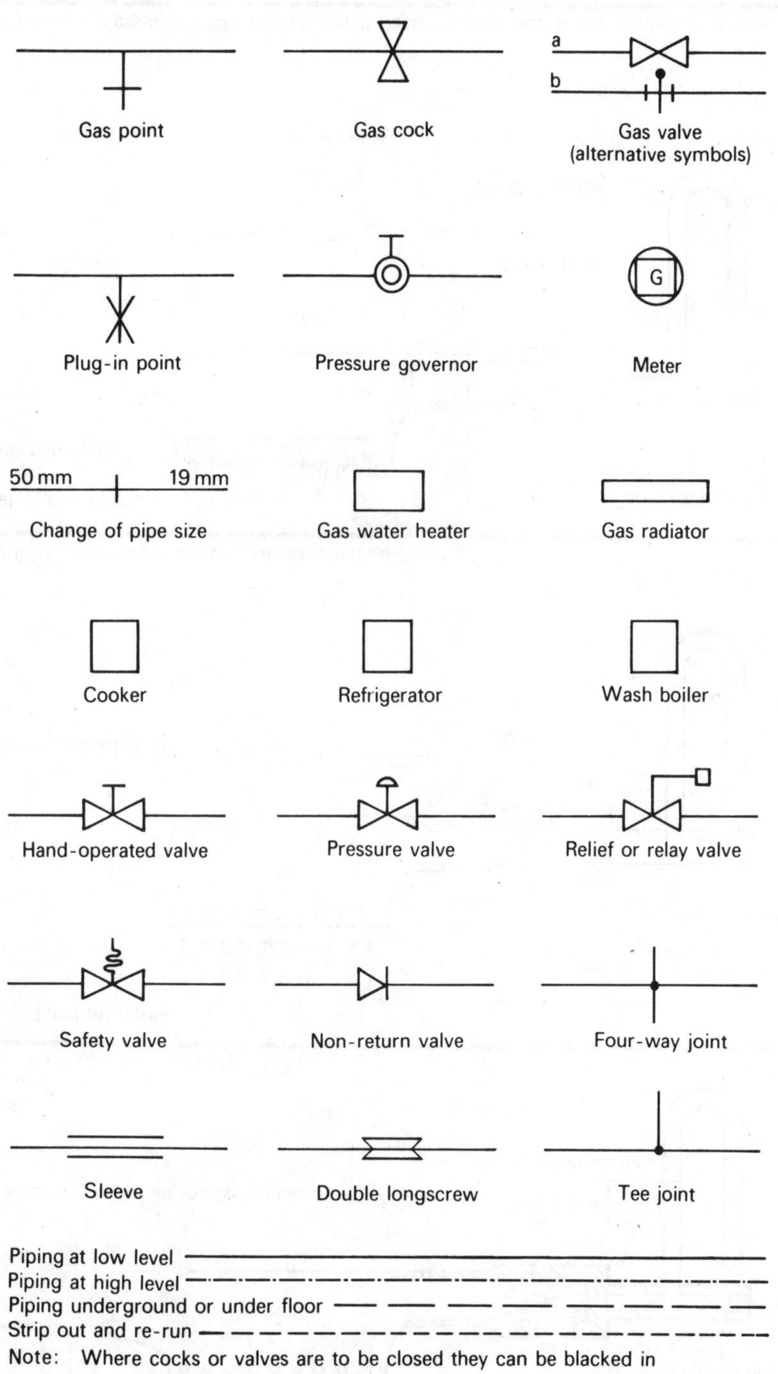
a
b
Gas point
Gas cock
Gas valve
(alternative symbols)
G
Plug-in point
Pressure governor
Meter
50 mm
19 mm
Change of pipe size
Gas water heater
Gas radiator
Cooker
Refrigerator
Wash boiler
Hand-operated valve
Pressure valve
Relief or relay valve
Safety valve
Non-return valve
Four-way joint
Sleeve
Double longscrew
Tee joint
Piping at low level
Piping at high level
Piping underground or under floor
Strip out and re-run
Note: Where cocks or valves are to be closed they can be blacked in

Fig.6.7 Graphical symbols

2. The position of control cocks or valves.
3. The routes and diameters of all installation pipework.

Note: Except where a pipe has to be exposed on the internal surfaces, or laid in prepared ducts or channels permitting access, pipes should be tested as the erection of the building proceeds.

Pipe positions

Gas pipes should not be fitted inside electrical intake chambers, transformer rooms, lift shafts, refrigerator chambers or inside the space in a cavity wall. When a gas pipe runs close to some other service, contact between them should be prevented by a spacing or insulation.

Diameter of pipes

The diameter of a gas pipe depends upon its length, frictional loss due to fittings, amount of gas required and the permissible pressure required at the appliance. It should be not less than 25 mm nominal bore for a single family dwelling. If the pipe supplies more than one meter, the capacity of the pipe should be at least equal to the capacity of a 25 mm diameter pipe for each primary meter supplied.

In the case of a block of flats where a single service pipe may supply a considerable number of dwellings, consideration should be given to the effect of a diversity factor. Table 6.3 gives the probable diameter of the main service pipe to small gas meters.

Table 6.3 Internal diameter of main service pipe

Internal diameter of pipe (mm)	Number of primary meters supplied
25	1
32	2
40	4–6
50	6 or more

For a more accurate assessment of the pipe size for an installation Table 6.4 should be used in conjunction with Tables 6.5 and 6.6.

Table 6.4 Discharge rates through steel pipes

Nominal bore (mm)	Length of pipe in metres									
	3	6	9	12	15	18	21	24	27	30
6	0.1									
8	0.1	0.3								
10	1.6	1.3	0.85							
15	3.0	2.1	1.71							
20	6.5	4.5	3.7	3.0	2.8	2.5				
25	13.3	9.4	7.6	6.8	6.0	5.4	5.0	4.8	4.5	
32	26.3	18.4	15.3	13.3	11.9	10.7	10.0	9.3	8.7	8.2
40	36.8	26.0	20.7	18.4	16.4	15.0	13.8	13.0	12.0	11.6
50	73.6	53.8	42.5	36.8	34.0	31.0	28.0	26.3	25.0	23.8
	Discharge (m^3/h)									

Table 6.5 gives the typical gas consumption at appliances and Table 6.6 the additions to the net pipe run for fittings.

Table 6.5 Gas consumption of appliances

Appliance	Consumption (m^3/h)
Refrigerator	0.10
Wash boiler	1.13
Sink water heater	2.30
Instantaneous multi-point water heater	5.70
Cooker	3.70
Warm-air heater	2.30

Table 6.6 Additions to pipe run

Nominal bore	Elbows	Tees	90° bends
15–25	0.6	0.6	0.3
32–40	1.0	1.0	0.3
50	1.6	1.6	0.6

Example 6.1. *Find the internal diameter of a gas pipe to supply two sink water heaters, one instantaneous multi-point water heater and two cookers. The net length of the pipe required is 20 m and there will be six 90° bends and four tees in the pipe run.*

Gas consumption (Table 6.5).

Sink water heaters = 2 x 2.3 = 4.6 m^3/h

Instantaneous multi-point water heaters = 5.7 m^3/h

Cookers = 2 x 3.7 = 7.4 m^3/h

Total 17.7 m^3/h

Resistance of pipe fittings assuming a 50 mm diameter pipe (see Table 6.6).

90° bends = 6 x 0.6 = 3.6 m

Tees = 4 x 1.6 = 6.4 m

Total 10.0 m

Total length of pipe = 20 + 10 = 30 m

Table 6.4 shows that a 50 mm diameter pipe 30 m long will discharge 23.80 m^3/h and is therefore adequate.

Domestic installations

The service pipe should be laid at least 375 mm below ground and if possible with a fall towards the main of 1 in 120. At the point of entry into the building a 76 mm diameter sleeve should be provided for the pipe. The service pipe should be set centrally in the sleeve by means of a bituminous or similar non-setting material filling the space between the pipe and the sleeve throughout the length of the sleeve. The space between the sleeve and the wall should be filled with cement mortar throughout the thickness of the wall. A service pipe, when laid with screwed joints, should be jointed to the main by a connector or coupling for the purpose of facilitating connection or disconnection.

The use of sharp bends or square elbows should be avoided and the pipe should not be restricted by bending or cutting. It is not usual practice to insert a service valve on the service pipe for domestic installations.

Figure 6.8 shows the method of laying the service pipe when it is possible to obtain a fall towards the main. Figure 6.9 shows the method of laying a service pipe when it is not possible to provide a fall from the entry into the building and the gas main. A condensate receiver should be fitted at the lowest point of the service pipe so that any water in the pipe may enter the receiver.

Natural gas does not contain moisture and therefore water from condensation of the gas will not be present. However, it may be necessary to fit a condensate receiver to prevent the service pipe from flooding from extraneous causes, such as water from a broken water main entering a broken gas main. The receiver is emptied by a pump connected to the suction pipe. When a condensate receiver is fitted, a notice to this effect should be mounted on the service pipe adjacent to the consumer's control cock.

The service pipe may be of mild steel to BS 1387, Grade B (medium) with screwed joints protected from corrosion with bituminous paint or hessian wrapping. A PVC pipe may be used from the main to the point of entry into the building, and steel pipe connected to this for entry into the building up to the control cock.

Position of meter

The gas safety regulations state the following:

1. No person shall install a meter on or under a stairway or in any other part of a building with two or more floors above the ground floor, where the stairway or that other part of the building provides the only means of escape in case of fire, unless the meter replaces an existing meter and sub-paragraph (a) or (b) of paragraph (2) below is complied with.
2. No person shall install a meter in any building with no more than one floor above the ground floor on or under a stairway or in any other part of the building where the stairway or that other part of the building provides the only means of escape in case of fire, unless:

   (*a*) The meter is (i) of fire resisting construction, or (ii) housed in a compartment with automatic self closing doors and which is of fire resisting construction, or
   (*b*) The pipe immediately upstream of the meter or, where a governor is adjacent to the meter, immediately upstream of that governor, incorporates a device designed to cut off automatically the flow of gas if the temperature of the device exceeds 95°C.

Figures 6.10, 6.11 and 6.12 show meter positions.

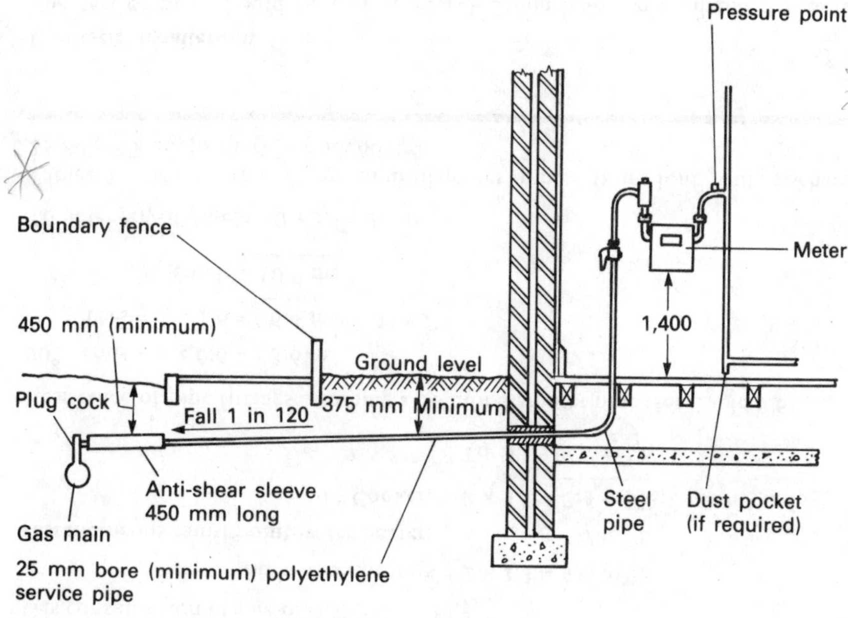

Pressure point
Boundary fence
Meter
450 mm (minimum)
1,400
Ground level
Plug cock
Fall 1 in 120
375 mm Minimum
Anti-shear sleeve
450 mm long
Steel
pipe
Dust pocket
(if required)
Gas main
25 mm bore (minimum) polyethylene
service pipe

Fig.6.8 Normal service pipe installation

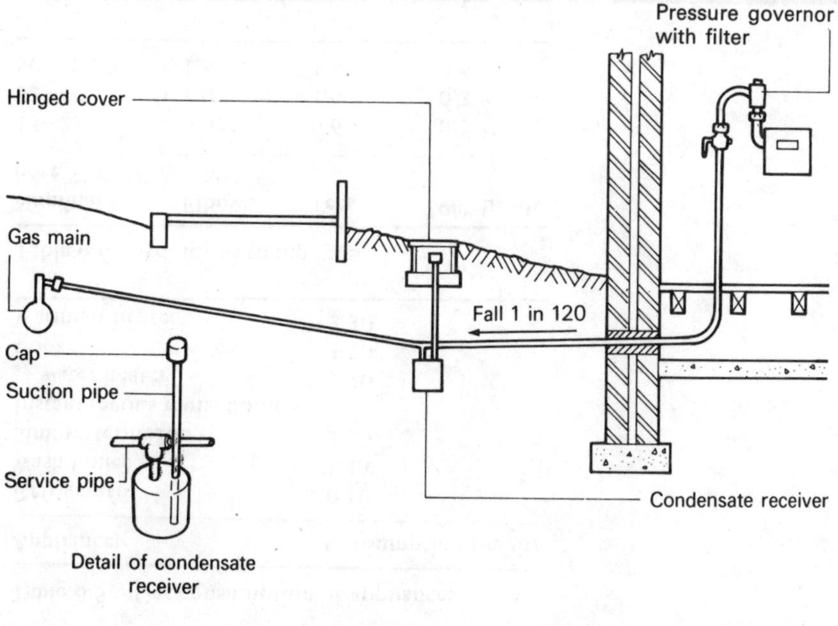

Pressure governor
with filter
Hinged cover
Gas main
Fall 1 in 120
Cap
Suction pipe
Service pipe
Condensate receiver
Detail of condensate
receiver

Fig.6.9 Service pipe with condensate receiver

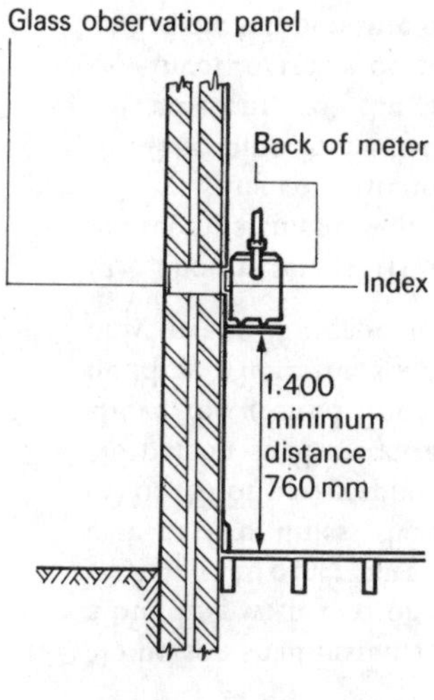

Glass observation panel
Back of meter
Index
1.400
minimum
distance
760 mm

Fig.6.10 Meter in dry, well-ventilated outbuilding

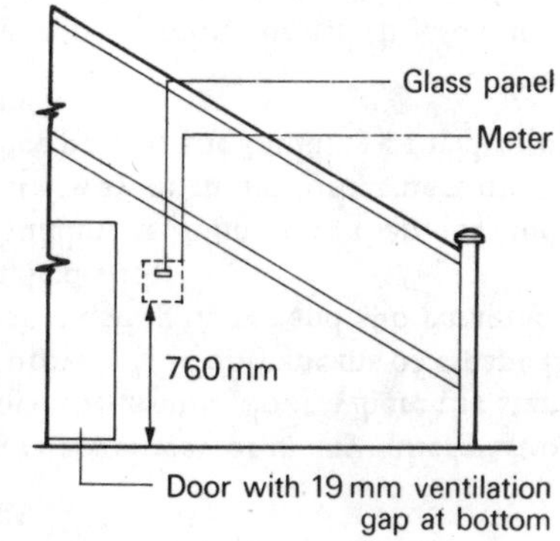

Glass panel
Meter
760 mm
Door with 19 mm ventilation
gap at bottom

Fig.6.11 Meter under staircase

Note: The polyethylene pipe and the steel pipe are joined together with a connector just below ground level.

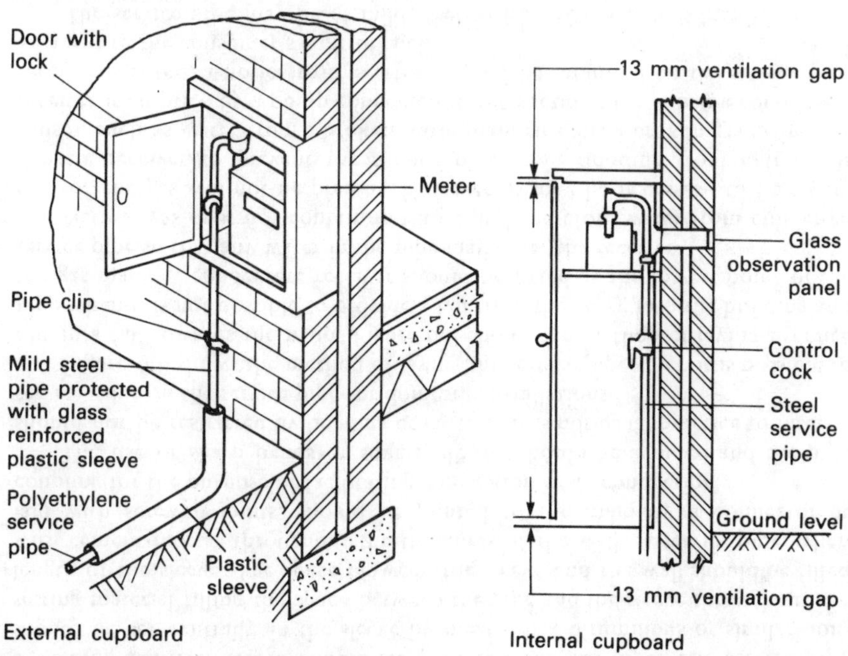

Door with
lock
Meter
Pipe clip
Mild steel
pipe protected
with glass
reinforced
plastic sleeve
Polyethylene
service
pipe
Plastic
sleeve
External cupboard
13 mm ventilation gap
Glass
observation
panel
Control
cock
Steel
service
pipe
Ground level
13 mm ventilation gap
Internal cupboard

Fig.6.12 Use of meter cupboard

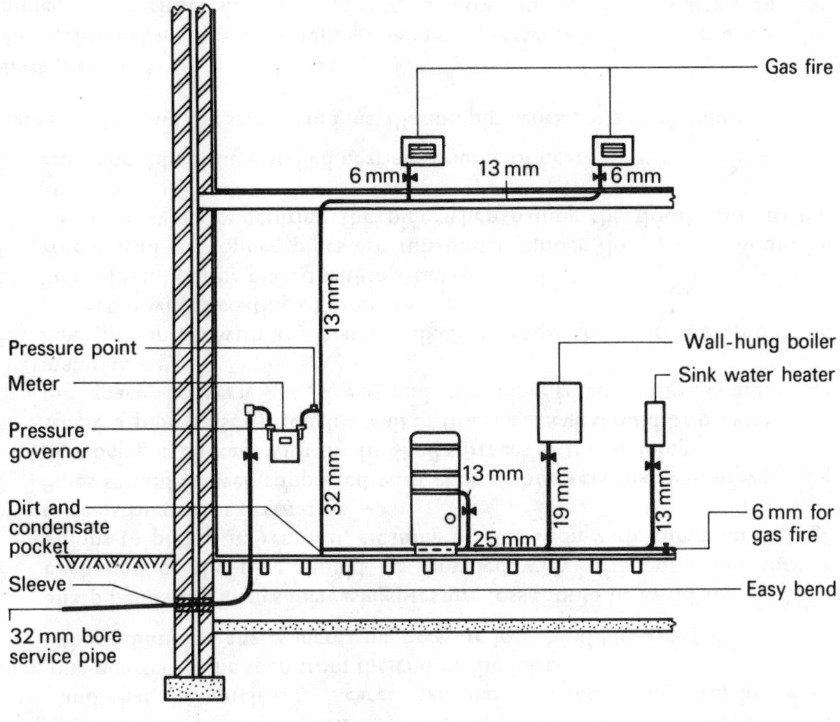

Gas fire
6 mm
13 mm
6 mm
13 mm
Pressure point
Meter
Pressure governor
Dirt and condensate pocket
Sleeve
32 mm bore service pipe
Wall-hung boiler
Sink water heater
32 mm
13 mm
19 mm
13 mm
25 mm
6 mm for gas fire
Easy bend

Fig.6.13 Gas installation for a small house

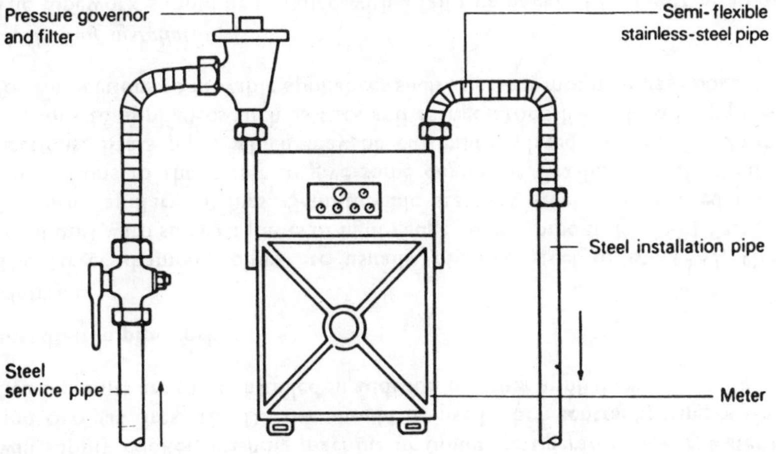

Pressure governor and filter
Semi-flexible stainless-steel pipe
Steel installation pipe
Steel service pipe
Meter

Fig. 6.14 Traditional meter installation

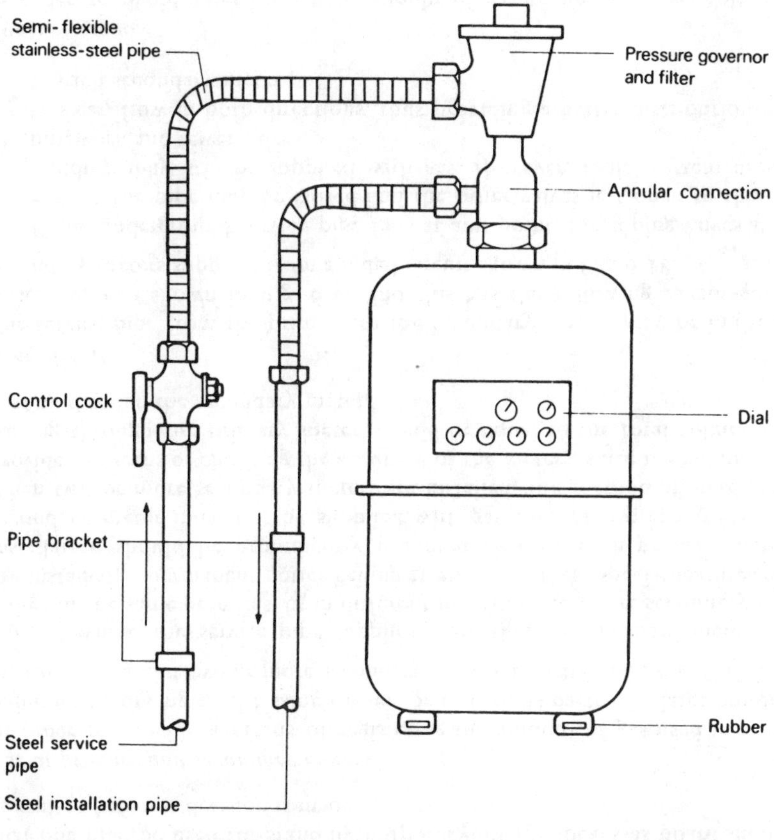

Semi-flexible stainless-steel pipe
Pressure governor and filter
Annular connection
Control cock
Dial
Pipe bracket
Rubber
Steel service pipe
Steel installation pipe

Fig. 6.15 Meter with single annular connection

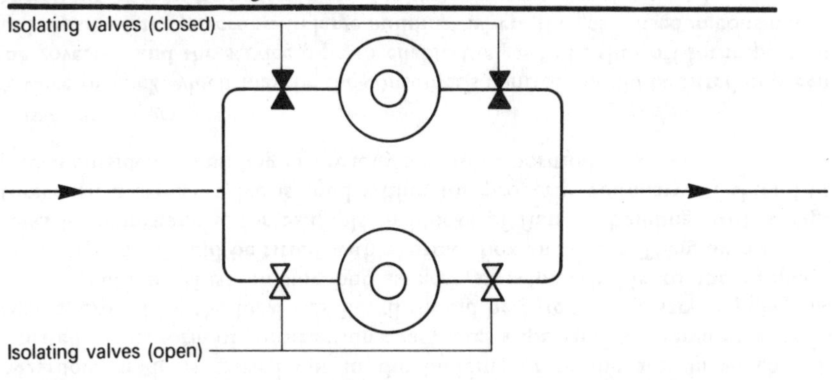

Isolating valves (closed)
Isolating valves (open)

Fig.6.16 Duplicated pressure governors

Types of meters

Domestic meters are available in two basic sizes, D1 and D2. The normal D1 size will supply cooker, washing machine or boiler, refrigerator, poker, water heater and two gas fires. The D2 size should be used when central heating or warm-air heating units are to be installed in addition to other appliances.

Installation pipework

Materials

The larger diameter pipes are usually of mild steel to BS 1387, Grade B (medium) with screwed joints or light-gauge copper pipe to BS 2871 Part I Table X, with capillary fittings. Semi-flexible stainless steel pipe is used for short connections to the meter to give some degree of flexibility at the meter connections. Brass pipe, which may be chromium plated, is used for short connections to appliances such as fires and refrigerators. Flexible steel pipe is used for connections to movable appliances such as castor-mounted gas cookers.

Method of installation

The pipework should be installed with a fall to a pocket fitted with a tap or plug to facilitate the removal of any condensation. A pressure test point should be fitted in a convenient position on or near to the outlet of the meter as possible. Pipes should be laid between and parallel with the floor joists and provided with good supports. Where this is not possible, they may be laid across the joists, provided that the depth of notch does not exceed one-sixth of the depth of the joists and that the notch is no nearer than one-sixth the span, or further away than one-quarter of the span from the end of the joists.

Other points and angles increasing pressure loss should be avoided.

1. Sharp bends and angles increasing pressure loss should be avoided.
2. Each end of the pipe should be provided with sufficient connectors or unions to permit its removal, cleaning or alteration with minimum damage to the structure or decoration.
3. Pipes should be well supported with clips or brackets that will prevent the pipe being in contact with the finished surfaces of the building.
4. Where a pipe passes through a wall or floor, a sleeve should be provided and the space between the sleeve and the pipe filled with incombustible material.
5. Pipes in contact with any material likely to cause their corrosion should be protected with a coating of bitumen.
6. Pipes should not be placed within a cavity.
7. Where lead or copper pipes are run under timber floors, care should be taken to avoid puncturing the pipe when nailing the floorboards to the joists.
8. Pipes should not be installed near any source of heat.

Figure 6.13 shows the service and installation pipework for a small house.

Meter pipework

The traditional method of installing the meter pipework is by an inlet and outlet connection at either side of the meter, as shown in Fig. 6.14. A single annular connection may also be used which permits a more compact installation (see Fig. 6.15).

Installations for large buildings

Service pipe and valve

Where the service pipe has an internal diameter of 50 mm or more, or where a hazardous trade is carried out in the building or in the area in which it is situated, or its form of construction constitutes a special risk, a service valve of a type approved by the local Gas Board should be fitted in the service pipe. The valve should be sited outside but as near as is practicable to the property boundary and should be fitted with a surface box and cover. Then, by reason of special circumstances, for example in blocks of flats or buildings with a large forecourt, a service valve is sited within the property boundary. It should be placed outside the building in a readily accessible position.

Service governors

A valve or cock which may be the consumer's control, should be fitted between the governor and the service pipe to enable the gas to be shut off for inspection and repair of the governor. In large buildings where the gas is used in continuous operations, it may be necessary to duplicate governors as shown in Fig. 6.16, so that one may be used for stand-by purposes, but the local Gas Board should be consulted before a decision is made.

Use of primary and secondary meters

In order to provide a means of registering the amount of gas used in separate buildings belonging to the same owner, or in various parts of a large building, a primary meter and two or more secondary meters are fitted (see Fig. 6.17).

Note: Where the service pipe supplies more than one primary meter, fitted either in the same premises, or in different premises and where secondary meters are installed, a permanent notice calling attention to these special features of the installation should be prominently mounted close to each primary meter. A secondary meter must not be supplied with gas through a prepayment meter. When two or more secondary meters are installed, this method of metering also provides a means of checking the accuracy of the meters, since the volume of gas passing through the primary meter should be equal to the total volume of gas passing through the secondary meters.

Ring circuit

The service pipe from the primary to the secondary meters may be in the form of a ring, also shown in Fig. 6.17 and this has the following advantages over secondary meters supplied from a radial circuit shown in Fig. 6.18:

1. If the underground service pipe fails at any point, the zoning valves at each end of the pipe may be closed and the other valves left open so that each building may still be supplied with gas. However, radial system uses less pipework and fewer valves.
2. Gas can flow in both directions, thus providing a better distribution of gas to each secondary meter.

Bypass to meter

In order to avoid interrupting the supply of gas to hospitals, schools, office blocks and factories when it is necessary to take out the meter for replacement, a system of pipes and valves (known as a bypass) should be fitted. The pressure governor should control both the main supply and the bypass supply and the

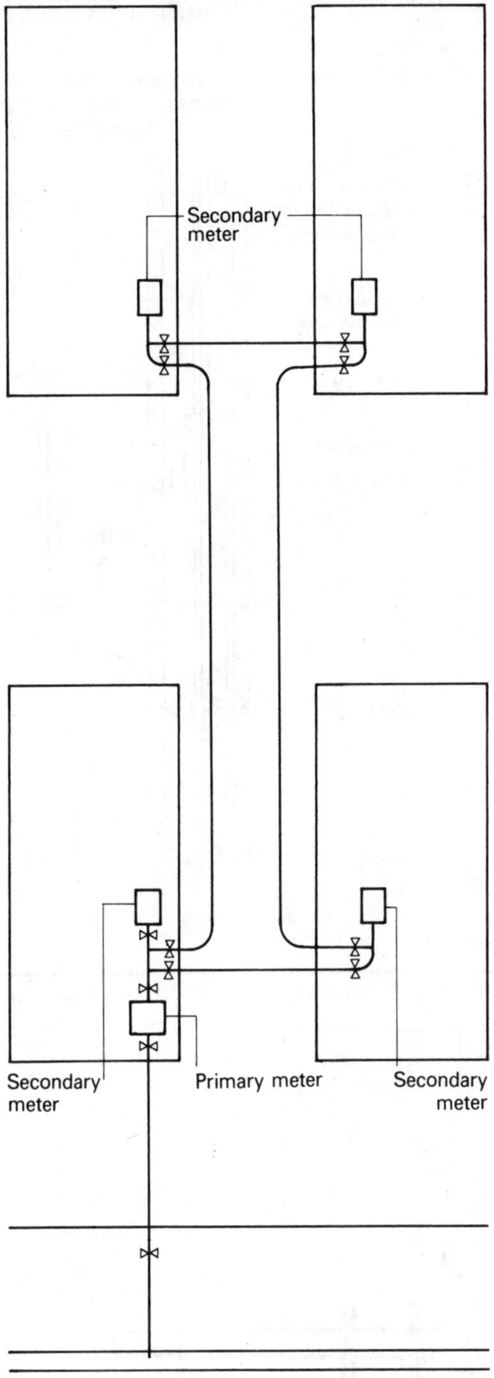
Secondary meter
Secondary meter
Primary meter
Secondary meter

Fig.6.17 Ring circuit with primary and secondary meters

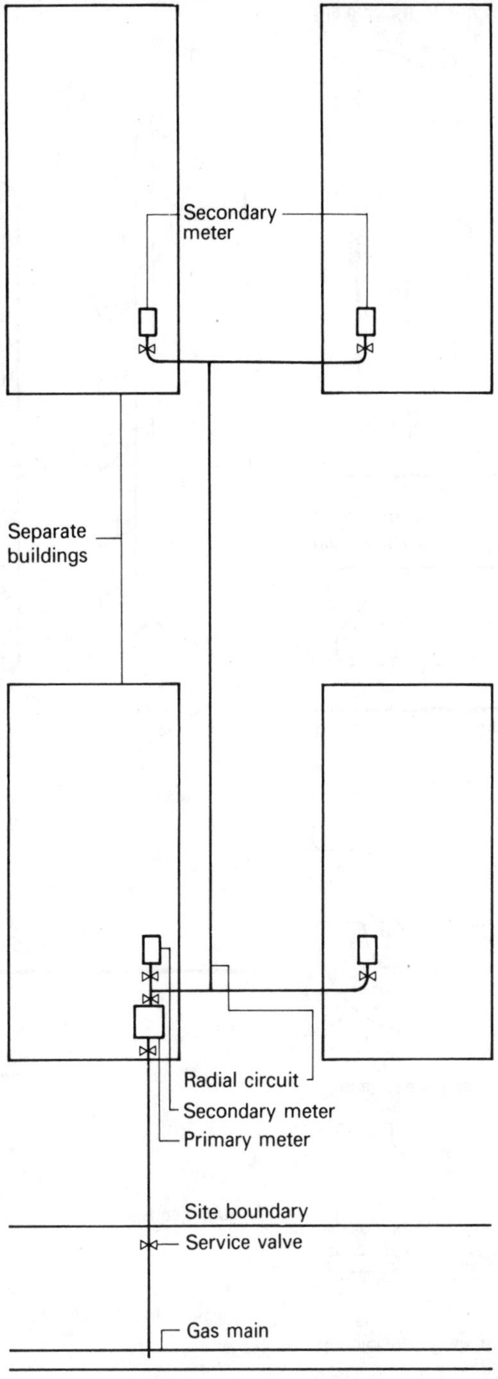
Secondary meter
Separate buildings
Radial circuit
Secondary meter
Primary meter
Site boundary
Service valve
Gas main

Fig.6.18 Radial circuit with primary and secondary meters

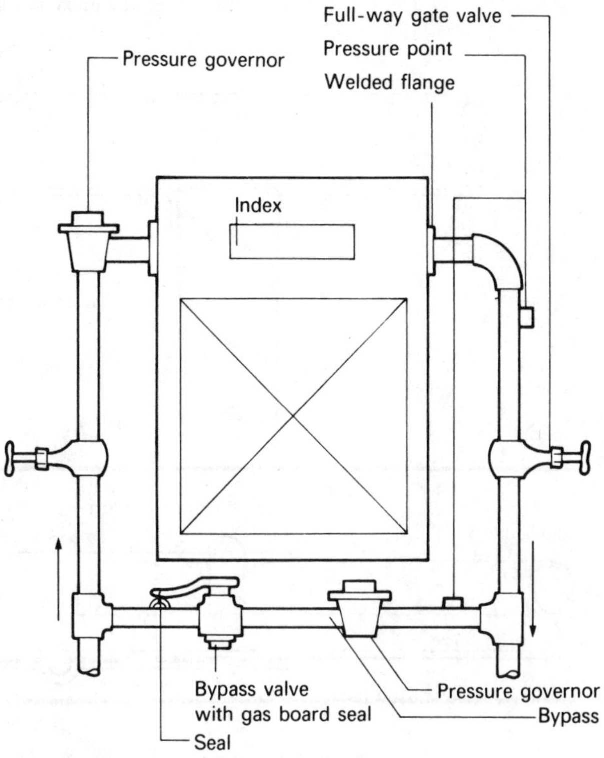
Full-way gate valve
Pressure point
Welded flange
Pressure governor
Index
Bypass valve
with gas board seal
Seal
Pressure governor
Bypass

Fig.6.19 Bypass to meter

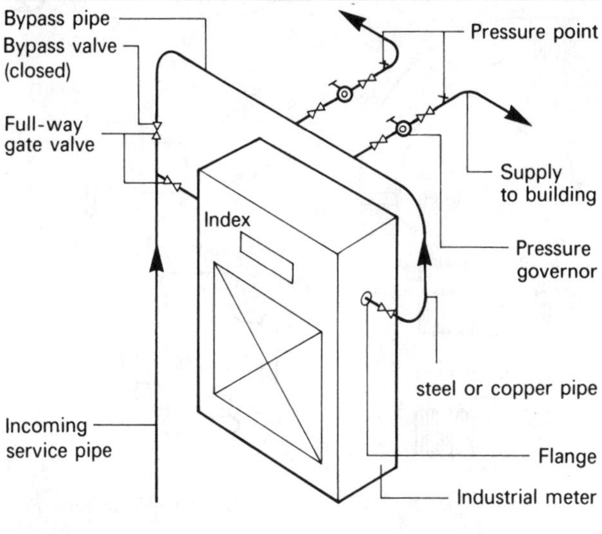
Bypass pipe
Bypass valve
(closed)
Full-way
gate valve
Pressure point
Supply
to building
Index
Pressure
governor
steel or copper pipe
Incoming
service pipe
Flange
Industrial meter

Fig.6.20 Bypass to meter

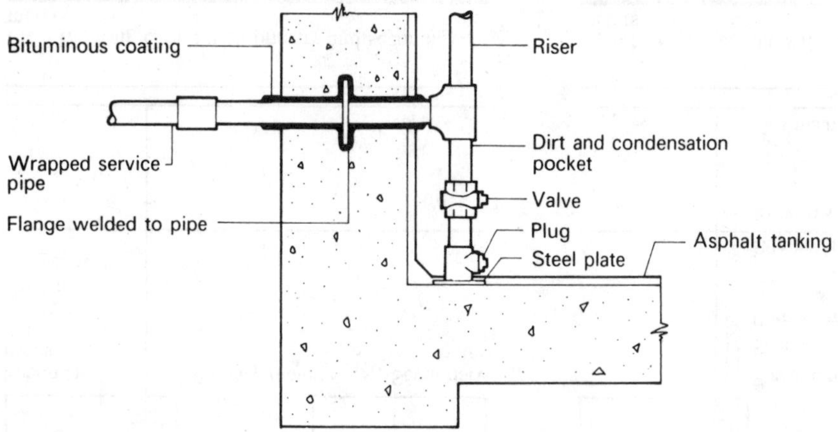
Bituminous coating
Riser
Wrapped service pipe
Dirt and condensation pocket
Flange welded to pipe
Valve
Plug
Steel plate
Asphalt tanking

Fig.6.21 Entry of service pipe through tanked basement

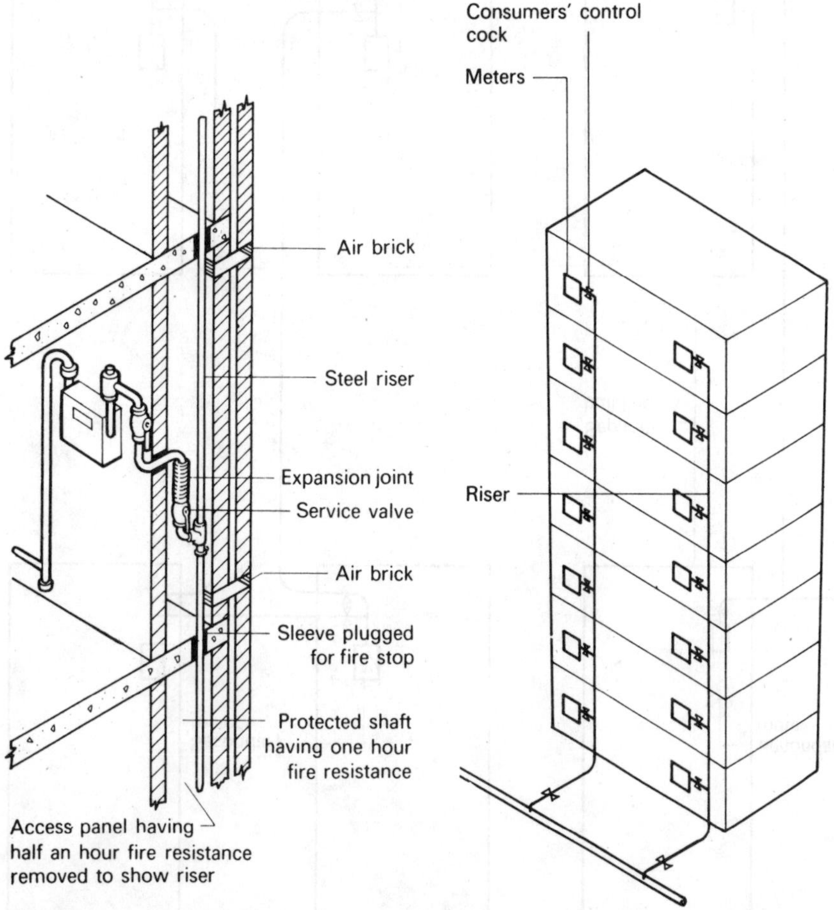
Consumers' control cock
Meters
Air brick
Steel riser
Expansion joint
Service valve
Riser
Air brick
Sleeve plugged for fire stop
Protected shaft having one hour fire resistance
Access panel having half an hour fire resistance removed to show riser

Fig.6.22 Supply to flats

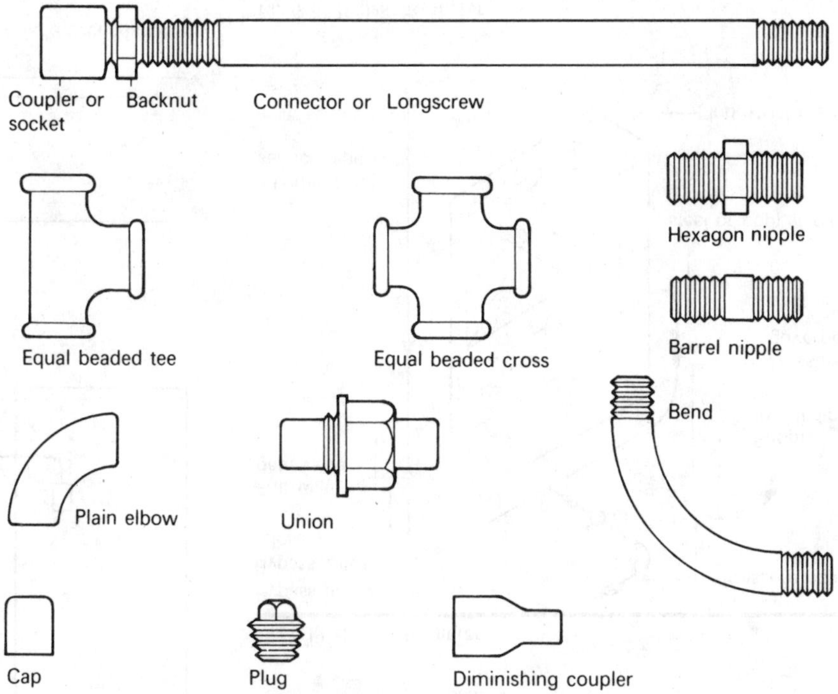
Coupler or socket
Backnut
Connector or Longscrew
Hexagon nipple
Equal beaded tee
Equal beaded cross
Barrel nipple
Bend
Plain elbow
Union
Cap
Plug
Diminishing coupler

Fig,.6.23 Fittings for steel pipes

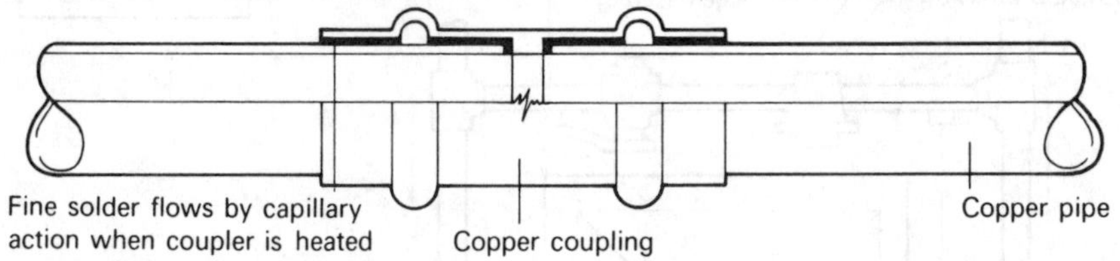
Fine solder flows by capillary action when coupler is heated
Copper coupling
Copper pipe

Fig.6.24 Soldered capillary joint

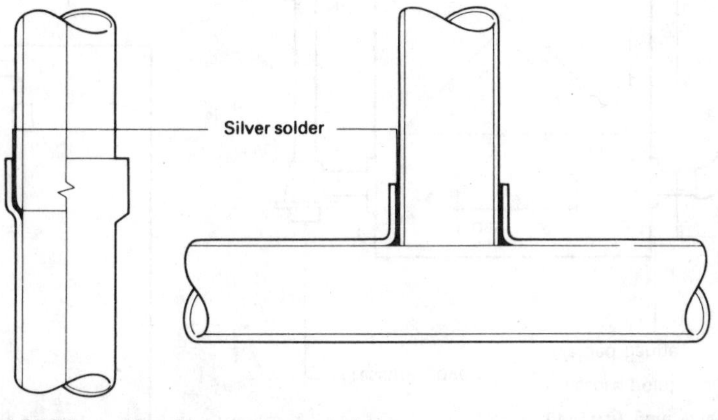
Silver solder

Fig.6.25 Silver soldered joints on copper pipe

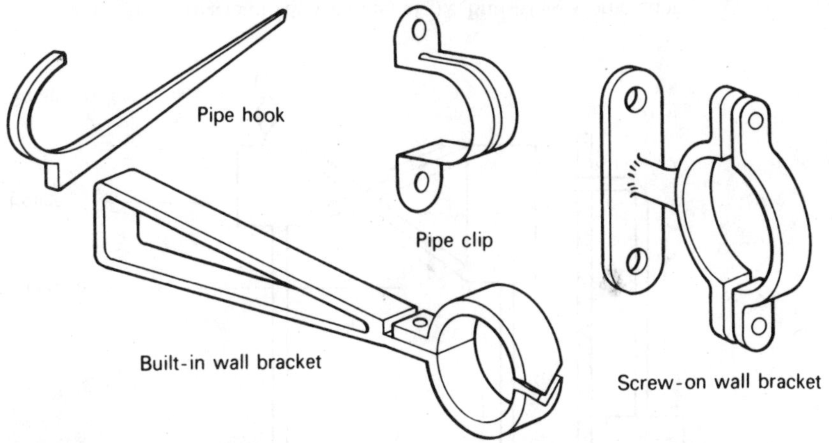

Pipe hook
Pipe clip
Screw-on wall bracket
Built-in wall bracket

Fig 6.26 Types of pipe fixings

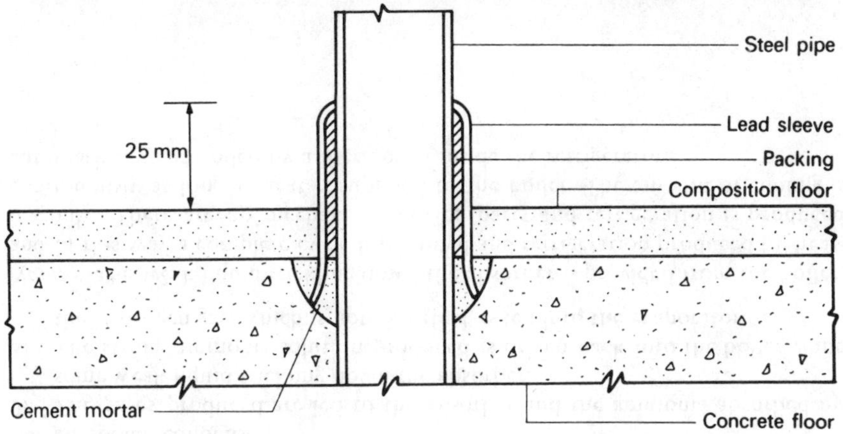

Steel pipe
Lead sleeve
Packing
Composition floor
25 mm
Cement mortar
Concrete floor

Fig.6.27 Protection of steel pipe

bypass valve should be sealed in a closed position in a manner approved by the Gas Board. Figures 6.19 and 6.20 show two methods of providing a bypass to a meter.

Large meters

Large meters for commercial or industrial gas supplies are fitted with flanged side connections suitable for 76 mm, 100 mm, 150 mm and 200 mm pipe connections. These flanged pipes are welded to the steel casing of the meter and the connections have sufficient strength to permit rigid steel or copper pipes to be connected direct to the meter, without the need for short lengths of flexible lead pipes which are required for small meters having soldered meter connections. The meter should be mounted level on a suitable base and should not be subjected to an internal gas pressure greater than that for which it was designed.

Meter housing

A ventilated, dry, incombustible compartment of adequate size for the housing of meters should be provided by the builder as part of the structure of the building. The compartment should be provided with a locked door to prevent unauthorised entry.

Blocks of flats

For blocks of flats the underground service pipe should be provided with a service valve fitted with a surface box and cover. The pipe should rise inside the building in a duct sited in a common corridor or landing, and where the riser has to pass through a wall or floor a sleeve should be provided. If the service pipe enters the flats through a tanked basement a flange should be welded on the pipe, and both the flange and the pipe coated with bituminous paint before being set in position and surrounded with concrete (see Fig. 6.21).

Tees for the gas supply to each floor should be fitted at suitable levels with connectors for accessibility. Meters should be installed in a dry, ventilated cupboard inside the flats, with the index visible from the outside so that the gas registered can be read from a common corridor or landing. Figure 6.22 shows the gas supply to blocks of flats.

Purging

Before a meter is fitted, the gas control cock or valve should be fully opened to allow gas to pass through and at the same time clear any dirt in the service pipe. After the meter is fitted, the installation pipe should be purged before connection of the appliances.

Types of pipe fitting and joints

Various types of fittings are obtainable for use with mild steel pipe and these are commonly made from malleable iron, with the exception of connectors and bends which are made from mild steel. Figure 6.23 shows some common types of fittings for steel pipes; fittings for use with copper pipes take similar forms but are of the soft soldered type shown in Fig. 6.24. Alternatively copper pipes may be jointed by means of silver solder as shown in Fig. 6.25.

Pipe fixing

Gas piping should be fixed so as to be secure and prevent sagging, the spacing of the pipe supports depends upon the type of material and the diameter of the pipe (see Chapter 12). Figure 6.26 shows a variety of pipe fixings.

Protection of pipes

Pipes that may come into contact with composition floors, walls and skirtings, particularly those of magnesium oxychloride, should be protected from corrosion by painting them with bituminous paint. Pipes which need to pass vertically through corrosive floors may be protected by use of a short lead sleeve, shown in Fig. 6.27.

Installation testing

Immediately on completion, whether or not gas is available, any new installation or extension of an existing installation should be tested under air pressure. In installations where the pipes are covered with concrete, etc., it is desirable to test each section of pipework as the work proceeds. Air is used because it is possible

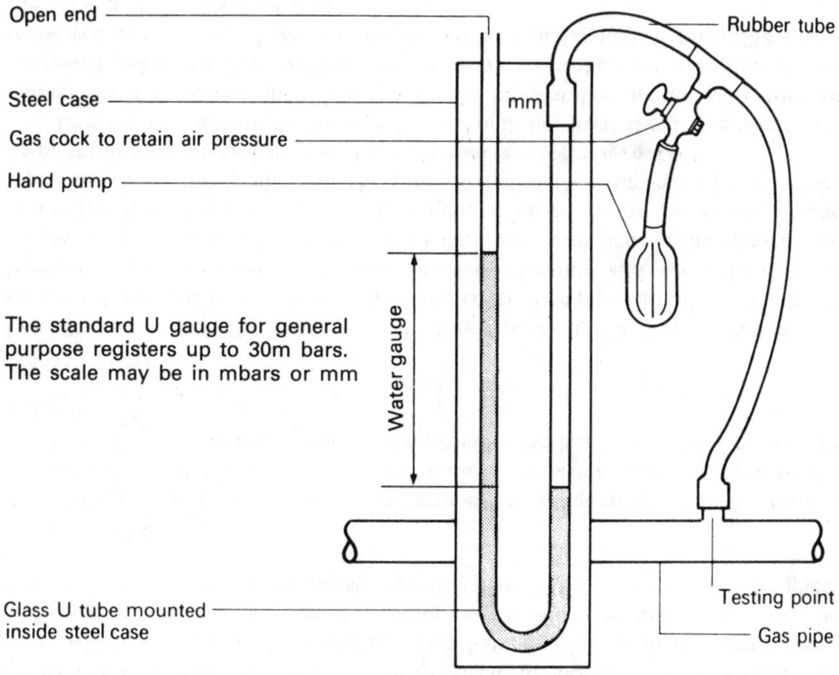
Open end
Rubber tube
mm
Steel case
Gas cock to retain air pressure
Hand pump
The standard U gauge for general
purpose registers up to 30m bars.
The scale may be in mbars or mm
Water gauge
Testing point
Glass U tube mounted
inside steel case
Gas pipe

Fig.6.28 U gauge for testing

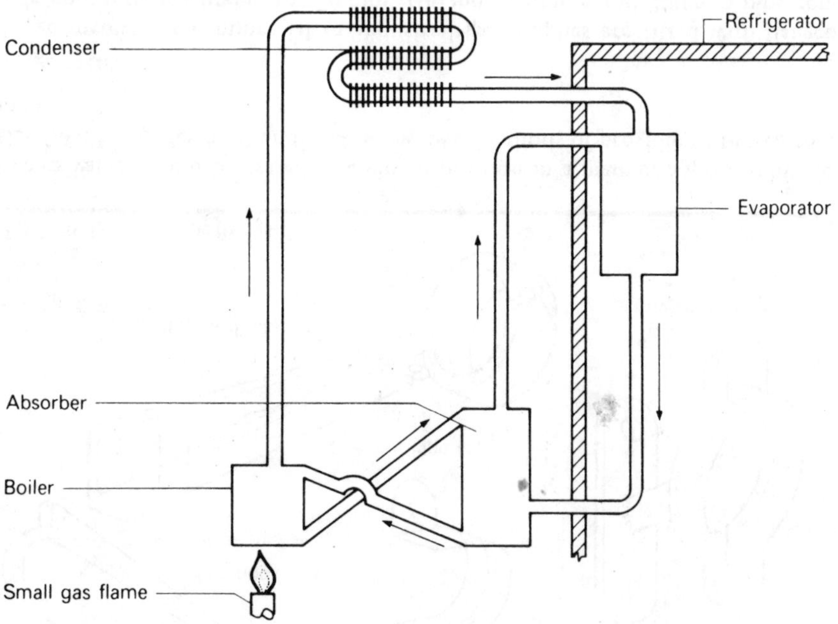
Refrigerator
Condenser
Evaporator
Absorber
Boiler
Small gas flame

Fig.6.29 The gas refrigerator (By Courtesy of the British Gas Corporation)

to test at a greater pressure than that to which the installation would normally be subjected to when containing gas.

Testing procedure (for soundness)

1. Close all taps at appliances, and open the testing point.
2. Connect U gauge to the pipework at the testing point.
3. Slowly pump air into the pipework until the U gauge registers a pressure of 20 mbars (the air pressure must never exceed 25 mbars).
4. Allow one minute for the temperature of the air in the pipework to stabilise.
5. Note any pressure loss during the next two minutes. If the distance between the water levels in each leg of the U gauge lessens a leak is indicated.
6. For a satisfactory test there must be no pressure drop during the two minute test period.

Note: A leak must be found by painting the joints with soapy water.

Purging the installation

After testing the installation for soundness air must be purged from the pipework as follows:

1. Ensure that there is good ventilation.
2. Turn off main gas cock and ensure all appliance taps are turned off.
3. Do not allow smoking or the operation of electrical switches.
4. Turn on the main gas cock and purge all pipes of air, commencing with the pipe furthest from the meter.
5. The pipework will be purged of air when the meter test dial has made five revolutions.

The gas refrigerator

The principle on which gas refrigeration is based is that when a liquid evaporates it extracts heat from its surroundings. Figure 6.29 shows the components of a gas refrigerator which operates as follows:

1. A solution of ammonia in water is heated in the boiler by a small gas flame and the ammonia gas driven off is condensed to liquid ammonia in the air-cooled condenser.
2. The gases produced are led to the absorber and the ammonia absorbed by some weak liquid trickling down the absorber.
3. The strong ammonia solution produced is driven back into the boiler while the hydrogen gas, which is not absorbed, is led into the evaporator.

Note: The weak liquid trickling down the absorber is provided from the boiler, and in this way a complete cycle is obtained and refrigeration produced by heating only. There are no mechanical moving parts and refrigeration is produced continuously as long as heat is supplied to the boiler. The amount of cooling is automatically controlled by a thermostat inside the refrigerator.

Chapter 7

Gas controls, safety devices, heating and flues

Gas cocks

The simplest and most common form of gas control is the gas cock. This usually consists of a tapered plug which fits into a tapered body, the two connecting surfaces being machined to provide a gas-tight seal. The two surfaces should be smeared with grease to act as a lubricant. The cock is restricted to a 90° movement between being fully closed or fully opened. Figure 7.1 shows the customer's main control cock which incorporates a union to facilitate the removal of the meter.

For safety reasons, a drop-fan cock may be used as shown in Fig. 7.2. The fan is hinged to the turning head of the plug and has lugs which are arranged to hold the fan upright when the gas is on. When the cock is in the OFF position, the fan falls and the lugs engage with a slot in the body of the cock. The plug in this OFF position cannot be turned to the ON position until the fan is deliberately held upright, thus preventing the cock from being turned on accidentally.

The plug-in safety cock shown in Fig. 7.3 ensures that the cock cannot be turned to the ON position without first connecting the outlet pipe to the cock. The end of the plug assembly has two lugs which engage inside notches on the outside of the cock plug. When the plug assembly is inserted and turned to engage the lugs, two pins on the plug cover also engage a groove, and a gas-tight connection between the two components is made by compressing the spring which forces the two conical surfaces together.

Pressure governor

Pressure control of gas may be achieved by use of a constant pressure governor fitted on the inlet pipe to the meter and to each appliance. The governor may be weight-loaded or spring-loaded, and the loading may be adjusted to provide the correct pressure at the appliance. A weight-loaded governor must always be fitted horizontally so that the weight acts vertically on the diaphragm. Spring-loaded governors are not restricted in this way, they may be fitted in any position and are therefore more popular than the weight-loaded types. Figure 7.4 shows a section of a spring-loaded, constant-pressure governor which operates as follows:

1. Gas enters the governor at inlet pressure and passes through valve A to the appliance and also through the bypass to space B between the two diaphragms.
2. The main diaphragm is loaded by the spring and the upward and downward forces acting upon the diaphragm are balanced. The upward and downward forces acting upon the compensating diaphragm are also balanced and this has a stabilising affect on the valve and counteracts any tendency to oscillation.
3. Any fluctuation of inlet pressure will inflate or deflate the main diaphragm, thus raising or lowering valve A and altering the resistance to the flow of the gas, and thus ensuring a constant pressure at the outlet to the appliance.

Note: The space above the main diaphragm is ventilated to atmosphere to allow unrestricted movement of the valve. The diaphragms may be made from rubber, leather or plastic. A filter may be fitted on the inlet side of the governor.

Thermostats

A thermostat is a device which opens or closes a gas valve in accordance with the temperature it senses. The action of the thermostat depends upon the expansion of metals, liquids or vapours when heated.

Rod-type thermostat

This type of thermostat is used for cookers and storage-type water heaters (see Fig. 7.5). The brass tube shown in the diagram encloses an Invar steel rod to which is attached a valve. When the gas is burning, the brass tube becomes hot and as it expands it carries with it the Invar rod which expands very little. This brings the valve closer to its seating, thus reducing the flow of gas. Should the air or water cool down the brass tube also cools down and contracts, thus moving the valve away from its seating and so allowing more gas to flow to the burner until once again the temperature setting is reached. The bypass permits a small amount of gas to flow when the valve is closed and maintains a small flame at the burner.

Vapour expansion thermostat

Many space heaters such as gas fires, are controlled by a vapour expansion-type thermostat (see Fig. 7.6). The coiled capillary tube, probe and bellows, are filled

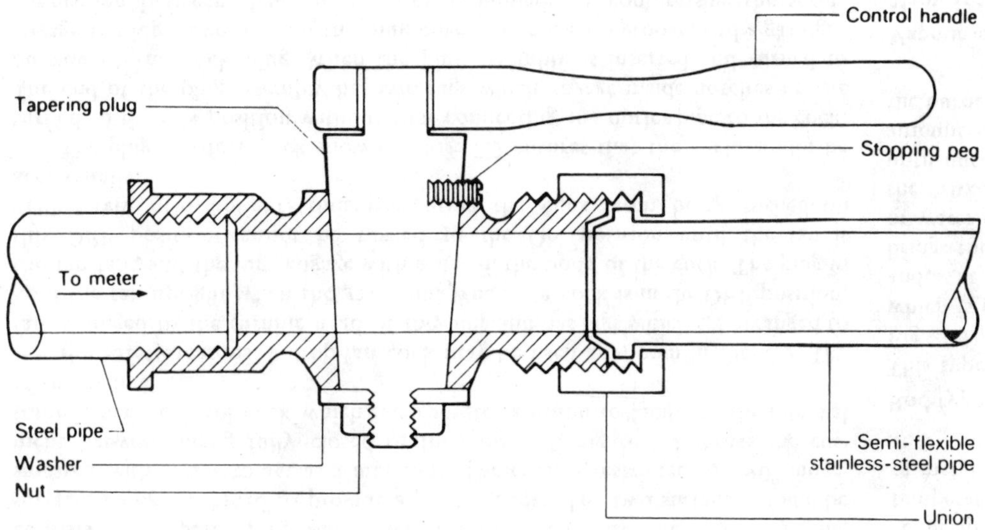

Control handle
Tapering plug
Stopping peg
To meter
Steel pipe
Washer
Nut
Semi-flexible stainless-steel pipe
Union

Fig.7.1 Customer's control cock

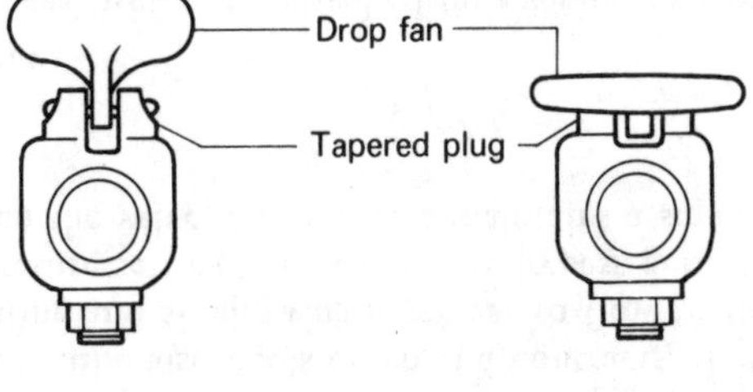

Drop fan
Tapered plug

Fig.7.2 Drop-fan safety cock

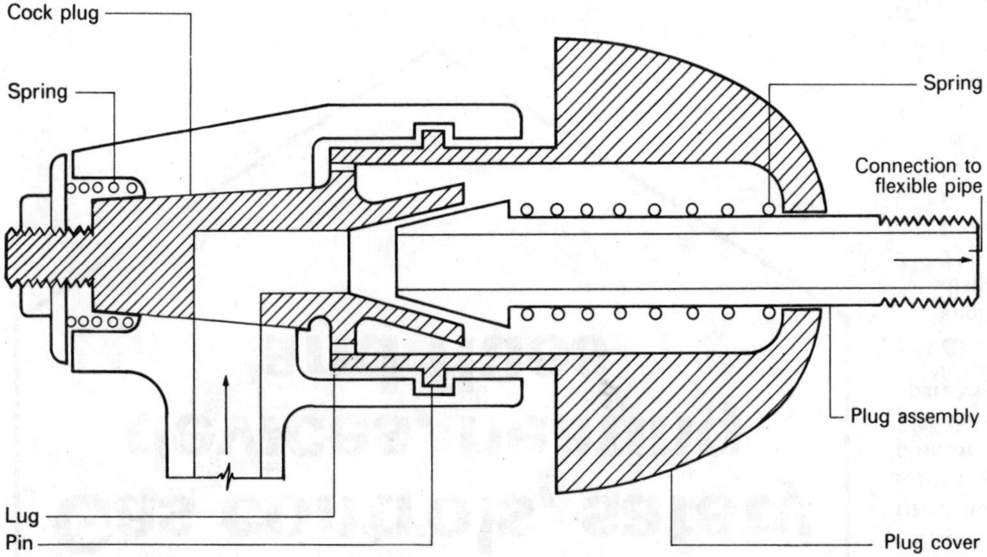

Cock plug
Spring
Spring
Connection to flexible pipe
Plug assembly
Lug
Pin
Plug cover

Fig.7.3 Plug-in safety cock

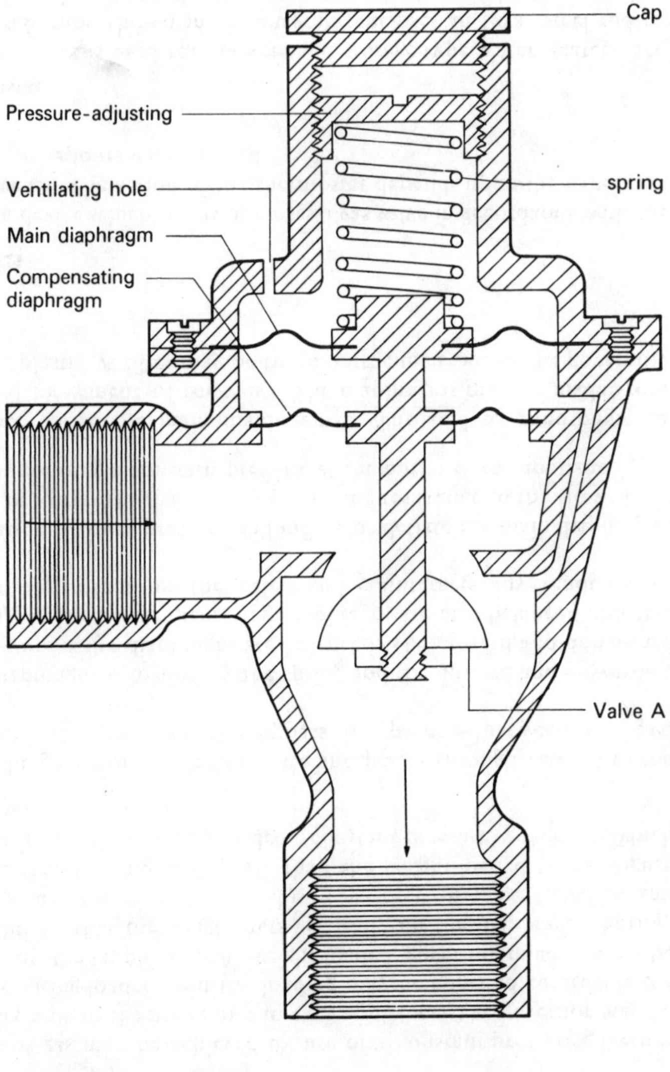

Cap
Pressure-adjusting
spring
Ventilating hole
Main diaphragm
Compensating diaphragm
Valve A

Fig.7.4 Spring-loaded pressure governor

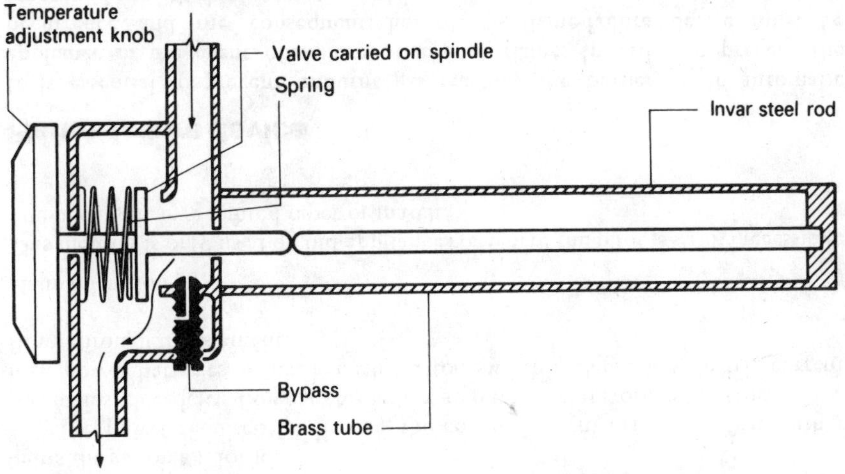

Temperature adjustment knob
Valve carried on spindle
Spring
Invar steel rod
Bypass
Brass tube

Fig.7.5 Rod-type thermostat

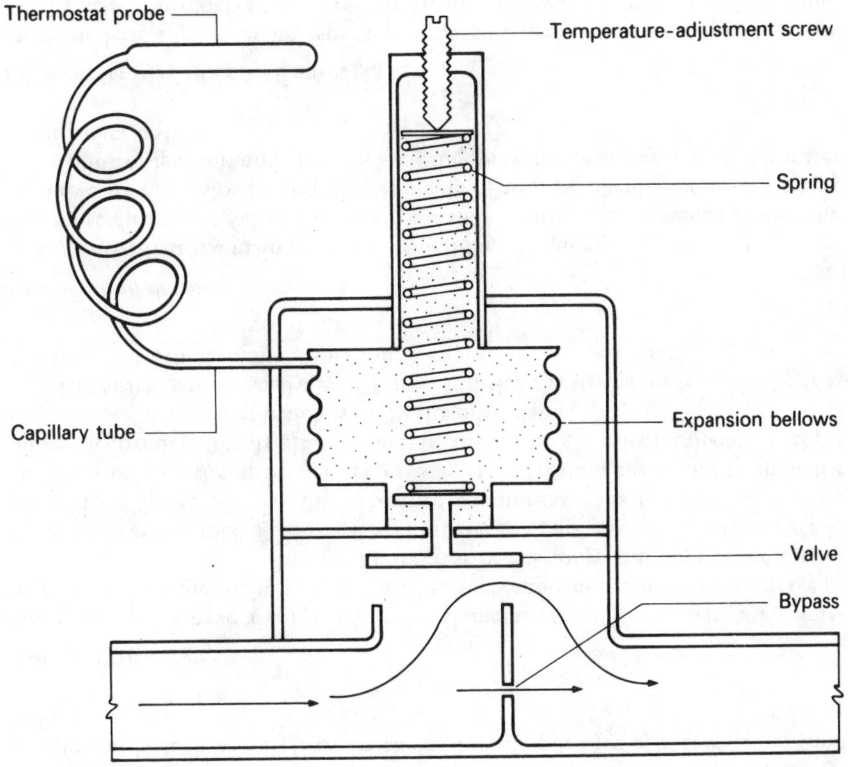

Thermostat probe
Temperature-adjustment screw
Spring
Capillary tube
Expansion bellows
Valve
Bypass

Fig.7.6 Vapour expansion thermostat

with ether and the probe is sited so as to sense the temperature of the air in the room. An increase in the air temperature causes the ether to rise up the tube to the expansion bellows, which in turn pushes the valve closer to its seating. The gas flow to the burner and the heat output to the room are therefore reduced. As the room temperature drops, the ether contracts, allowing the bellows to contract and the gas valve to open. The temperature-adjustment screw can be set to various temperatures which provide differing degrees of tension on the spring and therefore different control positions for the gas valve. The bypass again permits a small amount of gas to flow when the valve is closed and maintains a small flame at the burner.

ON/OFF thermostat and relay valve

The rod and vapour expansion-type thermostats are modulating in their action because they gradually change the flow of gas in response to temperature changes. Another thermostat, known as the ON/OFF type, is often used for central heating boilers and circulators (see Fig. 7.7).

The relay-valve system shown, operates as follows:

1. When the boiler or circulator is operated, the gas flows through to the burner because valves A and B are open and the gas pressures above and below the diaphragm are equal.
2. When the water reaches the required temperature, the brass casing expands sufficiently to draw valve A, connected to the Invar rod, to a closed position. The closing of valve A prevents gas from passing through the weep pipe to the underside of the diaphragm.
3. The gas pressure then builds up above the diaphragm in the relay valve and allows the valve assembly to fall under its own weight; valve B is thus closed.
4. When the water cools, valve A is again opened, allowing gas to flow to the underside of the diaphragm; valve B is again opened, which allows gas to again flow to the burner and be ignited by a pilot flame.

Note: The bypass allows gas below the diaphragm to escape through the weep pipe to the burner. This removes the pressure below the diaphragm and allows the valve assembly to fall under its own weight, aided by the gas pressure above the diaphragm.

Gas igniters

There are two types of gas igniters namely, spark and filament.

Spark igniters

Spark igniters are usually operated by mains electricity. In these igniters an electrical charge is built up in a capacitor until a trigger device allows it to discharge suddenly. The discharge current is stepped up by a transformer to a high voltage of 10 or 15 kV. This spark is strong enough to light the main burner or the gas of a pilot jet. After each spark the capacitor is discharged and the cycle repeated. This type of spark generator can be used to supply a number of burners. Figure 7.8 shows a schematic layout of a mains spark igniter. Spark ignition may also be powered by a battery and so may be used on appliances which have no mains electricity control.

Piezoelectric spark ignition (Fig. 7.9)

Such an igniter usually consists of two crystals each 12 mm long and 6 mm in diameter connected in parallel. Applying pressure on the crystals causes a large electrical voltage to build up until it overcomes the resistance of the spark gap and sparks are produced. The sparks are of sufficient strength to light the burner of appliances such as gas fires, or the gas of the pilot jet of water heaters and boilers.

Automatic ignition

Some appliances have a permanent pilot flame, i.e. the pilot is alight at all times whether or not the main burner is alight. Such appliances have the advantage of cheapness and simplicity but in the interest of economy the pilot flame should be as small as possible. To save gas still further the pilot may be lit automatically by a spark igniter. When the appliance thermostat calls for heat an electric current flows to the spark igniter causing it to create a spark and to light the pilot. Electronic controls also operate an automatic gas valve; it opens to let gas flow to the main burner which is lit by the pilot flame.

The spark gap is usually 3 to 5 mm and the electrodes must not be bent as this may crack the ceramic insulation.

Slow ignition devices

It is essential that the main gas burner should ignite quietly and safely. To ensure this most automatic valves incorporate a slow ignition device which allows the gas valve to open slowly and close quickly. The slow opening ensures a quiet, non-explosive ignition and the quick closing prevents over heating of the water in the boiler or heater.

Filament igniters (Figs 7.10 and 7.11)

Most modern appliances use spark igniters and filament igniters are therefore usually used on older appliances. The igniter consists of an electrically heated filament adjacent to a pilot jet. The filament consists of a small coil of wire which, due to the resistance of the flow of electricity, becomes red-hot and lights the gas on a pilot jet.

The power required for heating the coiled filament may come from either the mains electricity flowing through a step-down transformer or from 1.5 V high power batteries. When a push button switch is depressed electric current flows through the filament.

Manual ignition

This method is only used on old appliances or where the pilot is easily accessible. The pilot is lit by a lighted taper or match.

Flame-failure device

It is essential to prevent unburnt gas reaching the burner of an automatic appliance in the event of failure of a pilot flame. In order to prevent the occurrence and the consequent hazards, a flame-failure device must be incorporated with the automatic appliance.

Bimetal flame-failure device

In this type of safety device, the movement of the safety cut-out valve is effected by making use of the differential coefficient of expansion of brass and Invar steel. Two strips of the alloys are joined and the strips formed into a U shape with the brass which has the greater expansion on the outside. When the pilot flame is alight, heat causes the strip to bend inward and open the inlet valve, as shown in Fig. 7.12. If the pilot light is extinguished the strip cools and reverts to its former shape, thus closing the inlet valve as shown in Fig. 7.13. This device is normally used for water heaters.

Magnetically operated flame-failure safety device

This type of safety device makes use of the fact that a small amount of electric current is generated between a hot and cold body and this current is used to energise a magnet (see Fig. 7.14). The thermocouple is heated by a pilot flame which also ignites the burner flames and the magnet holds the safety cut-out valve in an open position and allows gas to flow to the burner.

To light the burner, the following procedure is carried out:

1. The starter button is pressed down to the full length of its travel. This first closes the main gas valve and sets the lower end of the valve against the electromagnet.
2. The pilot cock is opened and the pilot flame established.
3. The starter button is kept depressed for about 30 s, in which time the magnet should be sufficiently energised to hold open the gas valve.
4. The button is released and the open interrupter valve allows gas to flow to the burner which is ignited by the pilot.

Note: As long as the pilot flame remains alight, electric current is generated and the electromagnet will hold the safety cut-out valve open. If the pilot flame is extinguished, the thermocouple no longer produces electricity and the spring underneath the safety cut-out valve closes the valve, thus cutting off the gas supply to the burner. The device is normally used with boilers and air heaters.

Gas space heating

Boilers

Gas-fired, wall-hung boilers are popular because of saving in floor space. They can be obtained with either a conventional or a balanced flue. The heat exchanger of the boiler consists of a stainless steel or copper tube which holds approximately 0.7 litre of water, and this low water content permits a rapid thermal response. A typical wall-hung boiler has a heat output of 15 kW which is sufficient to heat domestic hot water and up to 20 m^2 of radiator surface, including the pipe runs.

Hearth-mounted boilers have heat outputs ranging from 8.8 kW for a small house to 1760 kW for a large building. The smaller boilers can be obtained with either a conventional or a balanced flue, but larger boilers are normally provided with a conventional flue. Figure 7.15 shows a small gas-fired, hearth-mounted boiler including the controls. These have been shown outside the boiler for clarity, but they are usually fitted inside the boiler casing.

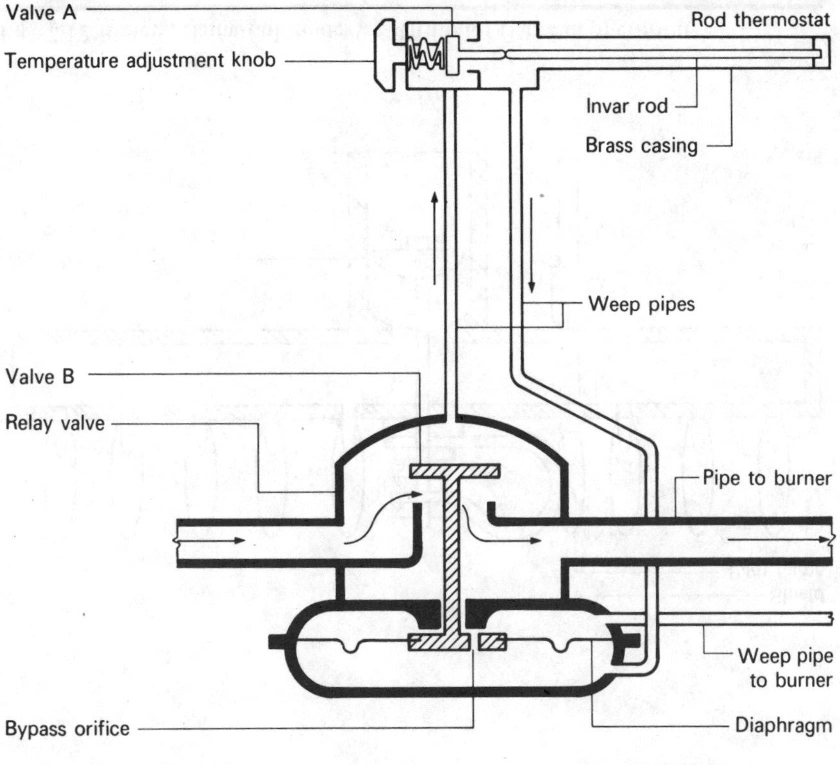
Valve A
Rod thermostat
Temperature adjustment knob
Invar rod
Brass casing
Weep pipes
Valve B
Relay valve
Pipe to burner
Weep pipe
to burner
Bypass orifice
Diaphragm

Fig. 7.7 ON/OFF thermostat and relay valve

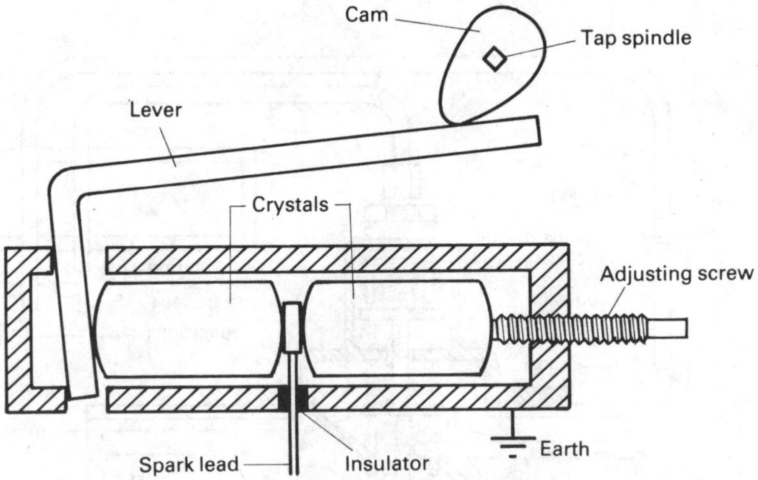
Cam
Tap spindle
Lever
Crystals
Adjusting screw
Earth
Spark lead
Insulator

When the gas tap is turned on the cam turns anti-clockwise. The lever is depressed and pressure is applied to the crystals. The voltage builds up until it overcomes the spark gap

Fig. 7.9 Piezoelectric spark ignition

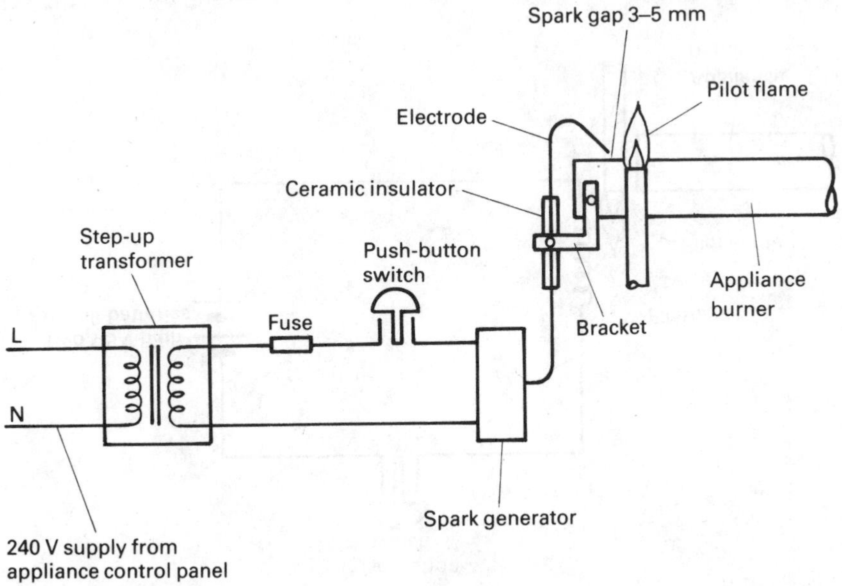
Spark gap 3–5 mm
Pilot flame
Electrode
Ceramic insulator
Step-up
transformer
Push-button
switch
Fuse
Appliance
burner
Bracket
L
N
Spark generator
240 V supply from
appliance control panel

Fig. 7.8 Mains spark igniter

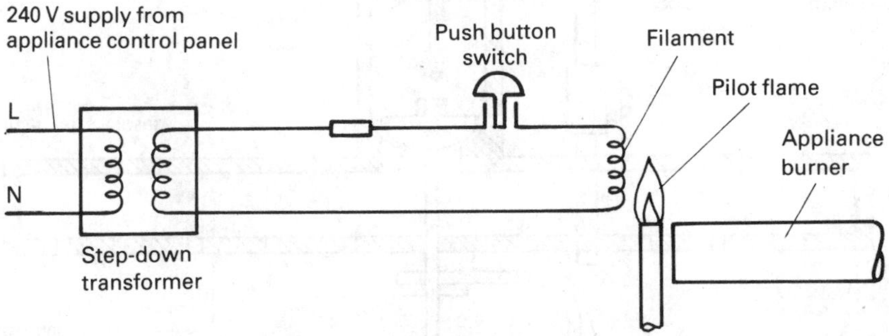
240 V supply from
appliance control panel
Push button
switch
Filament
Pilot flame
L
Appliance
burner
N
Step-down
transformer

Fig. 7.10 Mains filament igniter

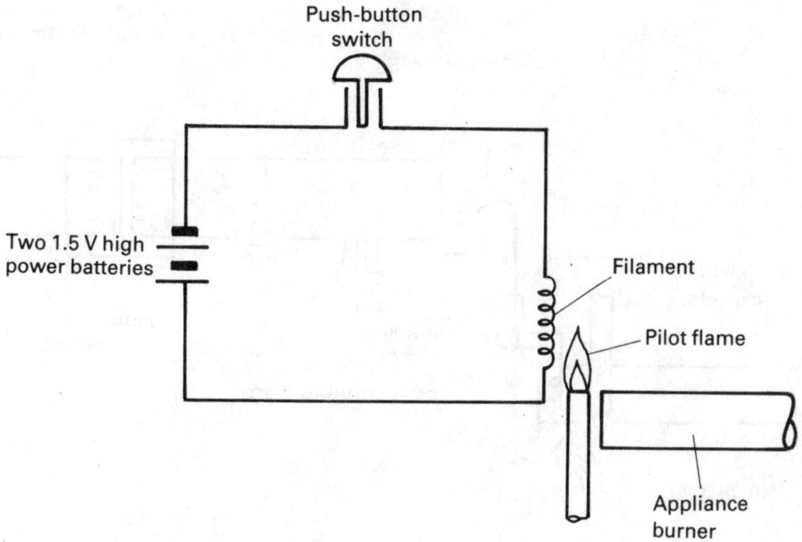
Push-button
switch
Two 1.5 V high
power batteries
Filament
Pilot flame
Appliance
burner

Fig. 7.11 Battery filament igniter

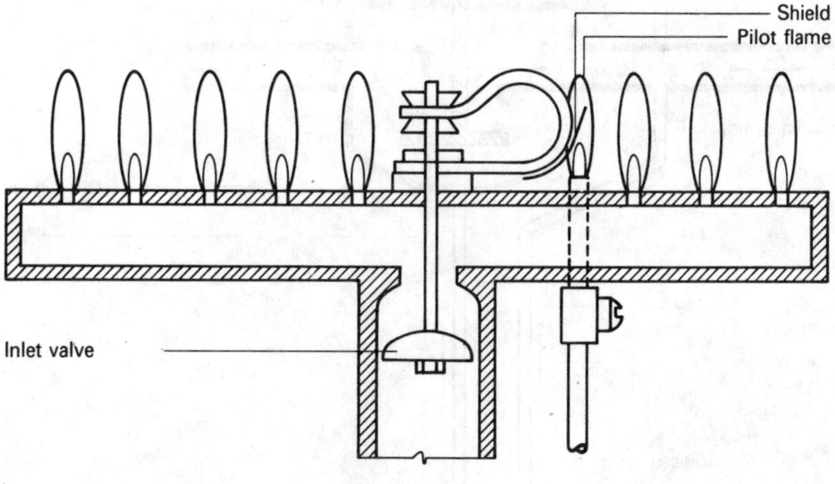
Shield
Pilot flame
Inlet valve

Fig. 7.12 Bimetal flame-failure device with pilot flame in operation

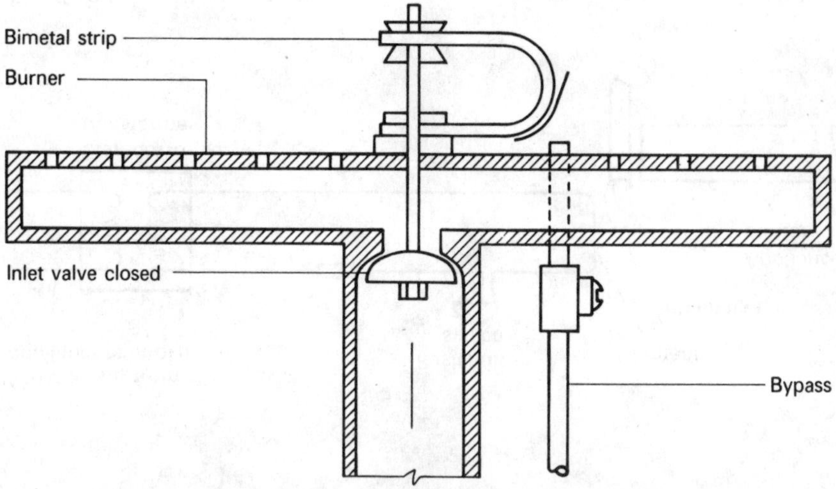
Bimetal strip
Burner
Inlet valve closed
Bypass

Fig. 7.13 Bimetal flame-failure device with pilot flame extinguished

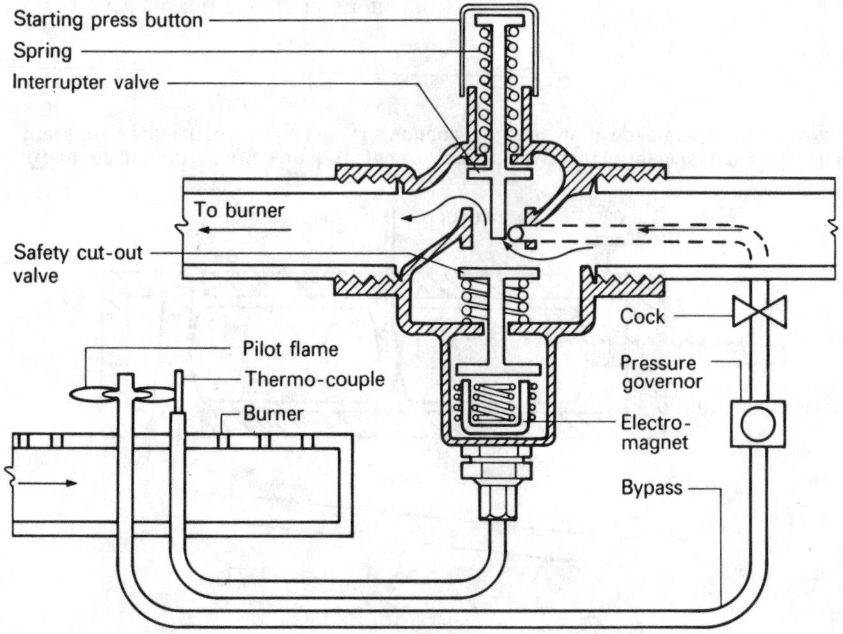
Starting press button
Spring
Interrupter valve
To burner
Safety cut-out
valve
Cock
Pilot flame
Thermo-couple
Burner
Pressure
governor
Electro-
magnet
Bypass

Fig. 7.14 Magnetically operated flame-failure safety device

Position

The boiler can be placed on the rooftop or in any other convenient position since there is no storage of fuel and removal of ashes are no problem. This is of great value since ground-floor and basement space is exceptionally valuable. A shorter conventional flue is also possible which saves on construction costs and space. There is also less pressure on the boiler. All types of boilers are fully automatic in operation and incorporate thermostatic and clock control, a flame-failure device which ensures a 100 per cent gas cut-off in the event of pilot failure, and a constant pressure governor.

Efficiencies are high – of the order of 75–80 per cent – and space requirements per unit of heat output is slightly less than for solid fuel and oil-fired boilers.

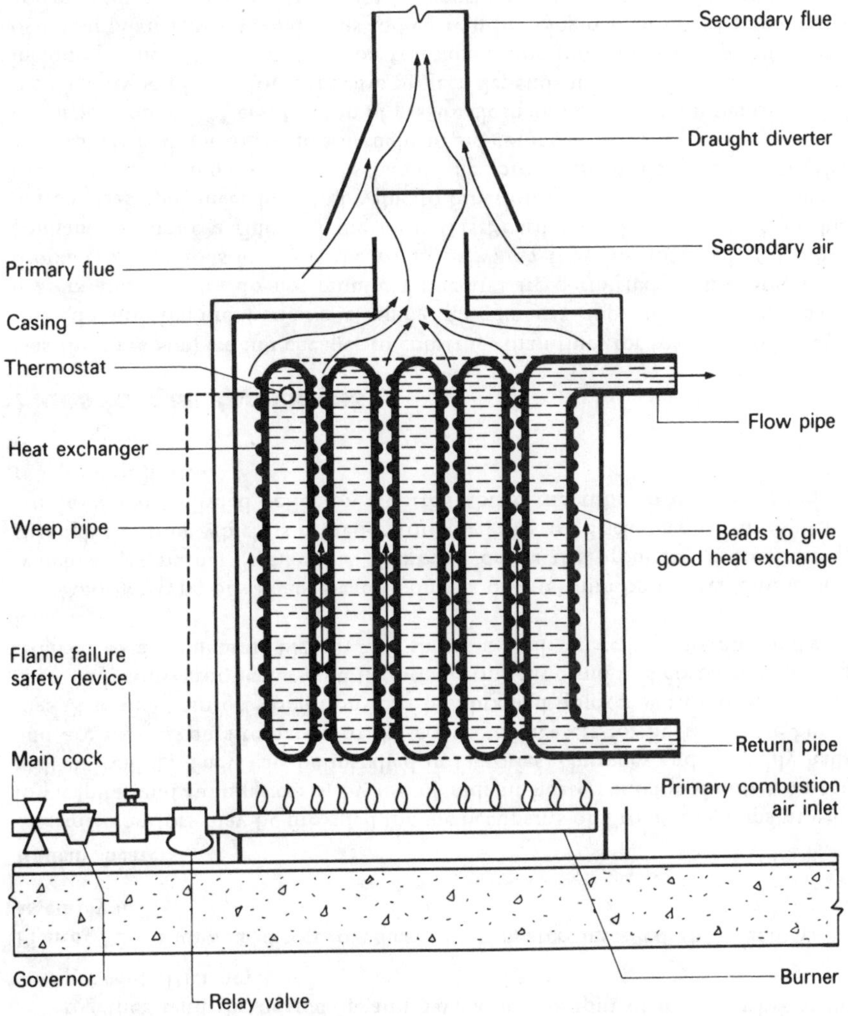

Secondary flue
Draught diverter
Secondary air
Primary flue
Casing
Thermostat
Flow pipe
Heat exchanger
Weep pipe
Beads to give good heat exchange
Flame failure safety device
Return pipe
Main cock
Primary combustion air inlet
Governor
Relay valve
Burner

Fig. 7.15 Floor-mounted gas boiler

Construction

Sectional cast-iron boilers are available with outputs of 8.8 kW and 440 kW. It is possible to erect the boilers in confined spaces and a single section can be readily replaced.

Welded plate steel boilers are normally fabricated in one piece, which reduces site work. Heat outputs of up to 180 kW are available. Welded tubular-steel boilers are available for larger installations, with heat outputs of up to 1760 kW. Steel boilers can withstand higher pressures than cast-iron boilers.

Fires

These are usually designed to provide both radiant and convective heating to the room; in addition they can incorporate a back boiler at the rear which can be easily fitted into a standard fireplace opening.

A typical fire has a heat output of 3 kW, and a gas-fired back boiler a heat output of 12 kW, which is sufficient to heat the normal domestic hot water and up to 18 m^2 of radiator surface, including the pipe runs. If domestic hot water is not required a back boiler will heat up to 20 m^2 of radiator surface, including the pipe runs. Fires can be either hearth- or wall-mounted. Figure 7.16 shows a section through a hearth-mounted gas fire providing both radiant and convective space heating.

Convector heaters

These are usually room-sealed combustion units, incorporating a balanced flue. The heat output from the convector is almost all by convection although some radiant heat is given off from the hot casing, usually about 10 per cent. They provide a simple, flexible space heating system and are usually cheaper to install than a boiler and radiator system. Figure 7.17 shows a natural convector and Fig. 7.18 a fan-assisted convector which gives a rapid thermal response Heaters are available with heat outputs of between 5 kW and 7 kW.

Warm-air units (see Fig. 7.19)

These are free-standing, self-contained packaged units which combine controlled heating, ventilation and air movement. They are provided with a silent fan and built-in noise attenuators which make them suitable for the space heating of houses, schools, churches and libraries. The warm air may be ducted from the unit to the various rooms, as shown in Fig. 7.20. The ducts must be well insulated to prevent heat loss. Units are available having heat outputs from 6 kW to 50 kW.

A warm-air space heating system provides a good distribution of heat and the air movement gives the required feeling of freshness. The system can be used in hot weather to circulate the air in the rooms. However, criticism of the system is that there is no radiant heating and a gas fire may be required to provide this, especially in the lounge.

Unit heaters for overhead use

These are suitable for factories, churches and assembly halls and two types are available:

1. Indirect, which have a flue to carry off the products of combustion and

have a heat output of up to 230 kW at 75 per cent efficiency.

2. Direct or flueless heater, in which the products of combustion are circulated together with the heated air and have a heat output of up to 41 kW at 90 per cent efficiency.

Figure 7.21 shows a direct overhead unit heater installed in a factory or assembly hall.

Radiant heaters

Thermal comfort may be provided for the occupants of a room at comparatively low capital and running costs, by use of radiant heaters mounted either on the wall or ceiling. They can be installed in factories, churches and assembly halls and are used as an alternative to the convector heaters for heating these buildings. Thermal comfort conditions can be obtained quicker with radiant heaters than with convector heaters and the pre-heating time and fuel costs are therefore reduced to a minimum. Figure 7.22 shows a wall- or ceiling-mounted radiant heater.

Another type of radiant heater consists of a 64 mm bore steel U tube into which a gas burner is fixed. A silent fan draws the products of combustion through the tube which is heated before the gases are discharged to atmosphere. An insulated polished reflector is fitted above the tube to ensure maximum radiation of heat.

Flues for gas appliances

Gas flues are simpler and cheaper to construct than flues for solid fuel or oil and they do not require periodic cleaning. Certain gas appliances such as space heaters and cookers do not require a flue and are permitted to discharge their products of combustion into the room in which they are installed. All other appliances require a flue and the local Gas Board should be consulted on this matter. Gas appliances have to be able to function without a flue, and the flue is therefore only required to discharge the products of combustion to the atmosphere and not to create a draught to aid combustion.

The products of combustion of gas are clean and comply with the Clean Air Act. The type of flue for a specific project depends upon several factors which include the height and type of construction of the building, the type and siting of the appliance, wind conditions and current building regulations. These factors are interacting and must therefore be considered as a whole. The products of combustion from gas appliances contain water vapour, and the placing and design of the flue should either prevent condensation or remove any water resulting from it.

Terms used

Draught diverter. A device designed for preventing downdraught or static conditions in the secondary flue of an appliance from interfering with the combustion of gas within the appliance. It also prevents excessive draught conditions by allowing the air in the room to mix with the products of combustion in the secondary flue and then cool down these hot gases. Figure 7.23 shows the operation of a draught diverter.

Duct. A tube or casing used for the passage of the products of combustion or air.

Excess air. Air in excess of that theoretically required for complete combustion of gas.

Flue. A tube or casing used for the passage of the products of combustion.

Primary flue. A length of flue prior to the draught diverter (see Fig. 7.24).

Secondary flue. This is the flue proper and is the flue between the draught diverter and the terminal (see Fig. 7.24).

Main flue. A flue used for carrying the products of combustion from two or more appliances (see Fig. 7.25).

Subsidiary flue. A flue connecting an appliance with the main flue (see Fig. 7.25).

Individual flue. A flue serving only one appliance.

Branched or shunt flue system. A flue system comprising a main flue into which the products of combustion from two or more appliances discharge, each by way of a vertical subsidiary flue (see Fig. 7.26).

Common flue system. A flue system taking the products of combustion from more than one appliance. It includes a branched flue system.

Room-sealed appliance. An appliance having the air inlet and flue outlets (except for the purpose of lighting) *sealed* from the room in which the appliance is installed. It includes a drying cabinet having an access door, with means of automatically closing the air inlet and flue outlet when the door is opened.

Balanced-flue appliance. An appliance designed to draw in combustion air from a point immediately adjacent to where it discharges its products of combustion. The inlet and outlet points are incorporated in a windproof terminal, which is sited outside the room in which the appliance is fitted (see Fig. 7.27 which shows the principles of operation of a balanced-flue appliance).

SE duct. A duct rising vertically which is open at the bottom and top, serving to bring combustion air to and take the products of combustion from room-sealed appliances to the external air (see Fig. 7.28).

U duct. A duct in the form of a U to one limb of which are fitted room-sealed appliances, while the other limb provides combustion air (see Fig. 7.29).

Terminal. A device fitted at the termination of the flue, designed to allow free passage of the products of combustion, to minimise downdraught and to prevent the entry of foreign matter which might cause restriction of the flue. Figure 7.30 shows various types of terminals.

Venting. The removal of the products of combustion from an appliance.

Principles of design of flues

A gas flue in a building may serve two purposes namely:

1. To remove the products of combustion from an appliance.
2. To assist in the ventilation of the room in which the appliance is installed.

The force causing the movement of the gases inside a flue is usually due to the difference in density between the hot gases in the flue and the cooler air inside the room. This force is small and it is therefore essential that bends and terminals used in the construction of the flue should offer low resistance to the flow of gases. The flue should also terminate in such a position in the open air that the effects of wind pressure will aid the updraught and not act adversely.

Note: Some gas flues use a fan to extract the products of combustion from the appliance, and therefore the force causing the movement of the gases in the flue

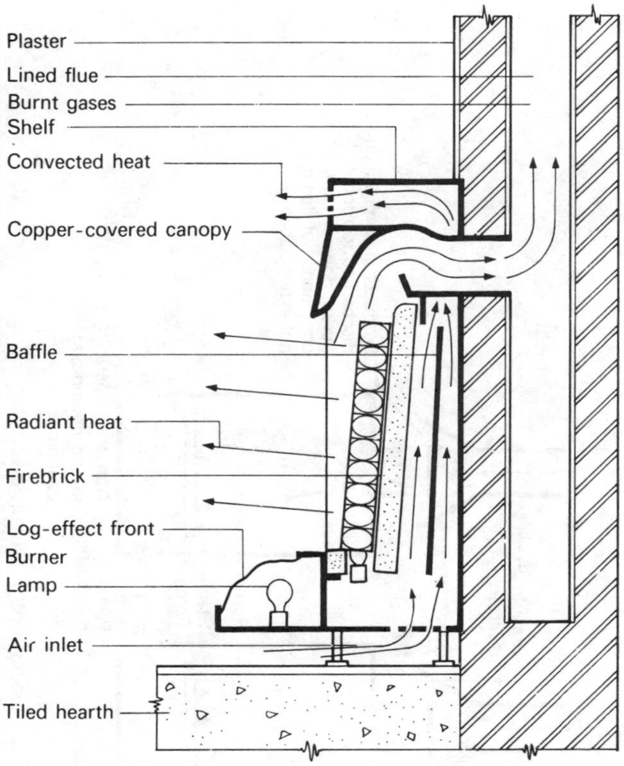
Plaster
Lined flue
Burnt gases
Shelf
Convected heat
Copper-covered canopy
Baffle
Radiant heat
Firebrick
Log-effect front
Burner
Lamp
Air inlet
Tiled hearth

Fig. 7.16 Built-in-type gas fire

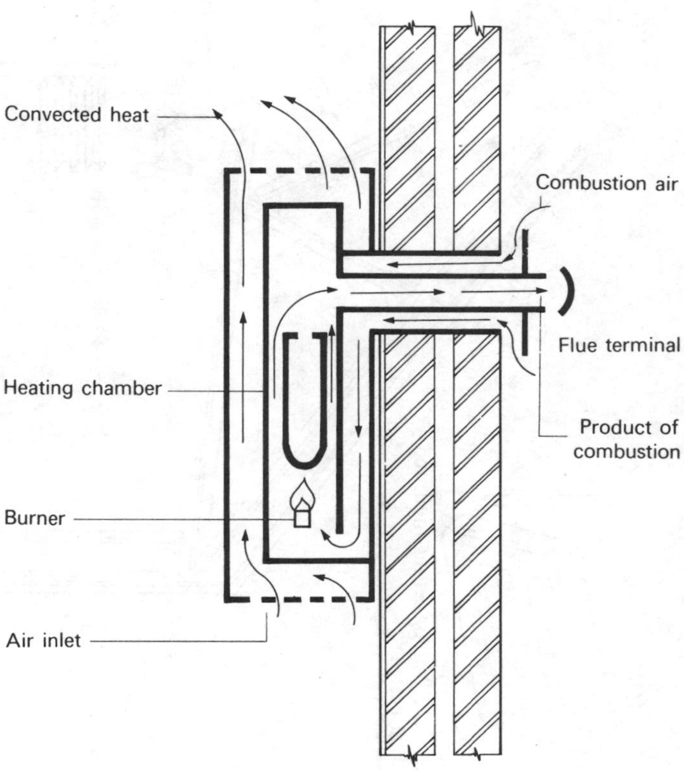
Convected heat
Combustion air
Flue terminal
Heating chamber
Product of
combustion
Burner
Air inlet

Fig. 7.17 Natural convector

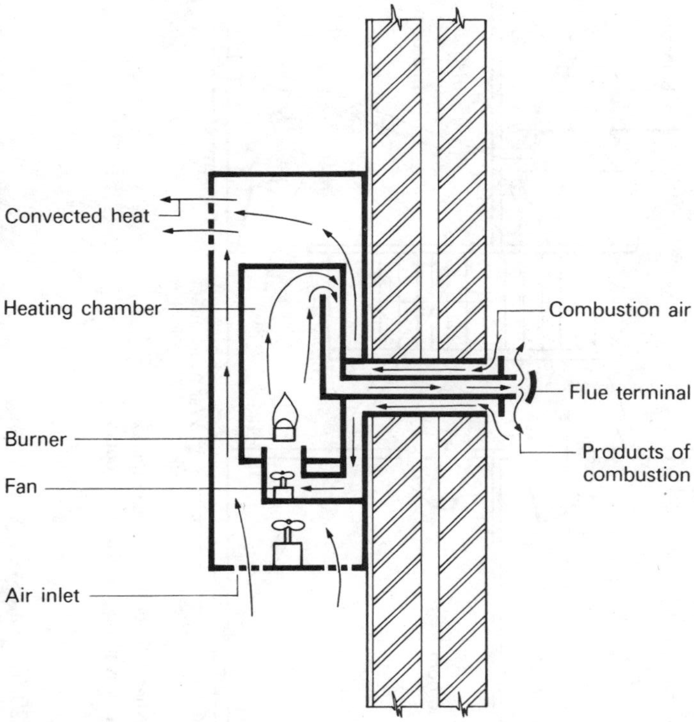
Convected heat
Heating chamber
Combustion air
Flue terminal
Burner
Products of
combustion
Fan
Air inlet

Fig. 7.18 Fan-assisted convector

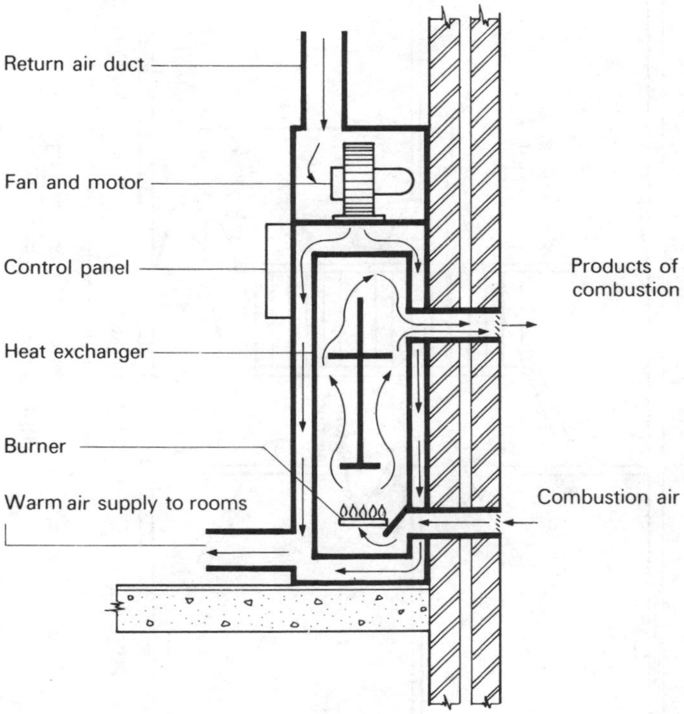
Return air duct
Fan and motor
Control panel
Products of
combustion
Heat exchanger
Burner
Combustion air
Warm air supply to rooms

Fig. 7.19 Gas warm-air unit

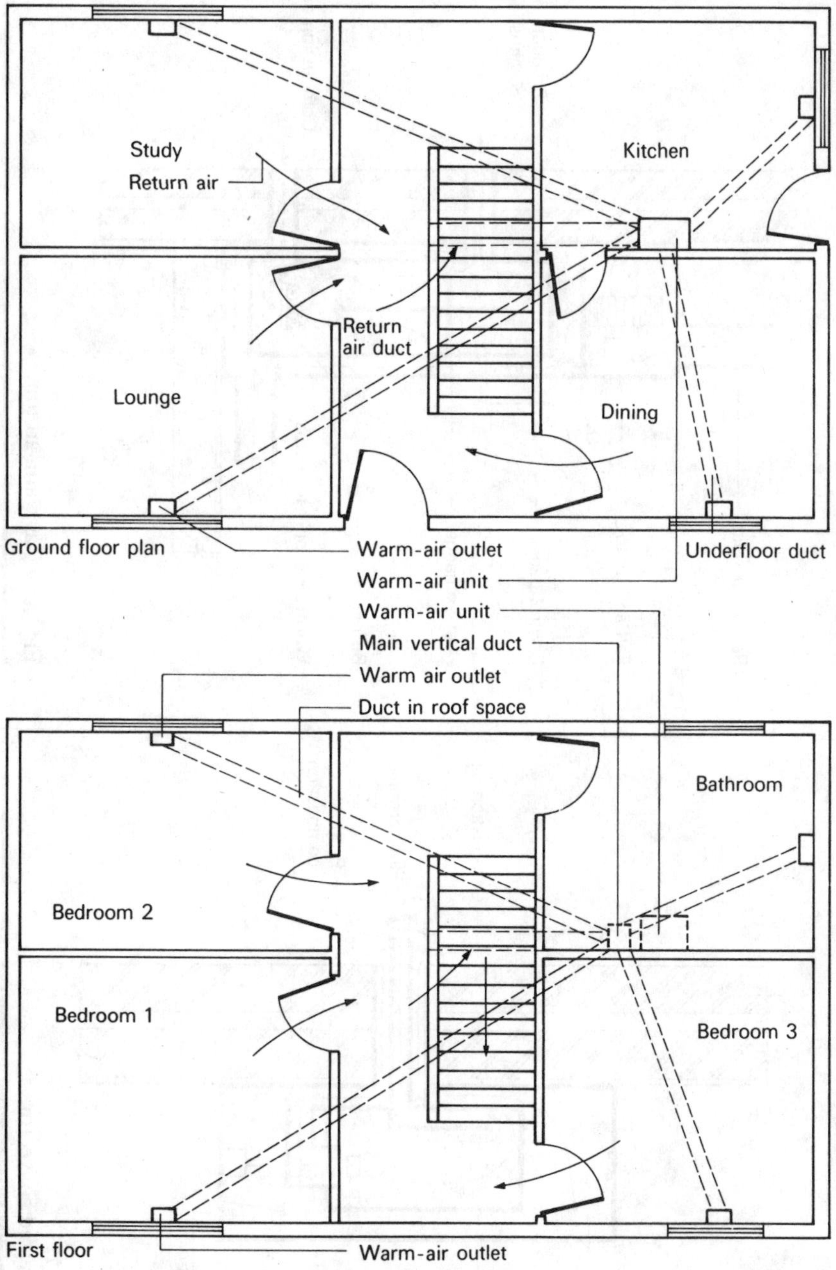

Study
Return air
Kitchen
Return air duct
Lounge
Dining
Ground floor plan
Warm-air outlet
Underfloor duct
Warm-air unit
Warm-air unit
Main vertical duct
Warm air outlet
Duct in roof space
Bathroom
Bedroom 2
Bedroom 1
Bedroom 3
First floor
Warm-air outlet

Note: When the doors are closed the air returns through grills in the doors

Fig. 7.20 Warm-air system

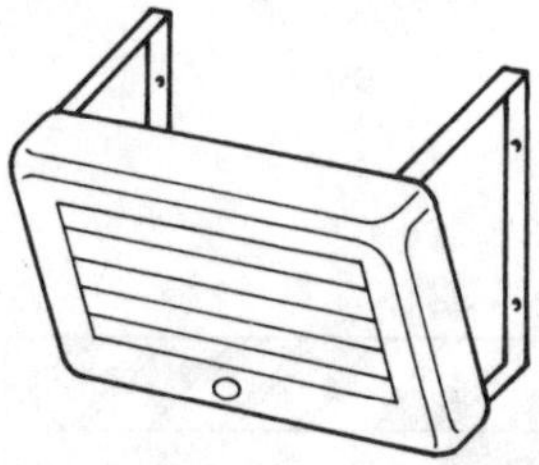

Fig. 7.22 Wall or ceiling-mounted radiant heater

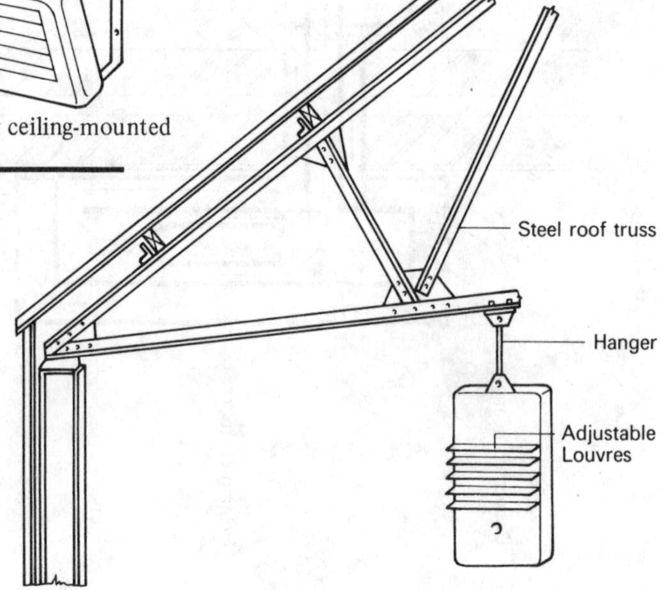

Steel roof truss
Hanger
Adjustable Louvres

Fig. 7.21 Direct-unit heater

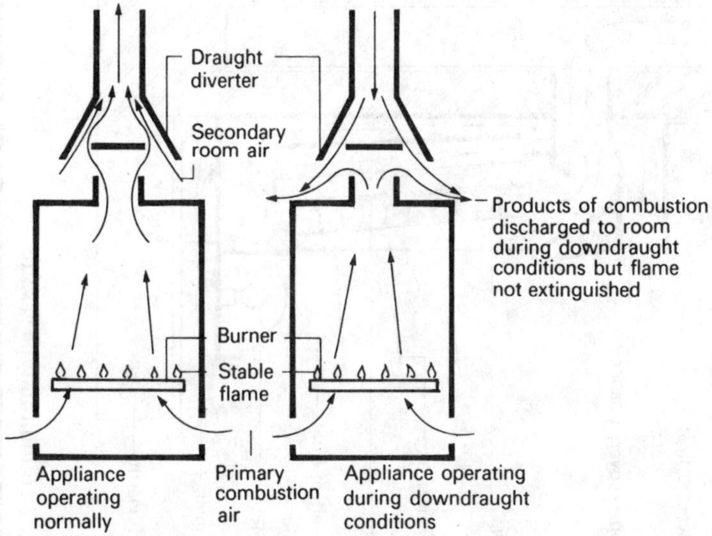

Draught diverter
Secondary room air
Products of combustion discharged to room during downdraught conditions but flame not extinguished
Burner
Stable flame
Appliance operating normally
Primary combustion air
Appliance operating during downdraught conditions

Fig. 7.23 Operation of draught diverter

is not due solely to the differences in density between the hot gases and the cooler air inside the room. However, this natural convection will reduce the power required for the fan. The size of a flue for a gas appliance is dependent upon the kilowatt rating of the appliance and the ventilation standard of the room in which the appliance is installed.

Materials for flues

1. Asbestos-cement pipes and fittings suitable for internal and external work and should conform to BSS 567.
2. Steel or cast-iron pipes protected by a good-quality vitreous enamel.
3. Sheet aluminium (except where temperatures and condensation is liable to be excessive, e.g. a water-heater flue in a cold position).
4. Sheet copper protected by chromium plating after fabricating.
5. Stainless steel.
6. Precast concrete flue blocks (see Fig. 7.31) made of acid resisting cement and pointed by acid resisting cement mortar.
7. Brick flue, lined with precast concrete made of acid resisting cement and pointed with acid resisting cement mortar, or glazed stoneware pipes pointed with high-acid resisting mortar.

Planning of flues

All necessary information with regards to the installation of gas flues, should be made at the early design stage of the building. The local Gas Board should be consulted before any drawings and specifications are made and also during the execution of the work. The drawings should show the proposed positions of appliances and full details of the routes of all flues, from the point of connection to the appliances to the position of termination.

Flue installations must conform with the Buildings Regulations, 1992. The following points must be considered when planning a flue:

1. The flue should rise progressively towards its terminal.
2. The primary flue should be as short as possible.
3. The secondary flue should be of adequate height and kept inside the building so far as is possible.
4. In many appliances the draught diverter is an integral part of the appliance and permits easy disconnection of the flue from the appliance. Where an appliance is not fitted with a draught diverter, a disconnecting device should be fitted as near as is practicable to the appliance.
5. Routes which expose the flue to rapid cooling should be avoided, or thermal insulation afforded to the flue external surface.
6. Horizontal flues and fittings having sharp angles must be avoided.
7. The joints of asbestos-cement pipes and fittings should be made by caulking with slag wool for about 25 per cent of the depth of the socket. The remaining 75 per cent of the socket should be filled with a good-quality fire cement and the outside edge neatly chamfered. All socket joints should face upwards.
8. Exterior flue pipes should be supported by brackets fitted th oughout the height of the flue at intervals of not more than 2 m, preferabl; ınmediately below the pipe socket.
9. Where a flue passes through combustible material, it must be covered by a metal sleeve, with an annular spacing of 25 mm packed with incombustible material (see Fig. 7.32).

Terminal position

The terminal of a gas flue should be placed in such a position that the wind can blow freely across it, the best positions being above the ridge of a pitched roof or above the parapet wall of a flat roof. Figure 7.33 shows the requirements for flues passing through a flat roof.

High-wind-pressure regions must be avoided and therefore terminals must not be placed below the eaves, in a corner or adjacent to another pipe. However, a terminal can be placed above the level of the eaves (as shown in Fig. 7.24) or on the unobstructed portion of a wall.

Condensate and its removal

Some initial condensation in a gas flue will occur immediately after the appliance has been lit and it tends to persist in flues placed on the external walls or in flues attached to appliances of high efficiencies, especially if these are run for long periods at much less than their rated maximum output. Well-insulated internal flues are preferable to external flues which increase the degree of condensation. The flues should be built so that condensate can flow freely to a point where it can be released, preferably into a 25 mm bore lead or copper condensate pipe (see Fig. 7.34).

Large boilers

It is desirable to house large gas-fired boilers in separate boiler rooms, which may be part of the main structure or completely separate. A boiler room sited on the roof of the building requires a very short flue (see Fig. 7.35). Alternatively, the boiler may be installed at ground level and a fan-diluting flue used (see Figs. 7.36, 7.37 and 7.38). With this type of flue, a fan draws in fresh air which is mixed with the products of combustion from the appliance and discharged to the outside air.

It is essential to provide permanent air inlets to the boiler room to ensure a sufficient supply of air for the efficient operation of the boilers and these inlets should be:

1. At least twice as great in free area as the area of the primary flue pipe.
2. Located at least 300 mm above ground level and fitted with a grill.
3. Constructed of durable material.

Note: The provision of combustion air to gas appliances, is covered by the Building Regulations.

Shared flues

Gas appliances in multi-storey buildings may be connected to a shared flue in the form of a SE-duct, U duct or shunt duct. The use of a shared flue in a multi-storey building, saves considerably in space and installation costs over the use of individual flues. Both the SE-duct and the U duct require room-sealed, balanced-flue appliances, and the shunt duct requires a conventional flue appliance which has the advantage of ventilating the room in which the appliance is installed. For safety reasons, all appliances connected to shared flues must be provided with flame-failure devices.

The degree of dilution of the products of combustion in a shared flue is sufficient to ensure the satisfactory operation of all gas appliances connected to the system.

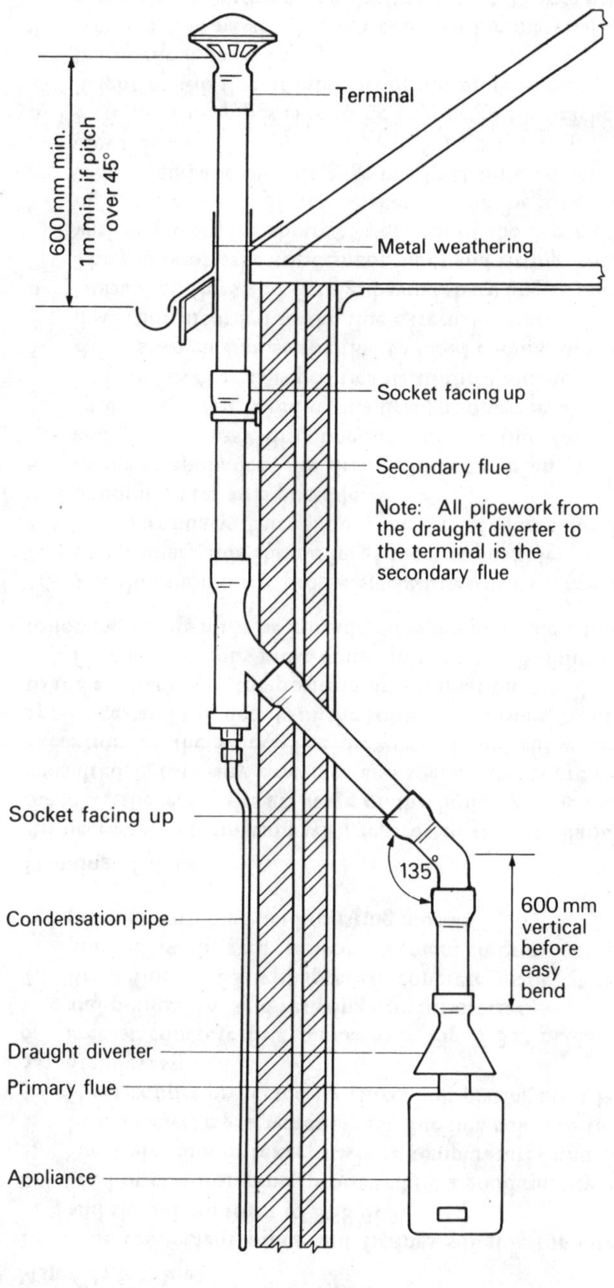
Terminal
600 mm min.
1m min. if pitch over 45°
Metal weathering
Socket facing up
Secondary flue
Note: All pipework from the draught diverter to the terminal is the secondary flue
Socket facing up
135°
600 mm vertical before easy bend
Condensation pipe
Draught diverter
Primary flue
Appliance

Fig. 7.24 Primary and secondary flue

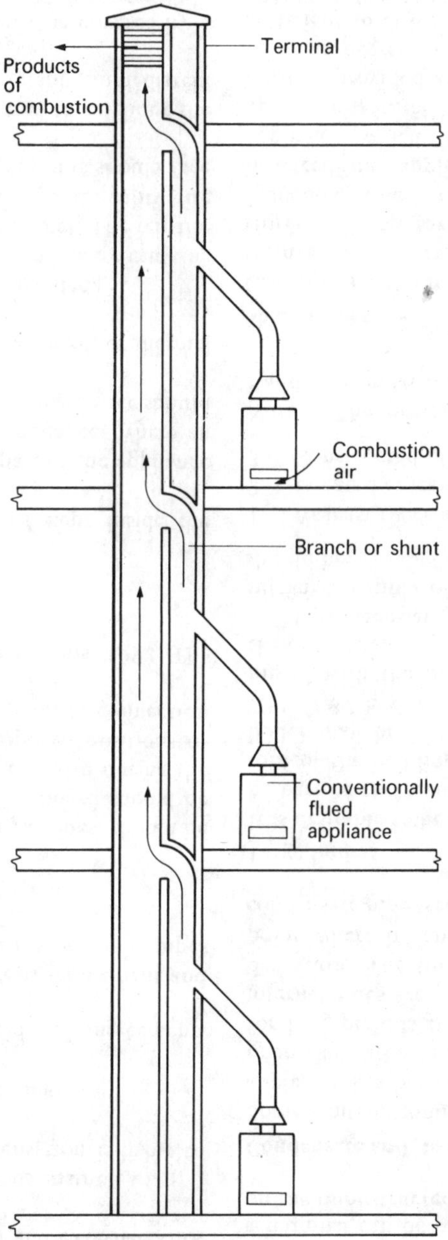
Terminal
Products of combustion
Combustion air
Branch or shunt
Conventionally flued appliance

Fig. 7.26 Branch and shunt flue

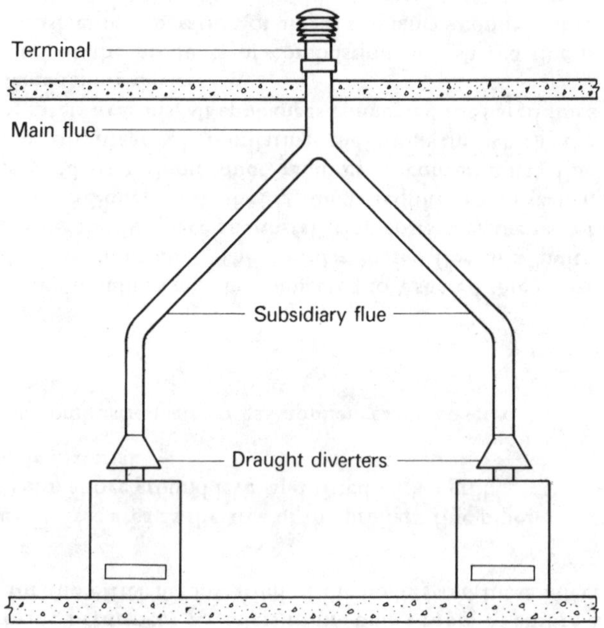
Terminal
Main flue
Subsidiary flue
Draught diverters

Fig. 7.25 Main and subsidiary flues

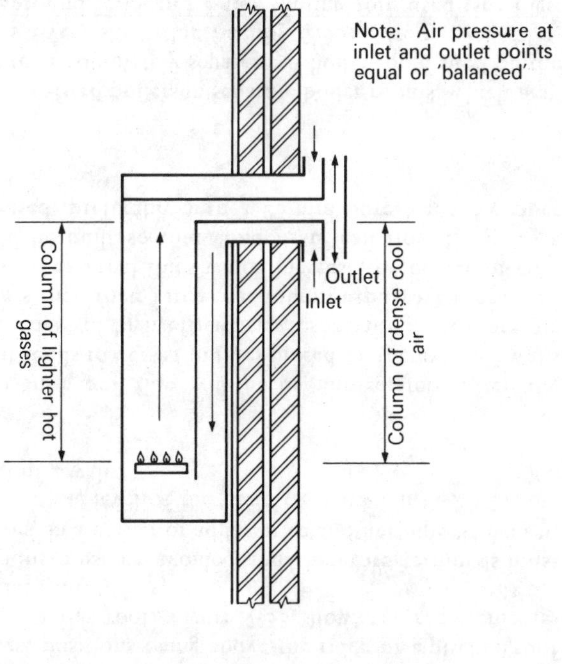
Note: Air pressure at inlet and outlet points equal or 'balanced'
Outlet
Inlet
Column of lighter hot gases
Column of dense cool air

Fig. 7.27 Principle of operation of balanced-flue appliance

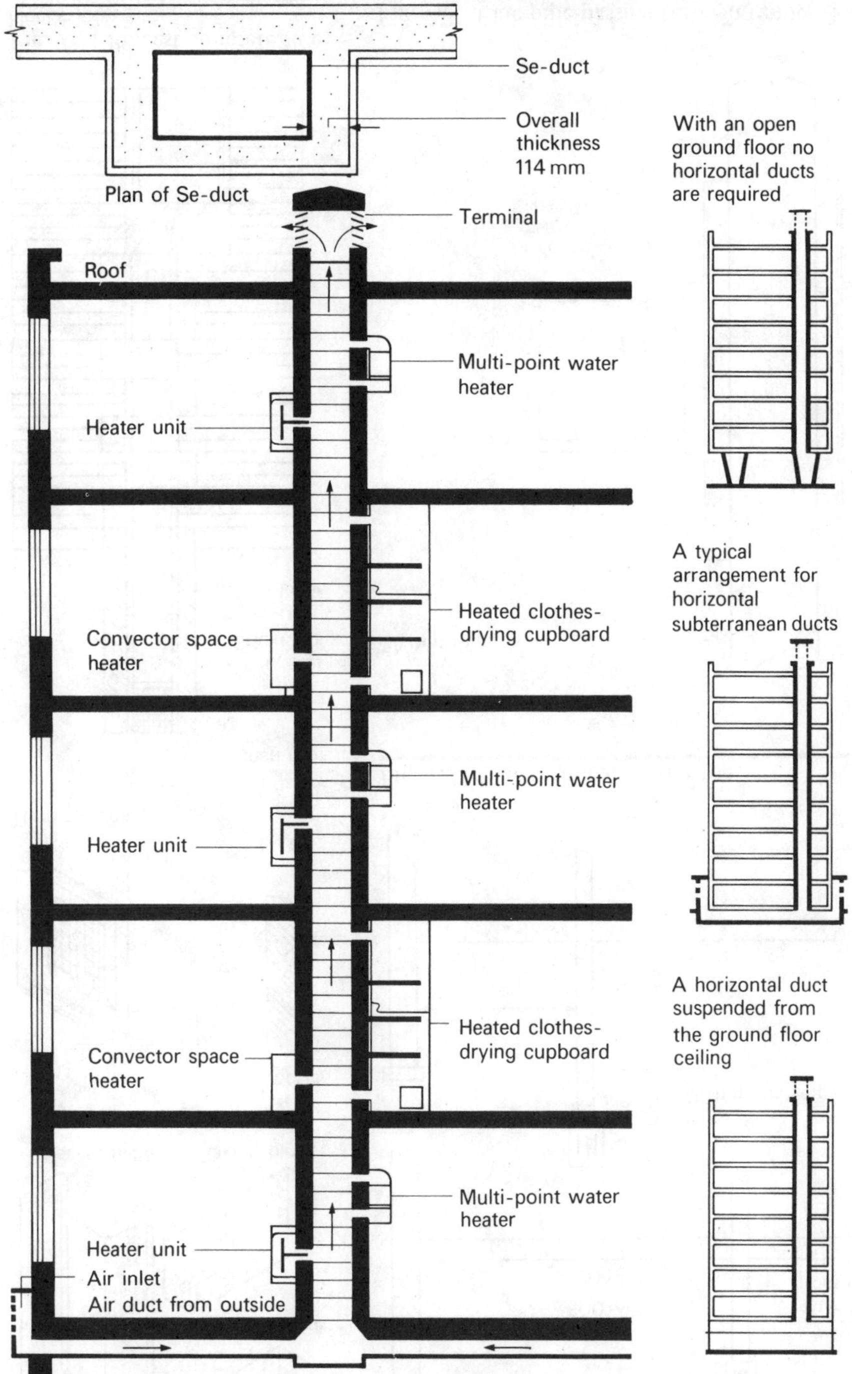
Se-duct
Overall thickness 114 mm
Plan of Se-duct
Terminal
Roof
Multi-point water heater
Heater unit
Convector space heater
Heated clothes-drying cupboard
Multi-point water heater
Heater unit
Convector space heater
Heated clothes-drying cupboard
Multi-point water heater
Heater unit
Air inlet
Air duct from outside
With an open ground floor no horizontal ducts are required
A typical arrangement for horizontal subterranean ducts
A horizontal duct suspended from the ground floor ceiling

A typical SE-duct installation showing how the various sealed appliances can be arranged. The flanking drawings show alternative air inlet arrangements.

Fig. 7.28 The Se-duct (Courtesy of the British Gas Corporation)

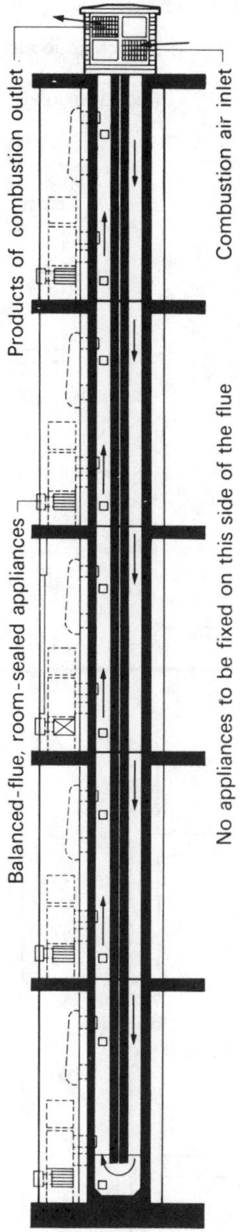
Products of combustion outlet
Combustion air inlet
Balanced-flue, room-sealed appliances
No appliances to be fixed on this side of the flue

Fig. 7.29 The U duct (Courtsey of the British Gas Corporation)

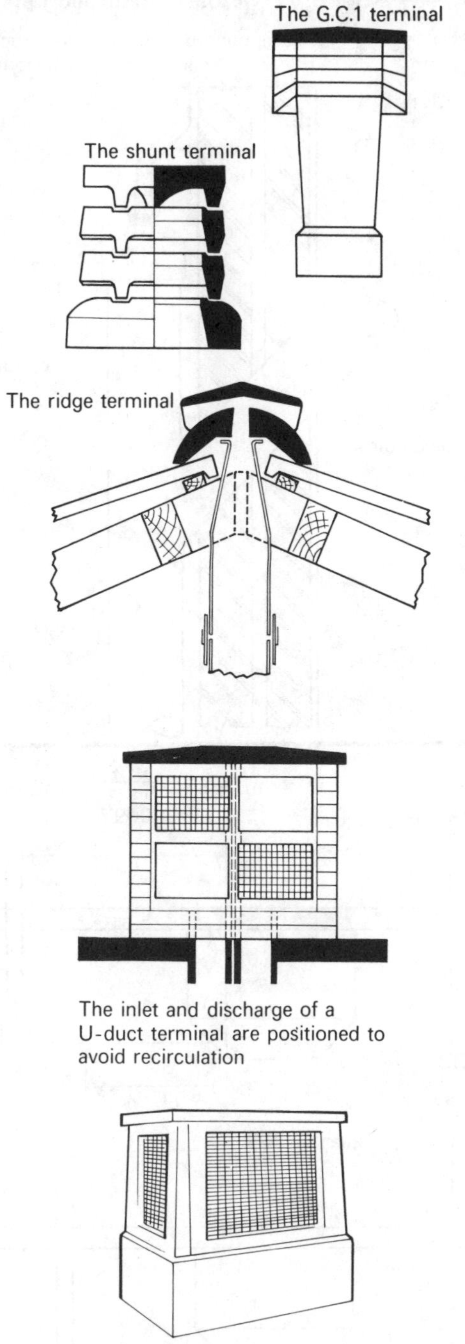
The G.C.1 terminal
The shunt terminal
The ridge terminal
The inlet and discharge of a U-duct terminal are positioned to avoid recirculation

A typical SE-duct terminal on the roof of a modern block of flats

Fig. 7.30 Types of terminals (Courtesy of the British Gas Corporation)

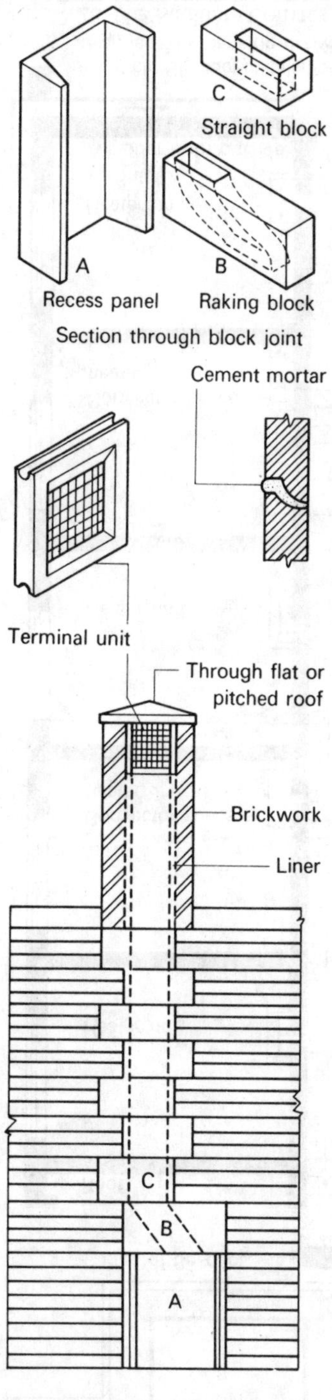
C
Straight block
A
Recess panel
B
Raking block
Section through block joint
Cement mortar
Terminal unit
Through flat or
pitched roof
Brickwork
Liner
C
B
A

Fig. 7.31 Precast concrete flue blocks

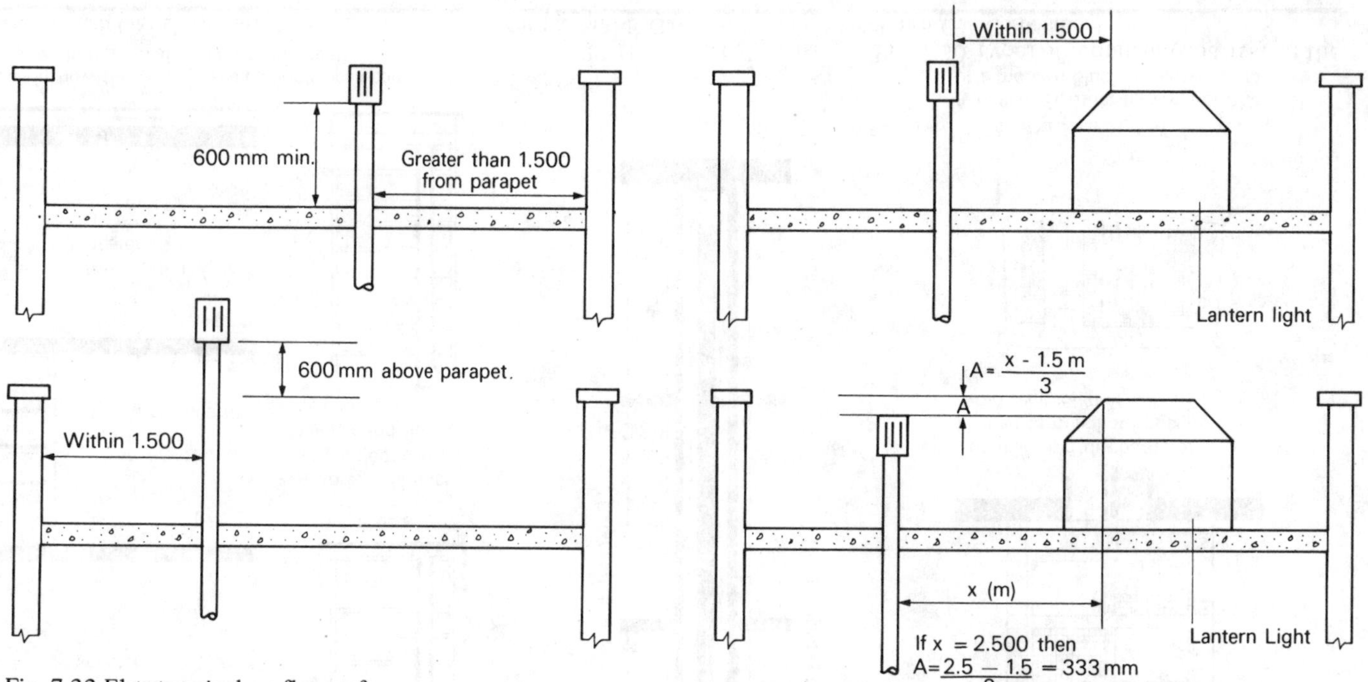
600 mm min.
Greater than 1.500
from parapet
600 mm above parapet.
Within 1.500
Within 1.500
Lantern light
A = x - 1.5 m / 3
A
x (m)
If x = 2.500 then
A = 2.5 − 1.5 / 3 = 333 mm
Lantern Light

Fig. 7.33 Flue terminal on flat roof

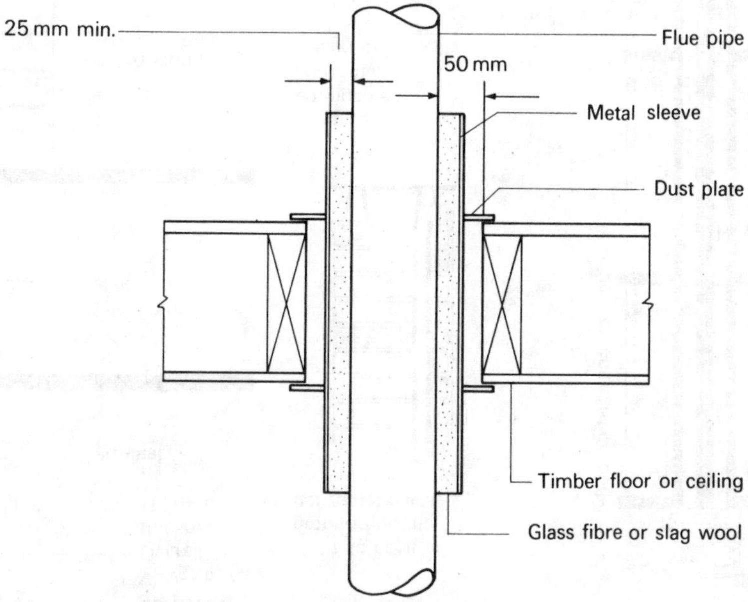
25 mm min.
Flue pipe
50 mm
Metal sleeve
Dust plate
Timber floor or ceiling
Glass fibre or slag wool

Fig. 7.32 Flue pipe passing through timber floor or ceiling

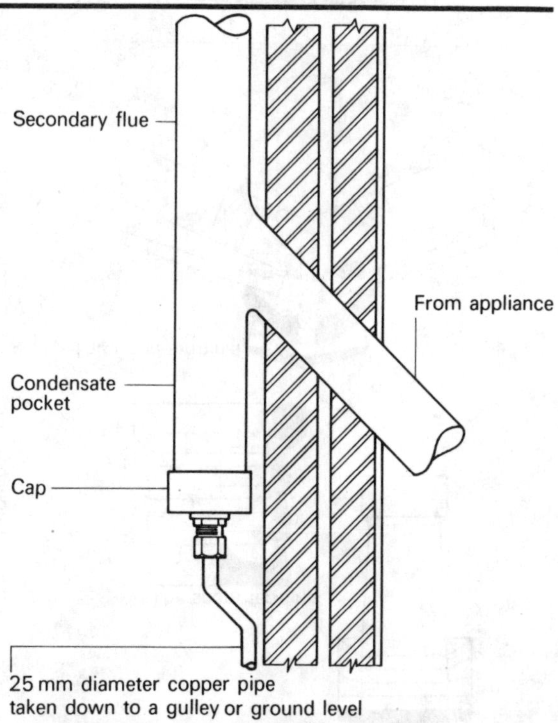
Secondary flue
From appliance
Condensate
pocket
Cap
25 mm diameter copper pipe
taken down to a gulley or ground level

Fig. 7.34 Condensate removal

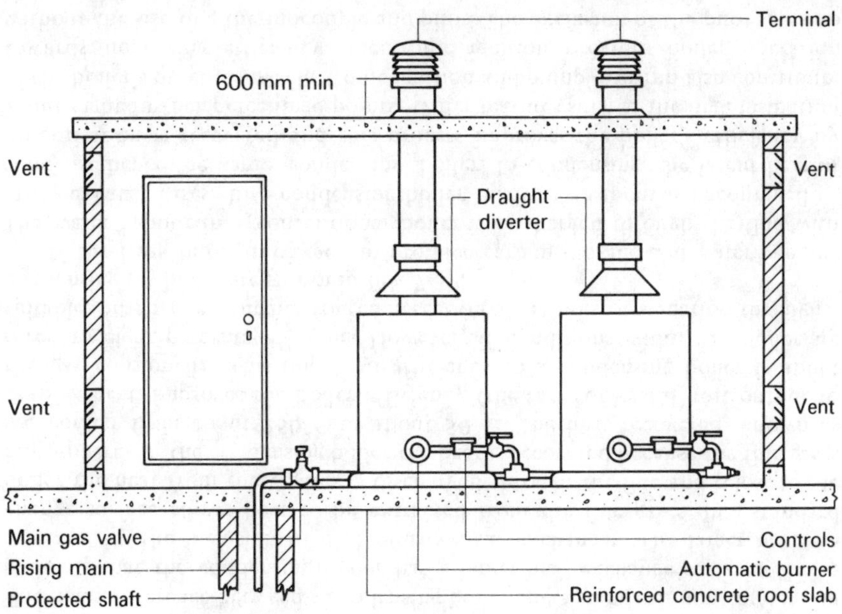
Terminal
600 mm min
Vent
Vent
Draught diverter
Vent
Vent
Main gas valve
Rising main
Protected shaft
Controls
Automatic burner
Reinforced concrete roof slab

Fig. 7.35 Rooftop boiler room

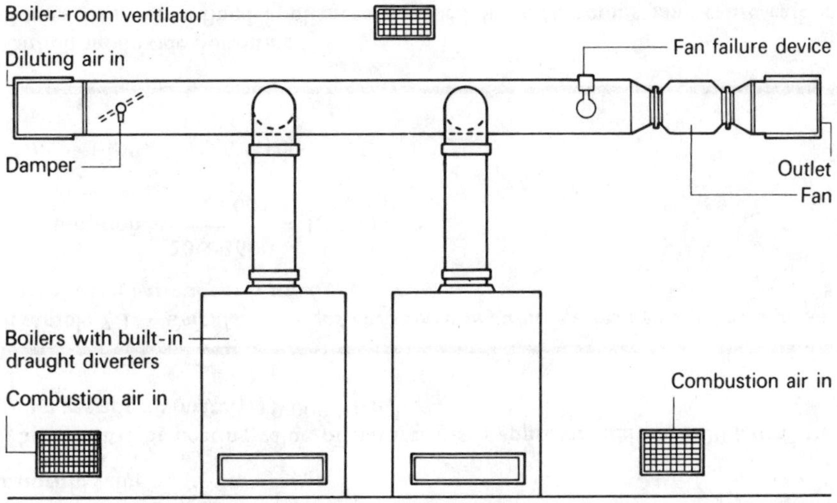
Boiler-room ventilator
Diluting air in
Fan failure device
Damper
Outlet
Fan
Boilers with built-in draught diverters
Combustion air in
Combustion air in

Fig. 7.36 Fan-diluted flue (Courtesy of Gas Council)

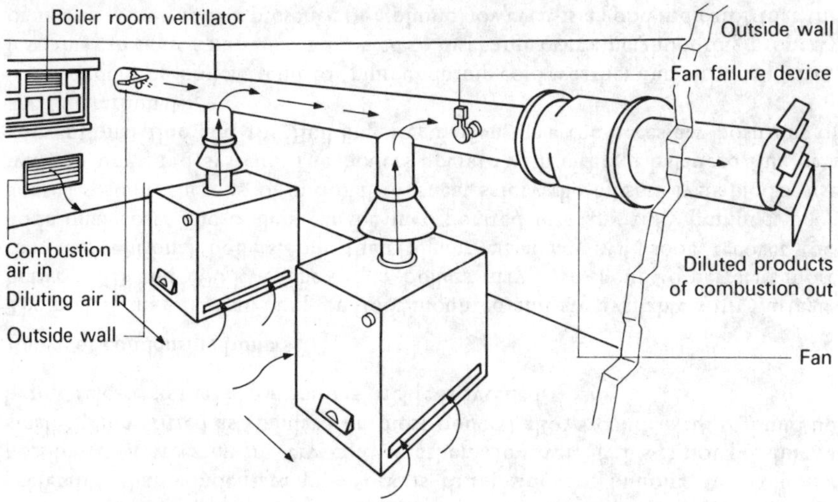
Boiler room ventilator
Outside wall
Fan failure device
Combustion air in
Diluting air in
Outside wall
Diluted products of combustion out
Fan

Fig. 7.37 Installation using two outside walls and boilers with draught diverters

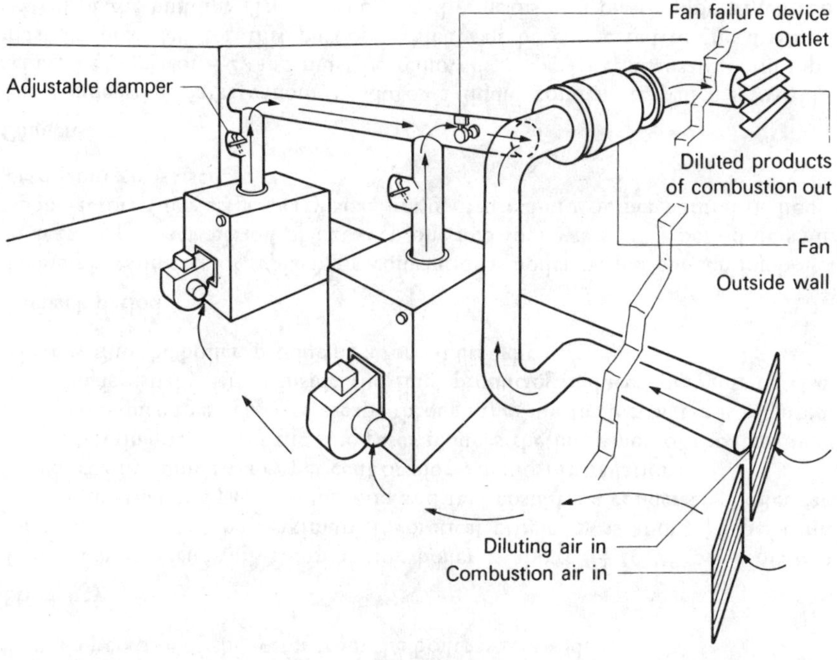
Fan failure device
Outlet
Adjustable damper
Diluted products of combustion out
Fan
Outside wall
Diluting air in
Combustion air in

Fig. 7.38 Installation using one outside wall and boilers with automatic gas burners

Calculation of gas consumption

The gas requirement from appliance may be found from the following formula:

$$\frac{\text{power}}{\text{calorific value}} = m^3/s$$

The calorific or heating value of natural gas is approximately 37 000 kJ/m^3 and town gas approximately 19 000 kJ/m^3.

Example 7.1. *Calculate the gas consumed in cubic metres per hour by a 20 kW boiler when natural gas is used.*

$$\text{Consumption} = \frac{20 \times 3600}{37\,000} = 1.946\ m^3/h$$

Note: 1 kW = 3.6 MJ/h

Carbon monoxide poisoning

Fumes from open-flued appliances installed in small rooms may cause carbon monoxide poisoning. The Building Regulations require that if an appliance is in a bath or shower room or a private garage, it must be a balanced-flued appliance. Since this appliance is room-sealed there is no risk (other than from the outside) of products of combustion entering the room.

The condensing gas boiler

The main difference between a condensing boiler and a standard, non-condensing boiler is that the condensing boiler has a larger heat exchange surface. This is achieved by the inclusion of a secondary heat exchanger. The larger heat exchange surface allows heat to be extracted from the flue gases. In a standard boiler the heat from these gases is discharged through the flue and is lost to the atmosphere. In the condensing boiler the heat is recovered because the flue gases are cooled from about 250°C to about 50°C. The heat recovered, known as *sensible* heat, improves the boiler efficiency (the ratio of useful heat output to the gas consumed). This improved efficiency in a condensing boiler is about seven to eight percentage points. However, if conditions within the boiler are suitable, heat from a further source is recovered. Details of a section through a condensing gas boiler are shown in Fig. 7.39.

Natural gas burns in oxygen to produce carbon dioxide and water vapour. The water vapour in a conventional boiler is discharged through the flue with other exhaust gases. In a condensing boiler this water vapour is encouraged to condense hence the name 'condensing' boiler. In condensing, the latent heat of vapourisation is recovered and this process improves the boiler's efficiency by about eight further percentage points. Other features such as the high insulation of the boiler's outer casing, electronic ignition and pump over-run also contribute towards improving efficiency. Electronic ignition provides boiler operation without the use of a thermocouple and pilot. The exclusion of the pilot saves on the running costs.

Boilers are now low content appliances. On boiler shut down, the circulating pump keeps running, extracting useful heat from the heat exchanger until the flow temperature of the water from the boiler is reduced.

Efficiency

The bench efficiency of a condensing boiler is about 94 to 95 per cent; in a conventional boiler the maximum theoretical efficiency is about 74 per cent. This means that the gas consumption and fuel costs for a condensing boiler can be reduced by some twenty per cent or more, in most installations.

Extracting heat from the flue gases reduces the buoyancy of the remaining gases. To induce these gases to escape through the flue an electric fan is required. This fan is usually fitted inside the flue, producing induced draught, or elsewhere within the boiler, producing a forced draught.

Payback period

Economic estimates of replacing a conventional boiler with a condensing boiler indicate a payback period of between one and four years. This period depends upon factors such as property size, hours of use and whether natural or liquid petroleum gas is used.

Condensate

The condensing water vapour produces a liquid condensate which is weakly acidic, i.e. a pH of 4.2, and must be removed. A special condensate drain pipe must be provided for this purpose which can be connected to the drainage system of the building. This can be done by means of a plastic pipe which can be connected to a waste pipe or a soakaway. The acidic nature of the condensate is seldom harmful to the drainage system because when diluted with the sink waste water, which is slightly alkaline, the overall effect is to neutralise the mixture.

Installation of boiler

The condensing boiler is as straightforward to install as a conventional boiler, but a condensate pipe must be connected to the internal waste system. It is preferable that condensate pipework is run inside the building to avoid the possibility of freezing in very cold weather. However, if this is not possible, a small siphon (fitted as standard on most boilers) allows condensate to build up before release so that a continuous trickle is avoided.

Range of condensing boilers

The boilers may be wall-hung or floor-mounted and are available with wattages between 10 and 600 kW. The larger boilers may cost about 20 per cent more than conventional boilers, but their higher efficiency will soon recover the additional cost. The boilers can be incorporated into any fully pumped, open vented system; there is an optional overheat safety thermostat available for use with an unvented system. The boilers operate with either a balanced flue or a conventional flue and the Building Regulations give the necessary positions of the flue terminals.

Condensing boilers tend to 'plume' during cold weather and although harmless, this can cause a nuisance if allowed to discharge over a neighbouring property or near a window. The presence of a plume however, is a good indicator that the boiler is working at or near maximum efficiency.

Summary

1. Condensing boilers can be cost effective at present fuel prices.
2. The existing radiator system can be used.
3. Conventional control systems can be used.
4. Installation is straightforward.
5. Use provides a saving in natural gas and operating cost.
6. Use provides a reduction in CO_2 emission which reduces the threat of global warming.
7. Condensate disposal is no problem.
8. Maintenance requirements are unchanged.
9. Condensate is not highly acidic (about the same as tomato juice).
10. Replacement of a conventional boiler with a condensate boiler is straightforward.
11. Comfort levels are maintained.

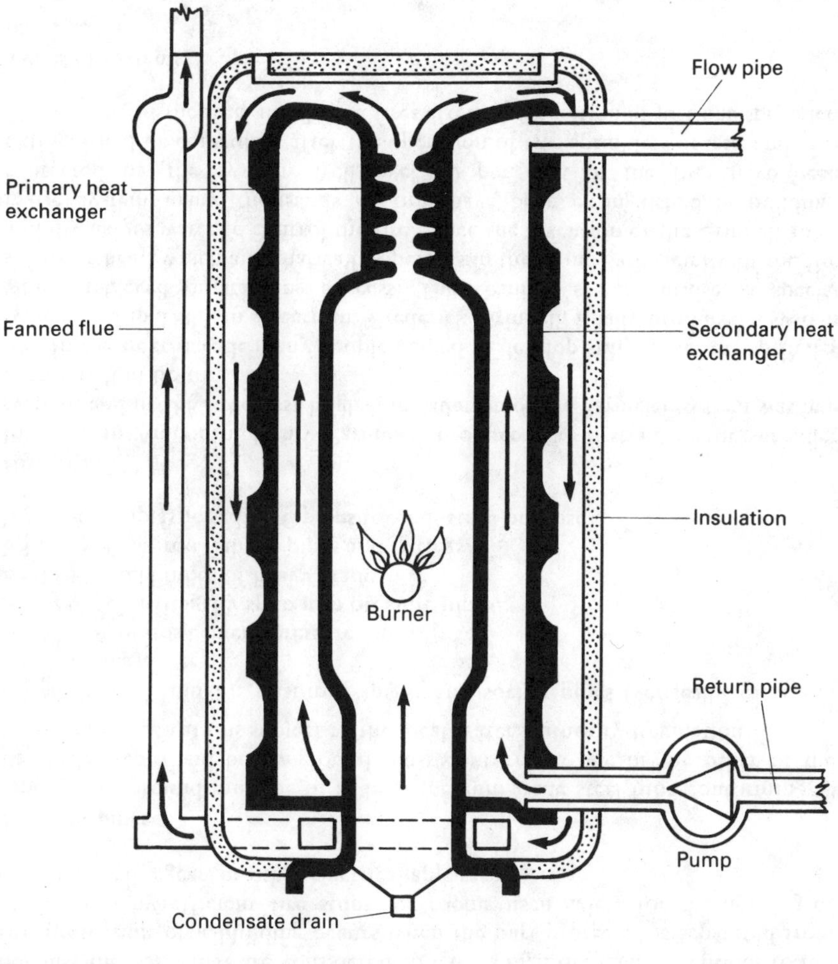
Flow pipe
Primary heat exchanger
Fanned flue
Secondary heat exchanger
Insulation
Burner
Return pipe
Pump
Condensate drain

Fig. 7.39 Section of a condensing boiler

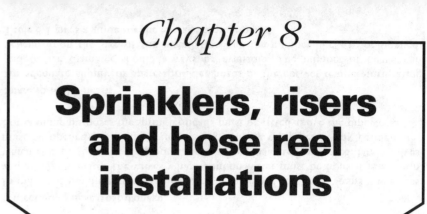

Chapter 8 Sprinklers, risers and hose reel installations

Fire-fighting equipment

Fire-fighting equipment for buildings may be divided into the following categories:

1. Sprinklers and other fixed water sprays.
2. Fixed foam, carbon dioxide and dry powder extinguishers.
3. Fixed wet or dry risers.
4. Portable extinguishers.
5. Fire doors, dampers and fire-resisting forms of construction.

Although fire detectors and alarms are not fire-fighting equipment, their installation will provide a warning of fire and reduce the time taken for the fire to be brought under control and extinguished. They will also reduce the risk of injury or loss of life by providing a warning of a fire.

Sprinkler systems

These are the most important and successful of the fire-fighting systems and fire insurers will allow a rebate of up to 70 per cent when an automatic sprinkler system is installed, which will offset the cost of the installation.

However, a sprinkler installation is a first-aid system for dealing with a fire in its early stages and cannot be relied upon to deal with a large fire which has started in, or spread from, an unprotected part of the building. It is essential, therefore, that a sprinkler installation should cover the whole of the building and not just the parts that are considered to have a high fire risk. Exceptions to full fire protection of a building occurs when the part protected is separated from other parts by efficient fire stops, or rooms used solely for the housing of electrical switchgear, or other electrical apparatus.

Types of pipes

Pipes above ground must be of at least medium-grade steel tube conforming to BS 1387. Pipes laid below ground must conform to one or the other of the following specifications subject to the local Water Authority Regulations:

BS 78 (Parts 1 and 2): Cast-iron spigot and socket pipes (vertically cast) and fittings.
BS 486: Asbestos-cement pressure pipes.
BS 1211: Centrifugally spun iron pressure pipes.
BS 1387: Steel tubes of heavy grade.
BS 2035: Cast-iron flanged pipes and fittings.
BS 3506: Unplasticised PVC pipes for industrial purposes.

Historical

In 1874 an American, Henry Parmelee, produced the first automatic sprinkler system, and in 1882 the first English installation of the Parmelee system was in a cotton mill in Bolton.

In the next decade many people helped to develop still further the sprinkler system and in 1883 an American, Frederick Grinnell, finally produced a system which achieved outstanding success. The Grinnell system utilises a special sprinkler head which automatically operates in the immediate vicinity of the fire and discharges water to control the fire. Since the invention of the Grinnell automatic system many thousands of fires have been extinguished in buildings protected by the system. Well over 50 per cent of the fires have been extinguished or controlled after the operation of two sprinklers or less, and over 38 per cent of the total number of fires after the operation of just one sprinkler.

Types of systems

Wet system

This is used in heated buildings where temperatures remain above 0 °C and there is therefore no risk of the water in the system freezing. As the name implies the 'wet' system pipework is constantly filled with water. Heat produced by an outbreak of fire causes the nearest sprinkler head or heads to open at their operating temperature. Immediately the water is discharged on to the fire, the flow of water activates a hydraulically operated alarm bell outside the building and arrangements can also be made to alert the local Fire Brigade. Figure 8.1 shows a typical wet sprinkler system where the water supply is taken directly from the water main.

Dry system

This is used in unheated buildings where the temperatures may be below 0 °C and therefore water in the system is liable to freeze. The installation pipework is charged with compressed air at a moderate pressure and the water is held back by a special differential valve which, when a sprinkler head opens and releases air in the pipes, lifts and allows water to enter the pipework.

Alternate wet and dry system

This is used in buildings where freezing is likely during the winter months. During the summer months the system is filled with water and operates as a 'wet' system. When winter approaches the pipework is emptied of water and charged with compressed air and operates as a 'dry' system. Figure 8.2 shows the installation of a dry, or an alternate wet and dry system. A pump has been included which is fitted on all except the smallest of systems. The purpose of the pump is to increase the velocity of the flow of water and to speed up the lifting of the differential valve, therefore reducing the time taken for the water to reach the fire after a sprinkler head has opened.

Pre-action system

This system is used where prior warning of system discharge is necessary, or where accidental discharge, due to damaged pipes or sprinkler heads, would be unacceptable.

The system consists of pipework charged with air at low pressure and an electrically operated valve used to retain the water. A system of heat-sensitive detectors respond to a fire and signals the valve to release water into the pipework, at the same time sounding an alarm bell. After a short delay, one or more of the sprinkler heads are affected by the heat and open to discharge water on to the fire.

Deluge system

This is employed for fire risks requiring total or zoned water coverage. The pipework in the system is not pressurised and all the water-discharge nozzles are open. The system can be put into operation either by manual or automatic detectors which open a control valve and allows the simultaneous discharge of water from all the open nozzles. The system is used for the protection of aircraft hangars.

Tail-end system

The tail-end or subsidiary system is used when a portion of a building is subject to damage by frost, while the rest of the building is adequately heated. The system is a variation of the wet system and the alternate wet and dry system. A differential air valve is connected to the pipework, and the pipework above the valve is charged with compressed air.

Protection against frost damage

In unheated buildings using the dry or alternate wet and dry systems, the valves and all piping below the valves which contain water must be housed in a room which is kept above freezing-point, or in a frost-insulated cupboard. Drain valves must be opened occasionally in order to remove the pipes of any condensate. For draining purposes the sprinkler heads must be fitted above the pipework.

Spare sprinkler heads

An adequate supply of spare sprinkler heads of the correct temperature rating and orifice diameter should always be available. The number of heads will depend upon the size of the installation and the number likely to be opened. Table 8.1 gives a guide for the average requirements.

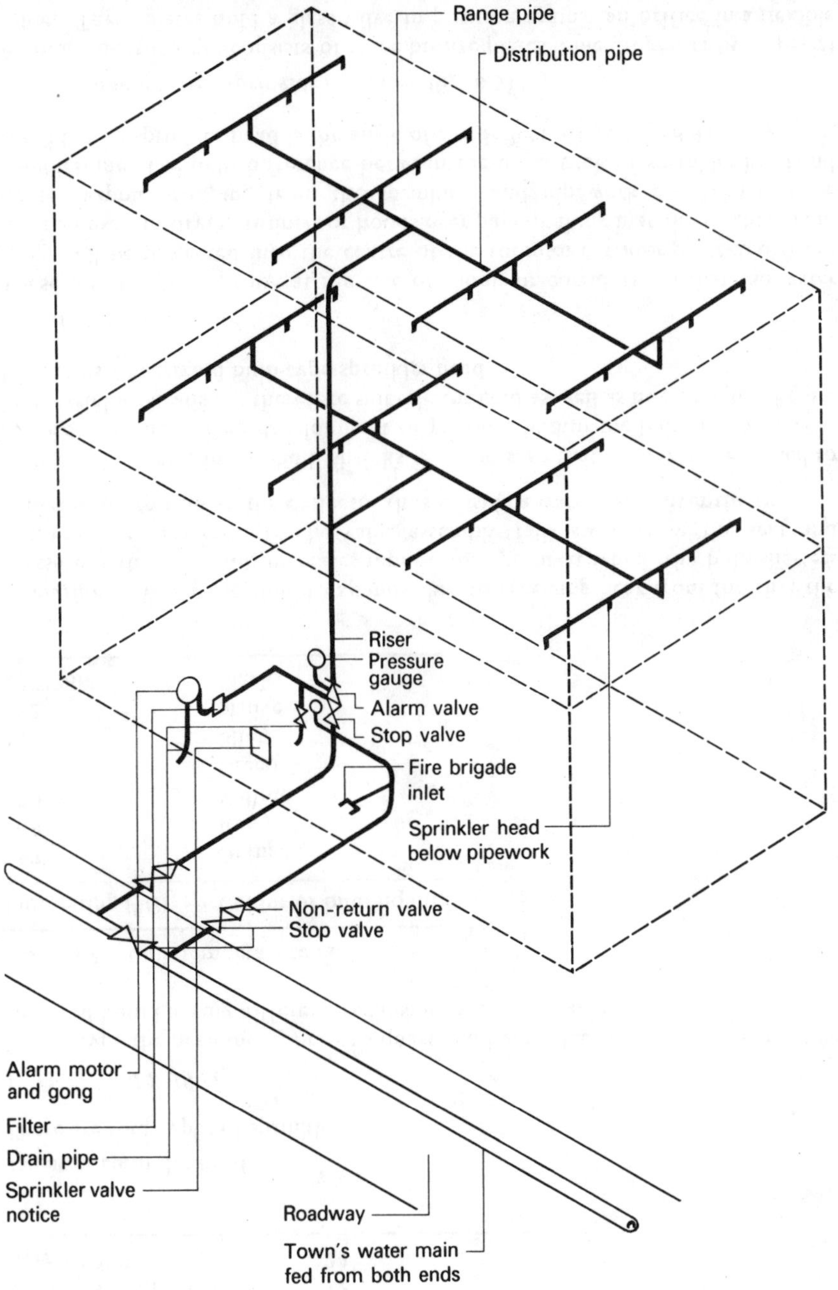
Range pipe
Distribution pipe
Riser
Pressure gauge
Alarm valve
Stop valve
Fire brigade inlet
Sprinkler head below pipework
Non-return valve
Stop valve
Alarm motor and gong
Filter
Drain pipe
Sprinkler valve notice
Roadway
Town's water main fed from both ends

Fig.8.1 Typical 'wet' sprinkler installation for a two-storey building

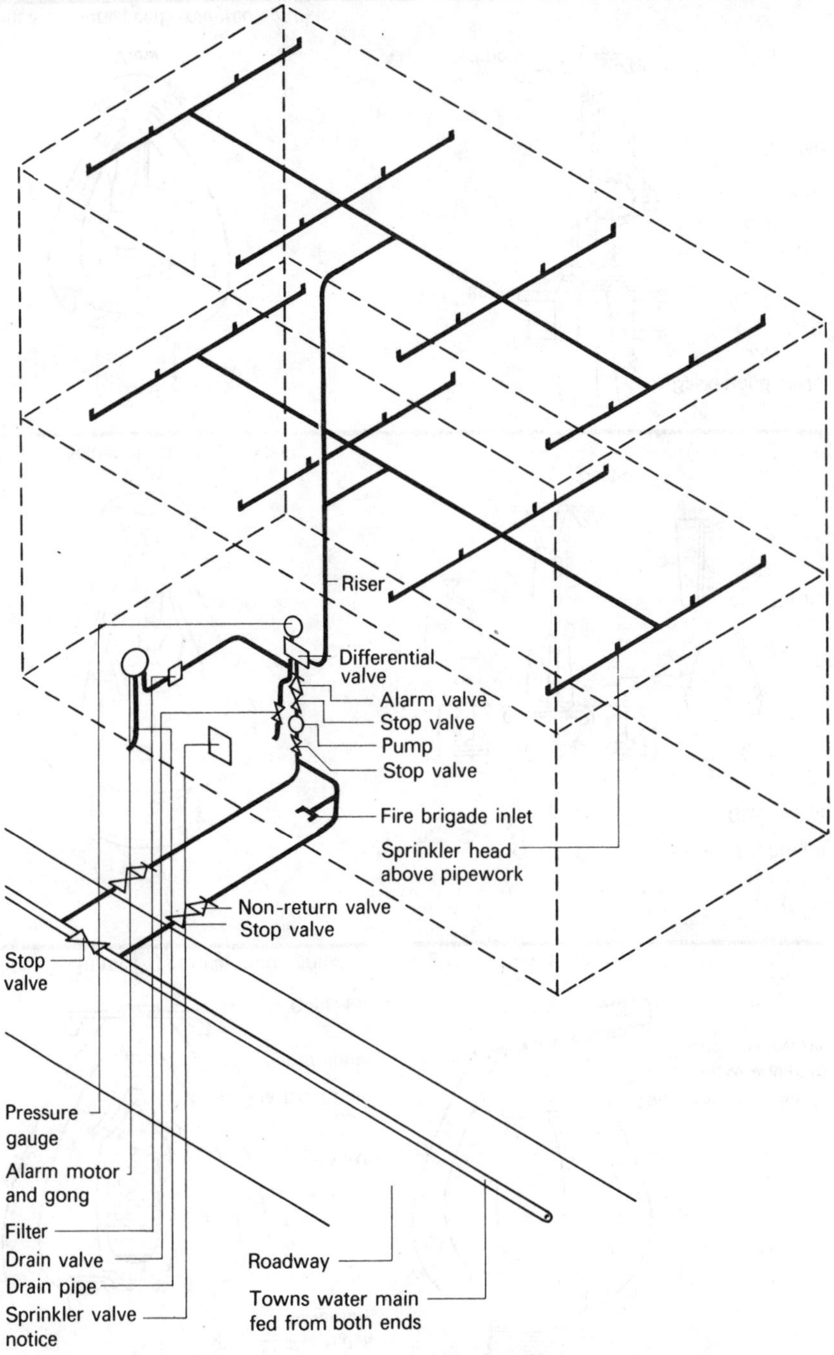
Riser
Differential valve
Alarm valve
Stop valve
Pump
Stop valve
Fire brigade inlet
Sprinkler head above pipework
Non-return valve
Stop valve
Stop valve
Pressure gauge
Alarm motor and gong
Filter
Drain valve
Drain pipe
Sprinkler valve notice
Roadway
Towns water main fed from both ends

Fig 8.2 Typical 'dry' or 'alternate wet and dry' sprinkler installation for a two-storey building

Table 8.1

Number of sprinkler heads installed	Number of spare sprinkler heads
300	6
Above 300 up to 1000	12
Over 1000	24

Types of sprinkler heads

There are four types of sprinkler heads:

1. *Quartzoid bulb type*

In this type the head incorporates a quartzoid bulb filled with a highly expansive coloured liquid having different expansion rates (see Table 8.2).

Table 8.2 Bulb temperature rating

Bulb rating (°C)	Colour of bulb liquid
57	Orange
68	Red
79	Yellow
93	Green
141	Blue
182	Mauve
227/288	Black

Operation: When the liquid expands due to receiving heat from the fire, the pressure within the bulb increases rapidly to a point at which the bulb shatters (bulb-temperature rating). The valve assembly falls away from the seat and allows water to flow to the deflector thus causing a water spray over the fire.

Note: The liquid in the bulb shrinks as it cools, so that even if it is exposed to intense cold there is no development of pressure within the bulb; the quartzoid bulb sprinkler heads are therefore suitable for cold as well as hot climates. Figure 8.3 shows a quartzoid bulb-type sprinkler head.

2. *Side-wall types*

These are designed for use at the side of rooms or corridors, so that the water spray will be projected into the centre of the room or corridor protected. They are also used in drying tunnels or hoods over papermaking machines, where condensed vapour dripping from the sprinklers and pipework overhead may be troublesome. The only difference between the usual type of sprinkler head and the side-wall sprinkler head is the angle of the deflector (see Fig. 8.4).

3. *The soldered-strut sprinkler head* (see Fig. 8.5)

A soldered-strut type consists of three bronze plates joined together by a special solder. These plates hold a glass valve in position against an orifice in a flexible

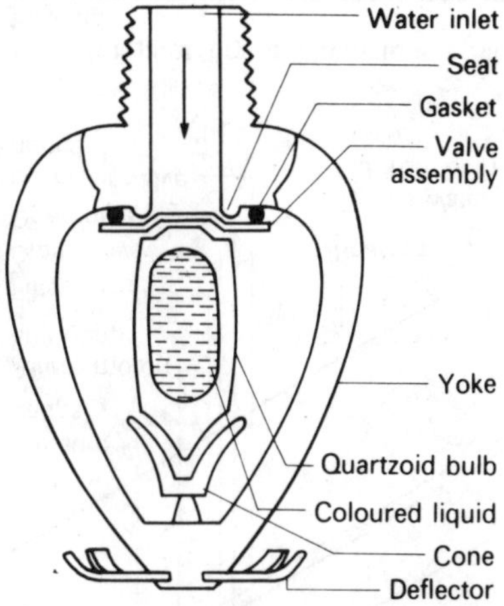
Water inlet
Seat
Gasket
Valve assembly
Yoke
Quartzoid bulb
Coloured liquid
Cone
Deflector
Fig. 8.3 Grinnell-type quartzoid sprinkler

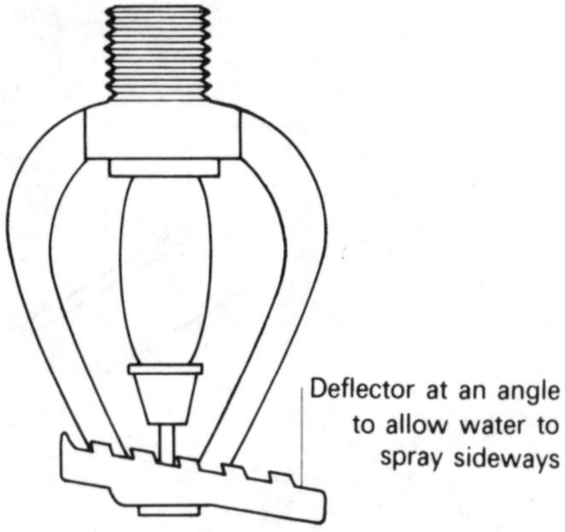
Deflector at an angle to allow water to spray sideways
Fig. 8.4 Side-wall sprinkler

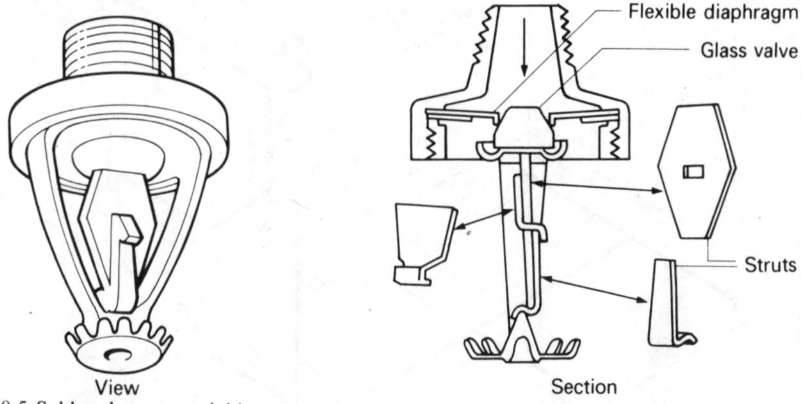
Flexible diaphragm
Glass valve
Struts
View
Section
Fig. 8.5 Soldered -strut sprinkler

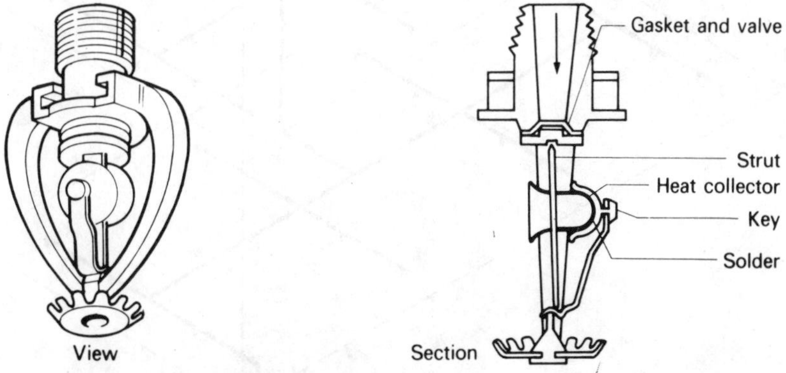
Gasket and valve
Strut
Heat collector
Key
Solder
View
Section
Fig. 8.6 'Duraspeed' soldered sprinkler

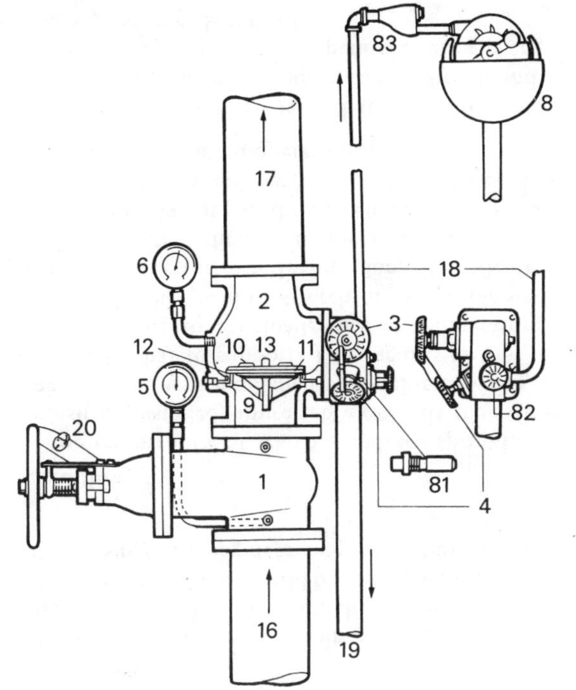

1. Installation stop valve
2. Alarm valve
3. Drain and test valve
4. 13 mm Test valve
5. Water supply pressure gauge
6. Installation pressure gauge
8. Alarm motor and gong
9. Combined valve seating and guide
10. Alarm valve clock
11. Composition disc
12. Annular groove in valve seating
13. Compensator
16. Water supply
17. Main feed to installation
18. Connection to alarm equipment
19. Drain pipe
20. Padlock and strap
81. Drip plug
82. Alarm stop valve
83. Strainer

Fig.8.7 Installation controlling valves - wet system (Courtesy of Mather and Platt Ltd)

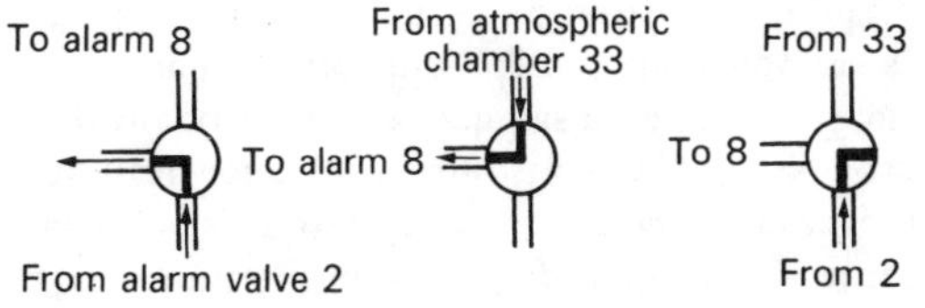
To alarm 8
From alarm valve 2
From atmospheric chamber 33
To alarm 8
From 33
To 8
From 2

Settings of three-way alarm cock

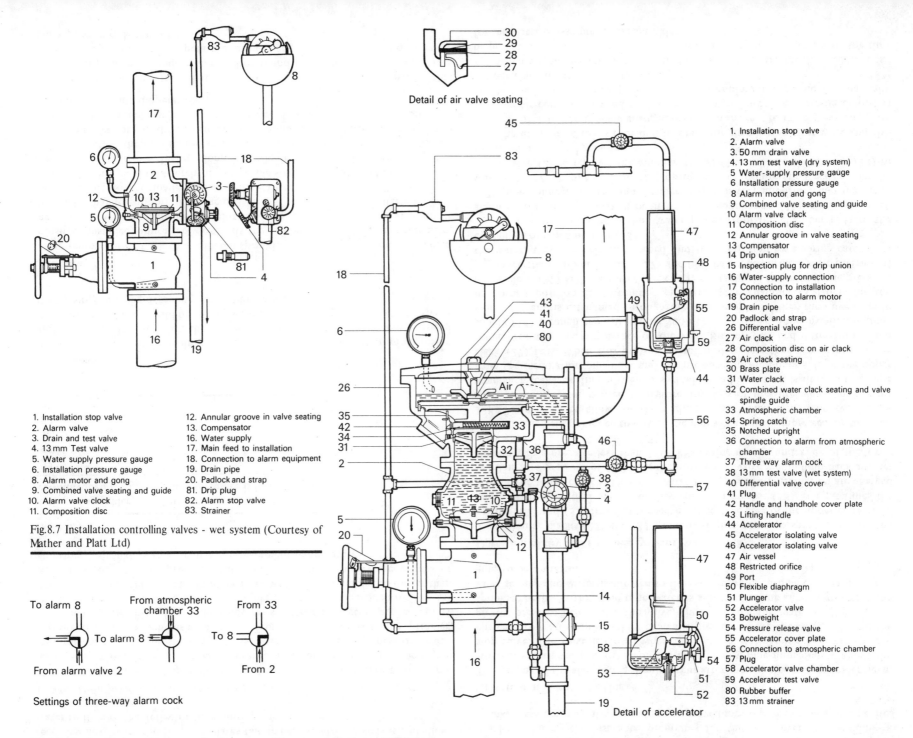

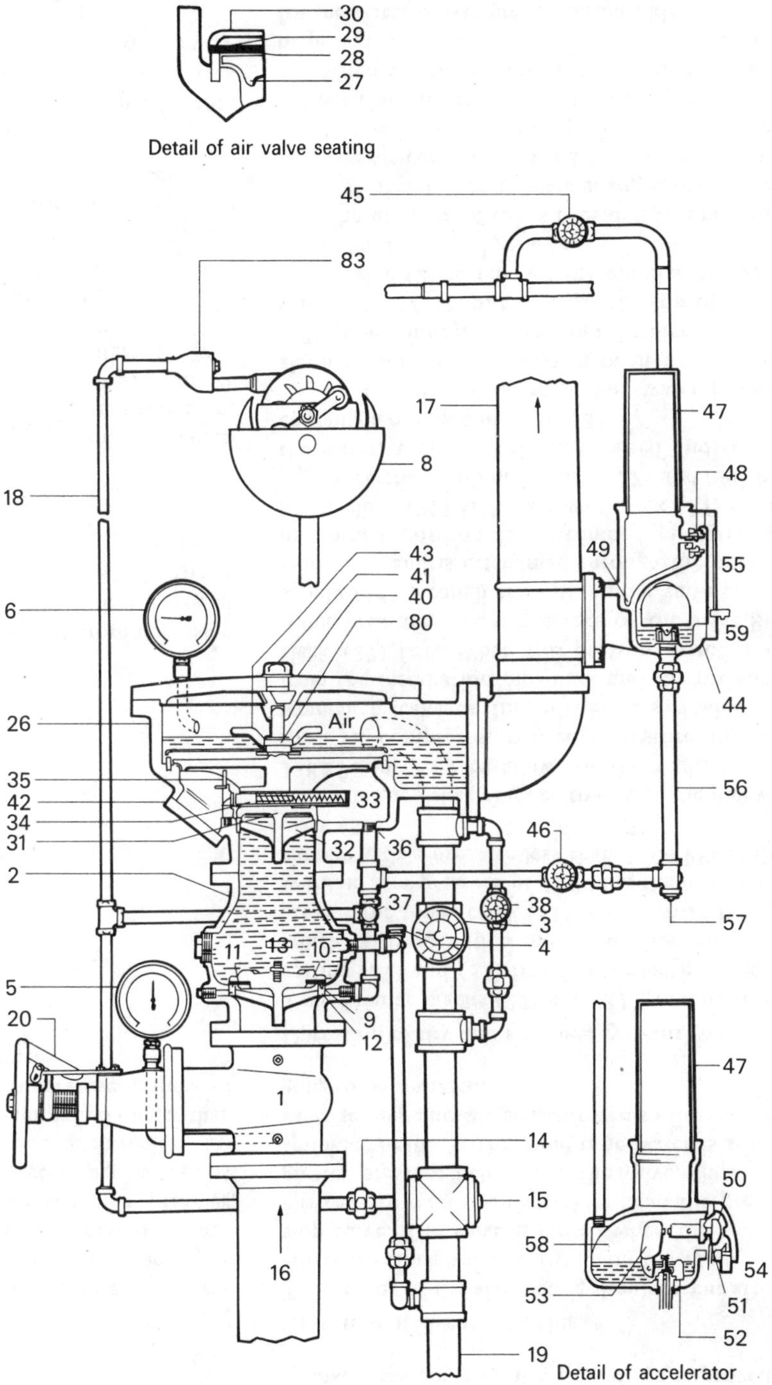
Detail of air valve seating
Air
Detail of accelerator

1. Installation stop valve
2. Alarm valve
3. 50 mm drain valve
4. 13 mm test valve (dry system)
5 Water-supply pressure gauge
6 Installation pressure gauge
8 Alarm motor and gong
9 Combined valve seating and guide
10 Alarm valve clack
11 Composition disc
12 Annular groove in valve seating
13 Compensator
14 Drip union
15 Inspection plug for drip union
16 Water-supply connection
17 Connection to installation
18 Connection to alarm motor
19 Drain pipe
20 Padlock and strap
26 Differential valve
27 Air clack
28 Composition disc on air clack
29 Air clack seating
30 Brass plate
31 Water clack
32 Combined water clack seating and valve spindle guide
33 Atmospheric chamber
34 Spring catch
35 Notched upright
36 Connection to alarm from atmospheric chamber
37 Three way alarm cock
38 13 mm test valve (wet system)
40 Differential valve cover
41 Plug
42 Handle and handhole cover plate
43 Lifting handle
44 Accelerator
45 Accelerator isolating valve
46 Accelerator isolating valve
47 Air vessel
48 Restricted orifice
49 Port
50 Flexible diaphragm
51 Plunger
52 Accelerator valve
53 Bobweight
54 Pressure release valve
55 Accelerator cover plate
56 Connection to atmospheric chamber
57 Plug
58 Accelerator valve chamber
59 Accelerator test valve
80 Rubber buffer
83 13 mm strainer

Fig.8.8 The installation valve set on the dry system (Courtesy of Mather and Platt Ltd)

diaphragm and seals the water outlet. When the fusible solder melts, due to receiving heat from the fire, the plates fall apart, the glass valve falls and allows water to flow to the deflector thus causing a water spray over the fire.

4. *The 'Duraspeed' soldered sprinkler head* (see Fig. 8.6)

The Grinnell 'Duraspeed' sprinkler has a much increased resistance to the harmful effects of atmospheric corrosion because the solder which secures the key to the heat collector is almost completely enclosed by the metal of these two parts. A protective film, which has no adverse effect on the sensitivity of the sprinkler, is applied to the thin edge of the solder as a further protection against atmospheric corrosion. In conditions which are likely to produce corrosion, the efficiency of the 'Duraspeed' sprinkler head will remain unimpaired longer than the soldered-strut-type sprinkler. Table 8.3 gives the temperature ratings of 'Duraspeed' sprinklers and the maximum ambient temperatures.

Table 8.3

Rating of sprinkler recommended (°C)	Temperature not to be exceeded where sprinkler is located (°C)
72	38
93	60
141	107
182	149
227	191

Maintenance

Sprinkler heads must be maintained in good condition, be free from corrosion and not covered with paint, distemper, dust or fluff. Where there is a risk of corrosion the heads should be coated with petroleum jelly and those defective – or suspected of being defective – should be replaced.

Cold flow

If the temperature at which a soldered sprinkler head is exposed when stored, in transit or installed, appreciably exceeds the safe value for which it is intended, the solder can be weakened to a point at which it begins to creep or yield under the load it has to support. This yield is at first minute, but once started it continues, owing to the load, even at temperatures which would be too low to initiate it. This yield or stretch of the fusible solder is called 'cold flow' and will continue, but for exactly how long is difficult to predict because it depends upon the extent the safe temperature was exceeded in the first place. If the safe temperature is not greatly exceeded, the continued yield of the solder can be so slow that many years may elapse before the cold flow becomes so advanced as to result in an unwanted operation.

Note: Manufacturers claim that 'cold flow' is a rare happening providing that the correctly rated sprinkler head is chosen and the ambient temperature does not rise abnormally. The temperature rating is engraved on the deflector of every 'Duraspeed' sprinkler head.

Escape from fire

Besides reducing the insurance premiums for the fire cover of a building, a sprinkler system may also provide an improved means of escape in case of a fire.

Operation of alarm gong (Fig. 8.7)

The ringing of the alarm gong should occur after the opening of a sprinkler head. An alarm-clock valve is fitted on the main supply pipe immediately above the stop valve. The valve is closed when the water is static, but when water is discharged from an open sprinkler head the pressure above the valve falls and the greater pressure below lifts the valve, thus allowing the water to flow to the sprinkler head. At the same time water is admitted to the annular groove in the valve seating and water is allowed to flow to the water turbine causing the alarm gong to be sounded.

Operation of dry and alternate systems (see Fig. 8.8)

Two circular clacks (27) and (31), mounted on a common spindle, rest on the seating (29) and (32) fitted in the valve casing. The area of the upper clack (27), which is fitted with a special quality composition disc (28), is eight times the area of the lower clack (31). Thus, where the pipes of an installation are charged with air at a pressure of 138 kPa, the differential valve would hold back a water supply with a pressure less than 1104 kPa acting on the underside of the lower clack (31).

It is the usual practice to charge the pipework with compressed air to about 138 kPa in order to allow for the slight leakage which cannot be avoided. A small air compressor is used to restore any loss of pressure which may result from such leakages. Immediately a sprinkler head opens, air is allowed to escape from the installation pipework and the pressure on the upper side of the upper clack (27) falls. When this pressure is reduced to less than one-eighth of the water pressure, the two clacks of the valve lift together and water flows from supply (16) through the differential valve chamber (33) to the installation pipework (17) and is discharged through the open sprinkler. When the valve is set in its normal position for automatic operation, the chamber (33) between the upper clack (27) and the lower clack (31) contains air at atmospheric pressure.

Once the combined clacks (27) and (31) have been lifted by the pressure of the water, they are held in a raised and open position by a spring catch (34) engaging a notched upright (35).

To operate the alarm, some water passes from the chamber (33) to the alarm motor and gong (8) through the pipe connection (36), three-way cock (37), pipe connection (18) and strainer (83). It is very important that the three-way cock (37) should be set in the appropriate position to allow water from the alarm valve (2) to pass via the annular groove (12) and the alarm cock (37) to the alarm motor and gong (8).

The drain valve (3) is provided to drain the installation after a fire and also when the installation is being changed over from the wet to the dry system.

The air compressor used to restore any loss of pressure is arranged to start automatically when the air pressure falls to a minimum value and automatically cut out when it has restored the pressure.

When a sprinkler operates, the air that escapes is very much more than can be provided by the compressor and therefore it has little or no effect on the time for the water to issue from the sprinkler.

Drencher systems

A drencher system provides a discharge of water over the external openings of a building, to prevent the transmission of fire from adjacent buildings. It comprises a system of pipework fitted on the outside of the building with discharge nozzles, known as 'drenchers', fixed at suitable intervals on roofs, under the eaves and over windows and doorways. In theatres, drenchers are fitted above the proscenium arch at the stage side for protection of the safety curtain.

Automatic drenchers are similar in construction to the quartzoid bulb sprinkler head and operate individually on the same principle. Non-automatic drenchers have open nozzles and normally operated valves, which when opened bring them all into operation simultaneously. The location of the valves for both systems should be clearly marked by a suitable notice.

The Mulsifyre system (see Fig. 8.9)

Oil fires can be extinguished by droplets of water travelling at high velocity, which bombard the surface of the oil to form an emulsion of oil and water that will not support combustion. The Mulsifyre system applies water on oil or paint fires in the form of a conical spray, consisting of droplets of water travelling at high velocity, thus extinguishing the fire by emulsification, cooling and smothering.

Operation: The conical spray of water, consisting of droplets at high velocity, is discharged through specially designed projector-mounted pipework. Quartzoid bulb detectors, mounted on independent pipework containing compressed air, are positioned so that wherever a fire occurs at least one detector will operate. If a fire occurs, the heat causes the bulb to shatter and the compressed air in the pipework to escape. This causes a rapid fall in air pressure on a diaphragm in the automatic deluge valve, which allows the valve to open and water to discharge through the projectors on to the fire.

An important feature of the system is the automatic alarm equipment: this is usually an alarm gong operated by a water motor which is driven by a small flow of water diverted at the installation control valves, when open.

The multi-jet sprinkler system (Fig. 8.10)

The multi-jet system is designed to provide an even distribution of water where the discharge from a standard sprinkler would be impeded by the presence of unavoidable obstructions.

Operation: When heat from a fire shatters the quartzoid bulb (B) in the automatic control valve (Fig. 8.11) the valve stem (E) slides down in guide (G) and allows valve (V) – seated in the central orifice diaphragm plate (D) – to open. Water flows through the orifice of the diaphragm plate and via the two outlets (Y) to all the distributions on the pipework. The automatic control valve is connected to the distribution pipe by unions (U) on either side, to facilitate the fitting of a new automatic control valve. The distributor (Fig. 8.12) has an orifice of 6 mm diameter which is normally sealed against the entry of dust and fumes by means of a light waxed-paper disc, which yields instantly to water pressure. A strainer (S) is fitted to prevent the orifice from being choked with foreign matter.

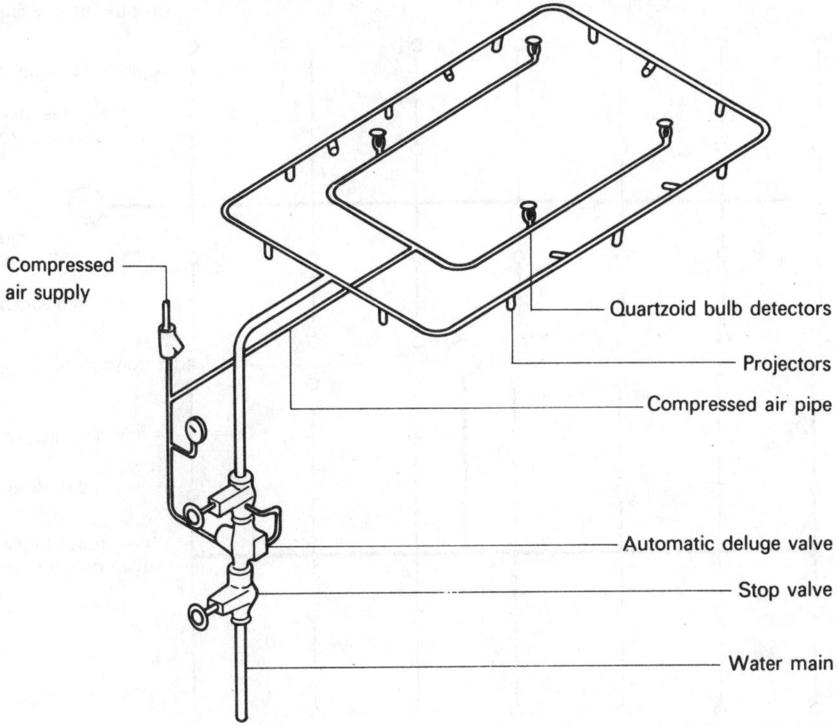
Compressed air supply
Quartzoid bulb detectors
Projectors
Compressed air pipe
Automatic deluge valve
Stop valve
Water main

Fig.8.9 The Mulsifyre system

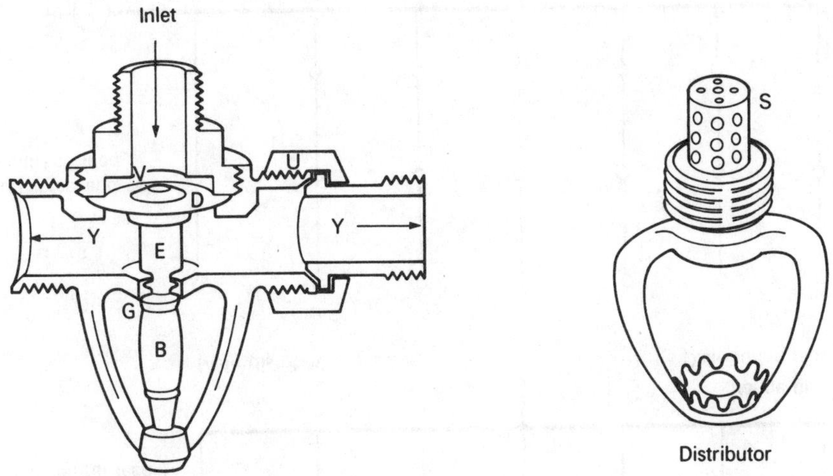
Inlet
V
U
D
Y
Y
E
G
B
S
Distributor

Fig.8.11, Fig.8.12 The multi-jet sprinkler system

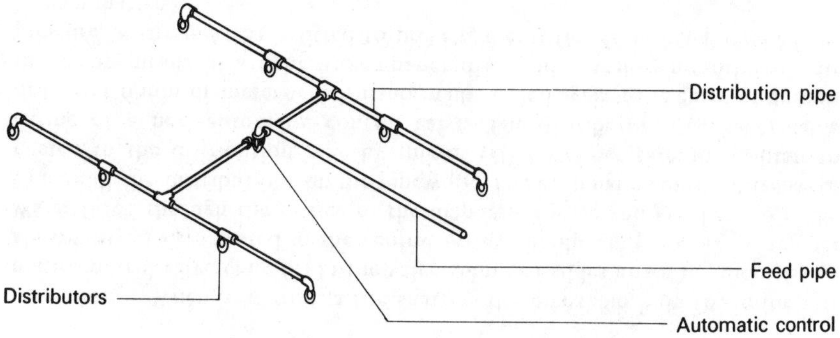
Distribution pipe
Feed pipe
Distributors
Automatic control

Fig.8.10 Installation of multi-jet system

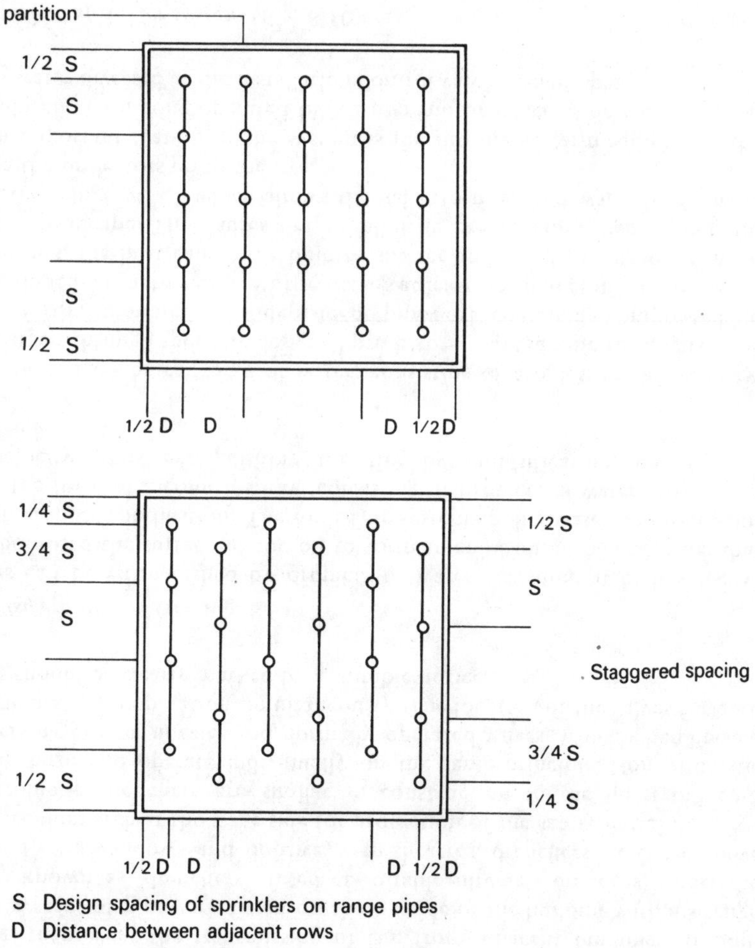
Wall or partition
1/2 S
S
S
1/2 S
1/2 D
D
D
1/2D
1/4 S
3/4 S
S
1/2 S
1/2 S
S
3/4 S
1/4 S
Staggered spacing
1/2 D
D
D
1/2D
S Design spacing of sprinklers on range pipes
D Distance between adjacent rows

Fig.8.13, Fig.8.14 Spacing arrangements

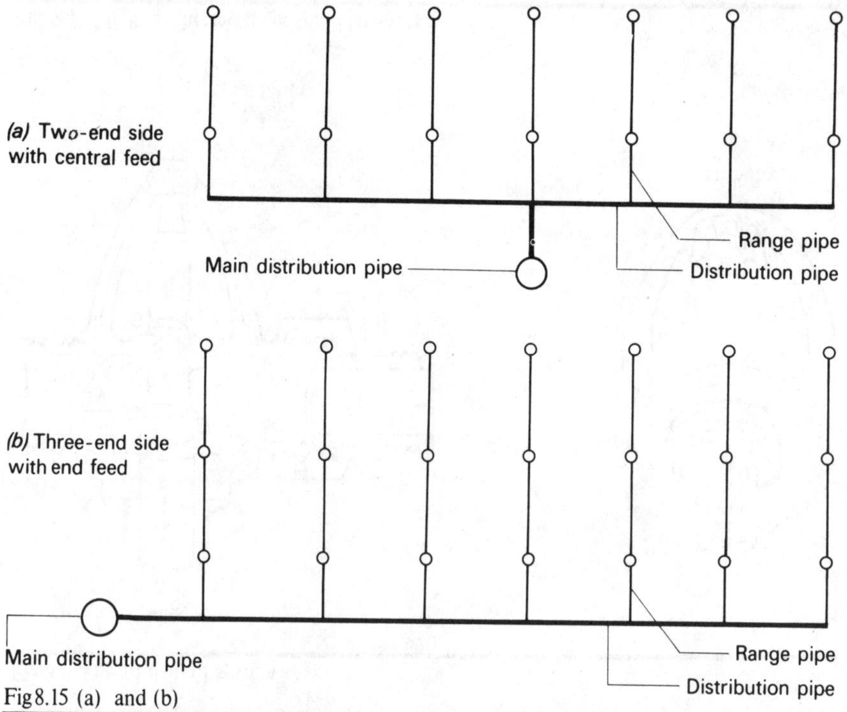
(a) Two-end side with central feed
Range pipe
Main distribution pipe
Distribution pipe
(b) Three-end side with end feed
Main distribution pipe
Range pipe
Distribution pipe

Fig 8.15 (a) and (b)

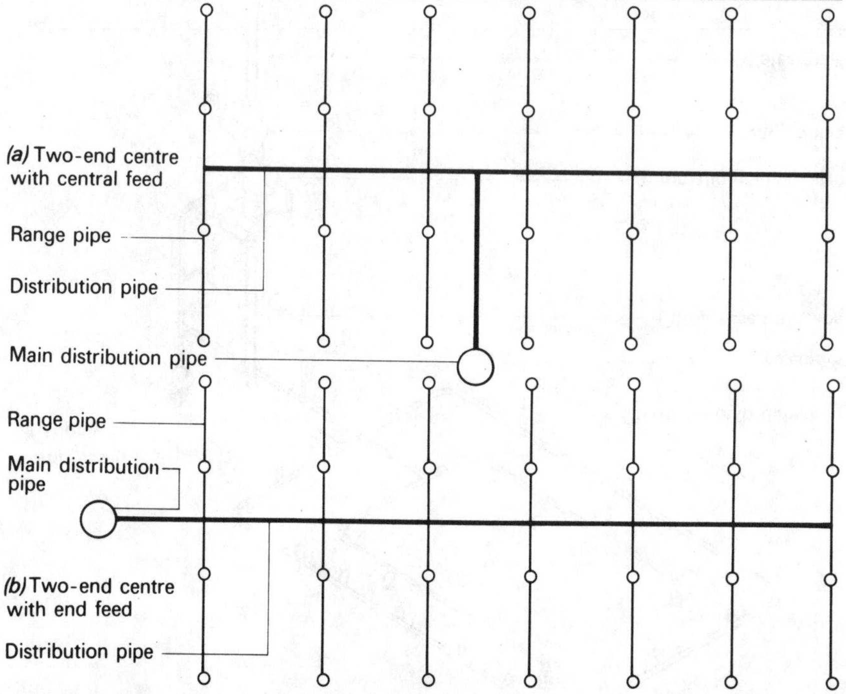
(a) Two-end centre with central feed
Range pipe
Distribution pipe
Main distribution pipe
Range pipe
Main distribution pipe
(b) Two-end centre with end feed
Distribution pipe

Fig 8.16 (a) and (b)

Classes of systems

Three classes of sprinkler systems have been developed to suit the following fire hazard classes of occupancy.

1. *Extra light hazard:* Non-industrial occupancies where the amount and combustibility of the contents is low.

2. *Ordinary hazard:* Commercial and industrial occupancies involving the handling, processing and storage of mainly ordinary combustible materials unlikely to develop intensely burning fires in the initial stages. Ordinary hazard occupancies have been divided into four groups according to the degree of fire hazard.

3. *Extra high hazard:* Commercial and industrial occupancies having abnormal fire loads. There are two types of these as follows:

(*a*) where the materials handled or processed are mainly of an extra hazardous nature likely to develop rapid and intensely burning fires;
(*b*) involving high piling of goods.

Note: The Rules of the Fire Offices' Committee for Automatic Sprinkler Installations, 29th edition, gives full details of fire hazard classification and types of occupancy, a summary is however given in Table 8.4.

Table 8.4 Classification of occupancies.

Extra light hazard Hospitals, hotels, libraries, museums, nursing homes, offices, prisons, schools, colleges
Ordinary hazard (Group I) Butchers, breweries, cement works, restaurants, cafés
Ordinary hazard (Group II) Bakeries, chemical works (ordinary), engineering works, laundries, garages, potteries, shops
Ordinary hazard (Group III) Aircraft factories (excluding hangars), boot and shoe factories, carpet factories, clothing factories, departmental stores, plastics factories, printing rooms, saw mills, tanneries, warehouses
Group III (special) Cotton mills, distilleries, film and television studios, match factories
Extra high hazard Celluloid works, foam plastics and foam rubber factories, paint and varnish factories, wood wool works, high piled storage risks, oil and flammable liquid hazard

Note: Some hazards in the ordinary hazard group having high fire load areas are placed in the extra high hazard class.

Spacing arrangements

Figure 8.13 shows a standard spacing of sprinkler heads where the distances between the heads is as follows:

S = design spacing of sprinklers on range pipes
D = distance between rows of sprinklers
{ max. 4.6 m extra light hazard; max. 4.0 m ordinary hazard; max. 3.7 m extra high hazard

$$S \times D = \begin{cases} 21\ \text{m}^2 \text{ or less, extra light hazard} \\ 12\ \text{m}^2 \text{ or less, ordinary hazard} \\ 9\ \text{m}^2 \text{ or less, extra high hazard} \end{cases}$$

Figure 8.14 shows the staggered arrangement for ordinary hazard systems where it is desirable to space sprinklers more than 4.0 m apart on the range pipes as follows:

S = design spacing of sprinklers on range pipes = max. 4.6 m

D = distance between adjacent rows of sprinklers = max. 4.0 m

Figures 8.15(*a*) and (*b*) shows the end side arrangements where there are range pipes on one side only of the distribution pipe. Figures 8.16(*a*) and (*b*), shows the end centre arrangements where there are range pipes on both sides of the distribution pipe.

Positioning of sprinklers

Sprinklers must be placed so that the deflectors are not more than 300 mm from non-fire-resisting ceilings and not more than 450 mm from fire-resisting ceilings. There must be a clear space of 300 mm below the level of a deflector within a radius of 600 mm from each sprinkler. In storage rooms, therefore, goods must not be stored within 300 mm of the level of the deflectors above them. Sprinklers must not be placed within 600 mm of columns or beams.

Water supplies for sprinklers

The efficiency of a sprinkler system depends largely upon the source of water supply, which should be adequate for the size and type of building protected. Acceptable sources of water supply are as follows:

1. Town main.
2. Elevated private reservoir.
3. Gravity tank.
4. Town main and automatic pump.
5. Town main, automatic pump, with either a pressure tank, gravity tank or elevated private reservoir.
6. Automatic pump drawing water from river or canal.

Water used for sprinklers must be free from suspension matter liable to cause accumulations in the system. Sea-water or brackish water is not normally allowed.

Town main (Figs. 8.1 and 8.2)
This must be capable of providing at all times the minimum pressure and flow

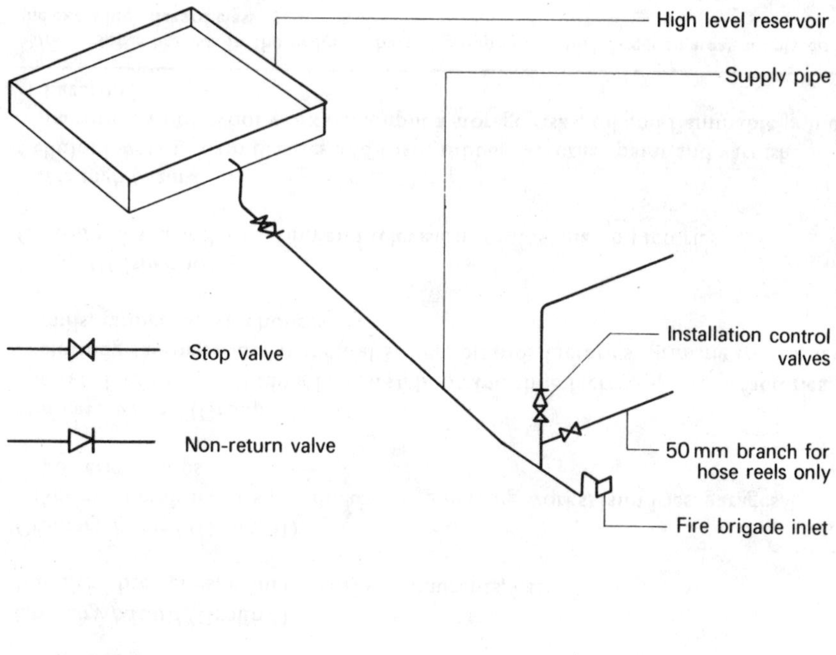
High level reservoir
Supply pipe
Stop valve
Non-return valve
Installation control valves
50 mm branch for hose reels only
Fire brigade inlet

Fig.8.17 Elevated private reservoir

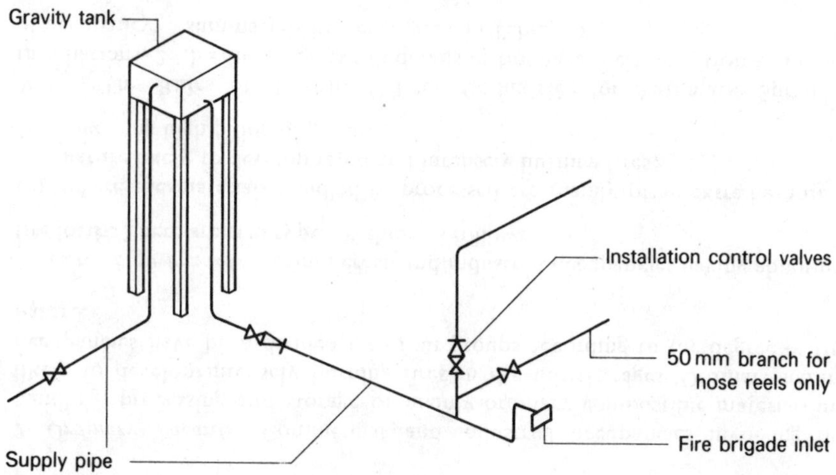
Gravity tank
Installation control valves
50 mm branch for hose reels only
Fire brigade inlet
Supply pipe

Fig.8.18 Gravity tank

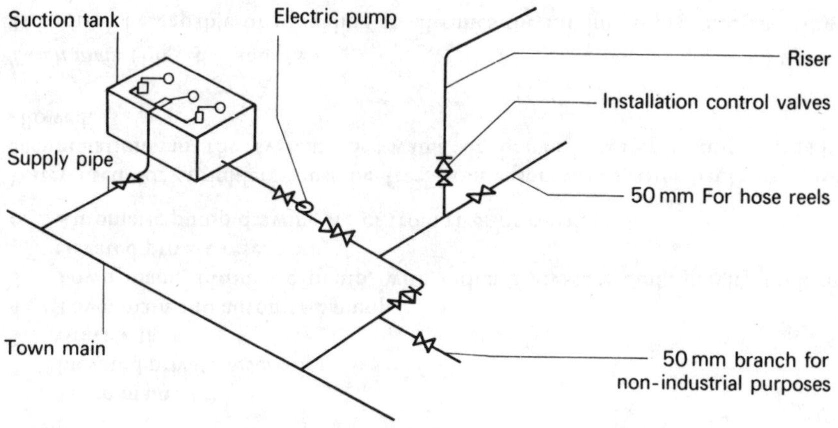
Suction tank
Electric pump
Riser
Installation control valves
Supply pipe
50 mm For hose reels
Town main
50 mm branch for non-industrial purposes

Fig.8.19 Town main and automatic pump

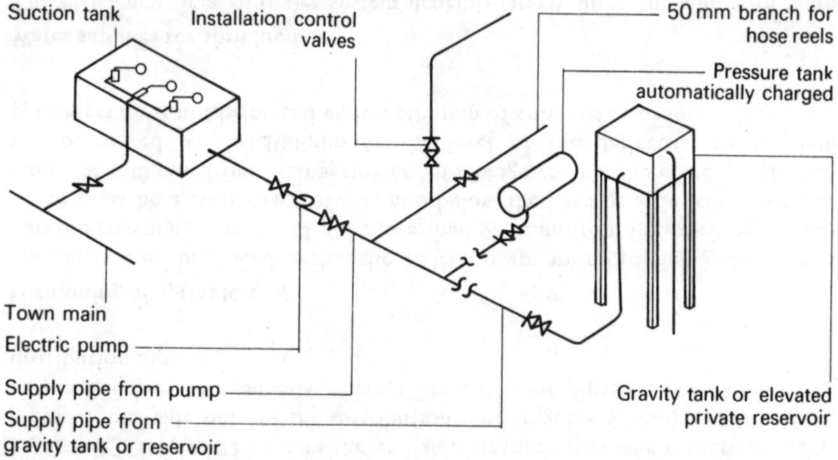
Suction tank
Installation control valves
50 mm branch for hose reels
Pressure tank automatically charged
Town main
Electric pump
Supply pipe from pump
Supply pipe from gravity tank or reservoir
Gravity tank or elevated private reservoir

Fig.8.20 Town main, automatic pump with either a pressure tank, gravity tank or elevated private reservoir

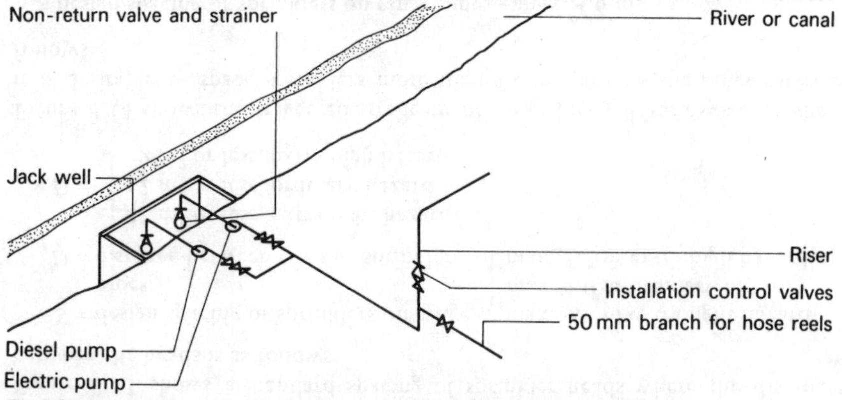
Non-return valve and strainer
River or canal
Jack well
Riser
Installation control valves
50 mm branch for hose reels
Diesel pump
Electric pump

Fig.8.21 Automatic pump supply drawing water from river or canal

rate to the sprinkler heads, specified by the Fire Offices' Committee Rules. Preferably, the main should be fed from both directions and the connection to the building duplicated. In special circumstances a single feed pipe may be used, providing there is a main stop valve on either side of the single feed pipe connection to the main.

Elevated private reservoir (Fig. 8.17)

This is a ground reservoir situated at a higher level than the building to be protected. The reservoir is under the sole control of the owner of the building protected by the sprinkler system. The reservoir must hold between 500 m^3 and 1875 m^3 of water.

Gravity tank (Fig. 8.18)

This is a purpose-made water vessel, erected on the site of the protected building, at such a height as to provide the required pressure and flow rate at the sprinkler heads. The supply to the tank should be such that it can be refilled in not less than 6 h. The tank must hold between 9 m^3 and 875 m^3 of water.

Town main and automatic pumps (Fig. 8.19)

If the local Water Authority permit direct pumping from the main, the latter should be capable of providing sufficient water at all times to match the rated output of the pump. Alternatively, a suction tank may be installed, having sufficient inflow of water to enable the pump to operate at full output, without emptying the tank. The suction tank must hold between 2.5 m^3 and 585 m^3 of water.

Town main (automatic pump, with either a pressure tank, gravity tank or elevated private reservoir (Fig. 8.20))

This provides a duplicated water supply so that in the event of the main failing and the suction tank emptying, there is another source of water supply from either a pressure tank, gravity tank or an elevated private reservoir.

A pressure tank, when used as the sole source of water supply, is only acceptable for extra light and ordinary (Group I) fire hazards. However, it is acceptable for the remaining groups of ordinary hazards providing it is only as one source of duplicated supply. The tank must hold between 7 m^3 and 23 m^3 of water.

Automatic pump (drawing water from river or canal (Fig. 8.21))

The automatic pumping arrangement of either two pumps (one of which is diesel-driven) or three pumps, any two of which, when connected in parallel, will provide the pressure and flow rate required and two of which are diesel-driven. A foot valve and strainer should be fitted on the end of the suction pipe, low enough to suit the lowest water level of the river or canal.

Flow rates and pressures

The minimum flow rates and water pressures for installations, are given in Table 8.5.

Table 8.5

Hazard class	Flow rate (litres/s)	Running pressure (kPa)
Extra light	3.75	220
Ordinary Group I	6.25	100
(light)	9.00	70
Ordinary Group II	12.10	140
(medium)	17.00	100
Ordinary Group III	18.3	170
(high)	22.5	140
Ordinary Group III	30.0	200
(special)	35.0	150
Extra high	Refer to Fire Offices' Committee Rules, which are too involved to include in the table	

Classification of fire risks

In order to provide the most efficient fire-extinguishing agent, fire risks are classified in four groups as follows:

Class A risk: Carbonaceous material such as wood, cloth and paper, where cooling by water is the most effective method of reducing the temperature of the burning material. Most fires are in this class.

Class B risk: Fires in inflammable liquids including petrol, oils, greases, paints, varnishes and fats, where the blanketing or smothering effect of agents, which exclude oxygen, is most effective.

Class C risk: Fires in inflammable gases such as acetylene, methane, propane, North Sea and natural gases, where the extinguishing by blanketing or smothering to prevent oxygen combining with the gas is the most effective.

Class D risk: Fires in inflammable metals such as uranium, zinc and aluminium, where the extinguishing or smothering effect of agents which exclude oxygen is most effective.

Class E risk: This is not a strictly separate classification, but includes all risks where the problem of extinguishing the fire is increased, due to the danger of an electric shock. The fire is in the presence of live electrical equipment or wiring and a non-conducting extinguishing agent is therefore required.

Extinguishing agents

Class A	Water or dry powder
Class B and C	Dry powder, foam, or carbon dioxide
Class D	Dry sand or dry powder
Class E	Dry powder, or carbon dioxide

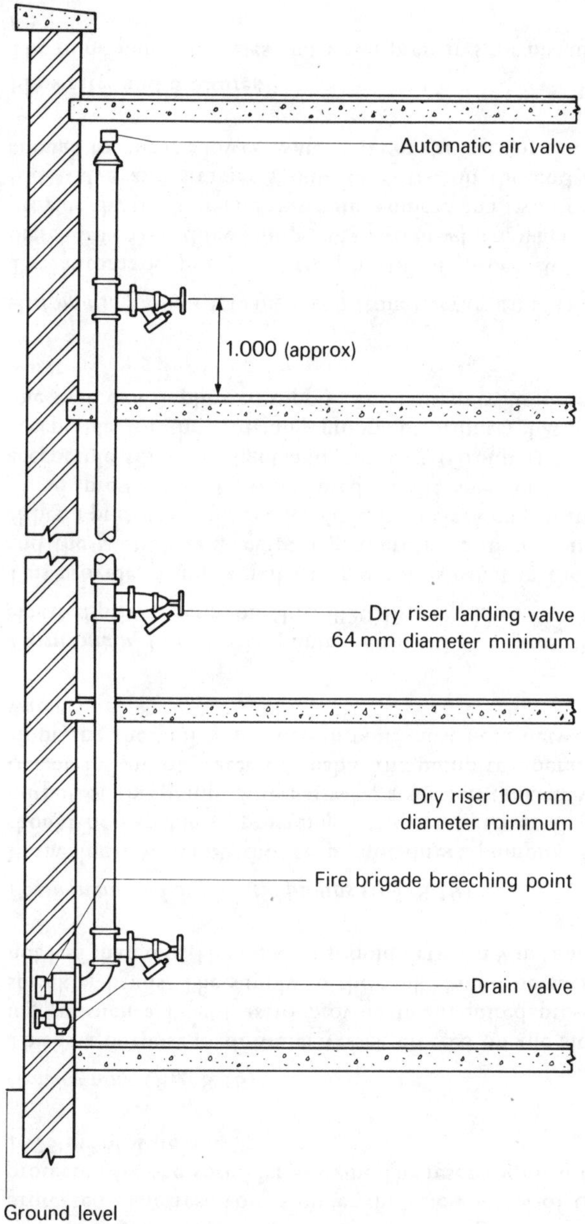
Automatic air valve
1.000 (approx)
Dry riser landing valve
64 mm diameter minimum
Dry riser 100 mm
diameter minimum
Fire brigade breeching point
Drain valve
Ground level

Fig.8.22 Dry riser

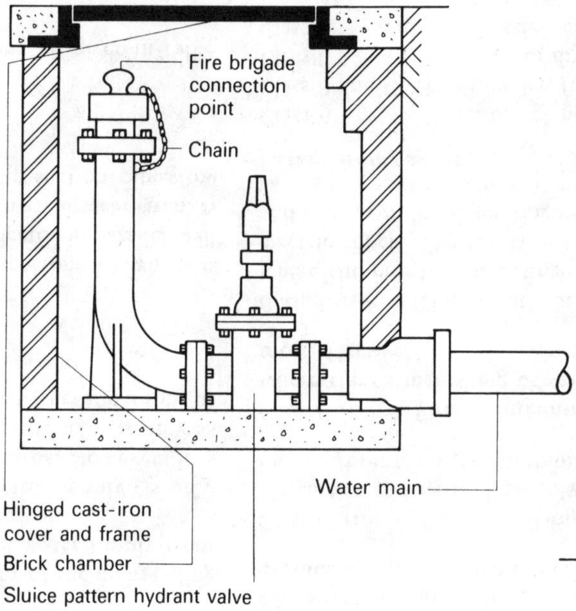
Fire brigade
connection
point
Chain
Water main
Hinged cast-iron
cover and frame
Brick chamber
Sluice pattern hydrant valve

Fig.8.23 Underground fire hydrant

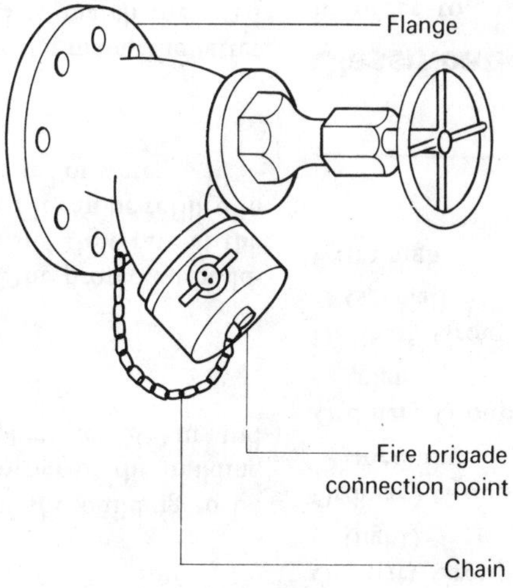
Flange
Fire brigade
connection point
Chain

Fig.8.24 Landing valve fire hydrant

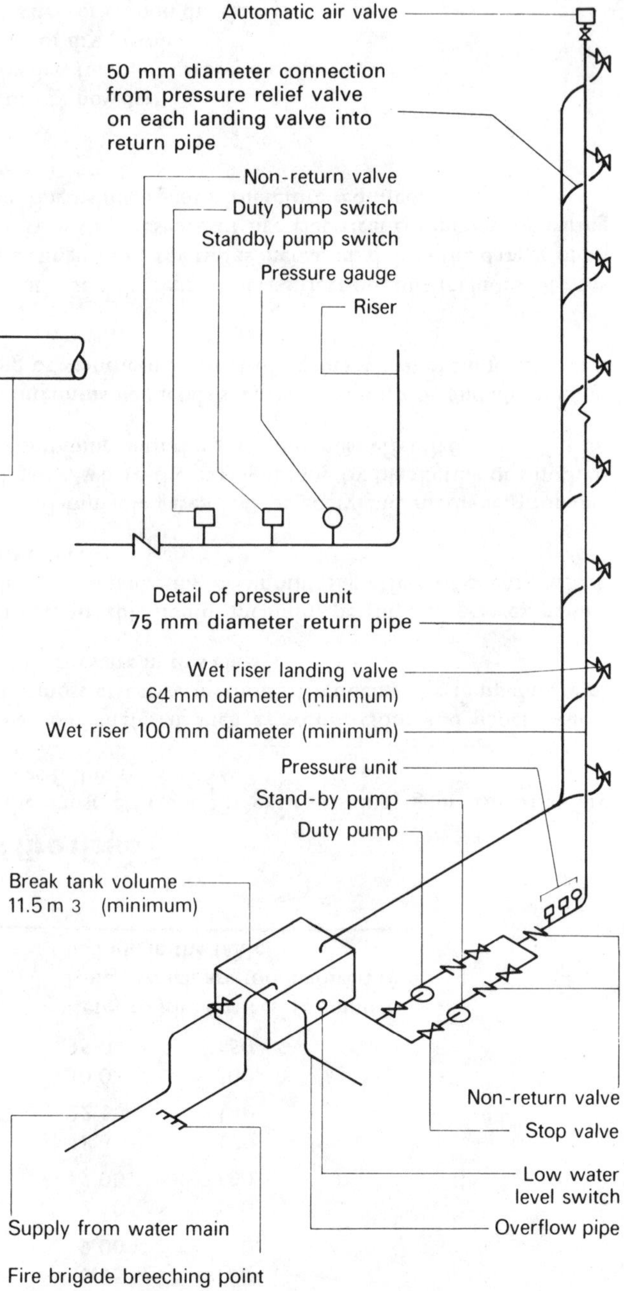
Automatic air valve
50 mm diameter connection
from pressure relief valve
on each landing valve into
return pipe
Non-return valve
Duty pump switch
Standby pump switch
Pressure gauge
Riser
Detail of pressure unit
75 mm diameter return pipe
Wet riser landing valve
64 mm diameter (minimum)
Wet riser 100 mm diameter (minimum)
Pressure unit
Stand-by pump
Duty pump
Break tank volume
11.5 m 3 (minimum)
Non-return valve
Stop valve
Low water
level switch
Overflow pipe
Supply from water main
Fire brigade breeching point

Fig.8.25 Wet riser

Testing of sprinkler installations

Weekly tests should be made to ascertain if the alarm apparatus is in order and if the air and water pressures are correct.

Quarterly tests should ascertain if each water supply is in order and the test should bring into operation any automatic pump. The sprinkler heads and pipework must be inspected and special attention must be paid to signs of corrosion and to ensure that sprinkler heads have not been painted.

Dry and wet risers

Multi-storey buildings are difficult to evacuate and can be extremely hazardous in the event of an outbreak of fire, unless fixed fire-fighting equipment has been installed.

The type of fire-fighting equipment for multi-storey buildings can take the form of either a wet or dry riser, depending upon the height of the building and the local Fire Authority Regulations.

Dry riser (Fig. 8.22)

As the name implies, the riser does not normally contain water, but is charged with water by the fire brigade during an outbreak of fire. The fire brigade connect the suction side of their pumps to a water main via an underground fire hydrant, or hydrants (see Fig. 8.23). The outlet side of the pumps are connected to the dry-riser inlet, or inlets at ground level and the pumps force water from the main into the riser. An automatic air valve at the top of the riser opens to allow air in the pipe to escape, but closes when the pipe is full of water. The firemen can now enter the building and connect their hose reels to landing valves fitted to the riser (see Fig. 8.24).

A dry riser, therefore, is merely an extension of the fireman's hose and should only be installed where prompt attention can be relied upon from the local fire brigade, or from trained fire-fighting personnel on the premises. The riser should be sited inside a ventilated lobby of a lobby approach staircase, or in a staircase enclosure.

Size

In buildings which do not exceed 45 m in height and where only one 64 mm diameter landing valve is provided on any floor, the internal diameter of the riser should be 100 mm. In buildings between 45 m and 60 m in height, and in buildings less than 45 m in which two 64 mm diameter landing valves are provided on each floor from the riser, the internal diameter should be 150 mm.

Note: Buildings above 60 m in height should be provided with a wet riser.

Inlets

The inlets to the riser should be sited on the external wall of the building at 760 mm above ground level and not more than 12 m from the riser. The inlets should be within 18 m of an access road suitable for the fire brigade pumping appliance.

A 100 mm diameter riser should be fitted with two inlets, and a 150 mm riser with four inlets, each inlet consisting of a 64 mm instantaneous male coupling and a back-pressure valve. The inlets should be protected by a cap secured with a suitable length of chain and a 25 mm drain valve should be fitted at the lowest point of the riser.

The inlets should be fitted inside a metal box, the door of which should be glazed with wired glass and its position indicated by 'Dry Riser Inlet' painted on the inner face of the glass in 50 mm block letters. The door should be fastened by a spring lock which can be opened from both outside or inside the building without the aid of a key, after the glass has been broken.

Construction of riser

The riser should be of galvanised steel piping, Class C, BS 1387 (red band) screwed and socketed. The fittings should be of steel or malleable iron galvanised and of steam quality. All changes in direction in the run of piping, should be made with standard bends and elbows must not be used.

Earthing

Dry risers should be electrically earthed.

Wet risers (Fig. 8.25)

Buildings exceeding twenty storeys or 60 m in height above ground level, whichever height is the least, should be provided with one or more wet rising mains to be used exclusively for fire-fighting purposes. As the name implies, the riser is always charged with water under pressure, fed by pumping sets from a break tank. Hydrants are connected to the riser on each floor and the pumps should be capable of providing a pressure of 410 kPa at the highest hydrant.

To protect the hosepipe connected to the riser, the hydrants on the lower floors of tall buildings should incorporate an orifice plate, so that when water is being discharged the outlet pressure is limited to 520 kPa.

A 75 mm diameter return pipe should be connected from the hydrants back to the supply source and the static water pressure with no flow of water should not exceed 690 kPa in the pipework.

The number of wet risers to be provided, the positions in which they are fitted, the outlets fitted to them and all similar details of construction and materials should conform to the same specifications as given for dry rising mains. The riser should only be installed in a heated building and should be electrically earthed.

Pumps

To provide an adequate supply of water to each riser at all times, duplicate pumps should be provided, one of which is for stand-by purposes. Each pump should be capable of delivering a minimum flow rate of 15 litre/s. The pumps should be connected in parallel, with their suctions permanently 'wet' when the tank is filled. The pumps may be run by electrical power, in which case a stand-by generator of sufficient capacity should be provided in case of mains failure. Alternatively, the stand-by pump may be driven by a petrol or a diesel engine. The pumps should be controlled to start automatically when a fall in pressure occurs in the riser exceeding 3 per cent of the normal static pressure, and to stop automatically when the normal pressure is re-established. The pumps may also be started by a flow of water when a fire brigade hose-reel is used, and therefore a pressure switch must be fitted in the pipeline on the delivery side of the pumps.

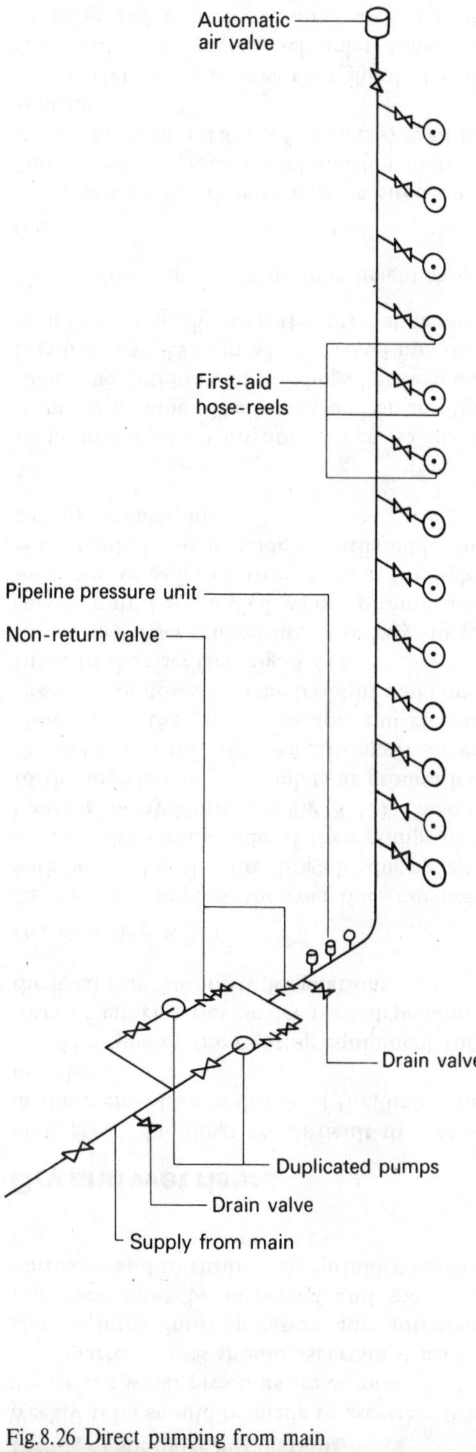
Automatic air valve
First-aid hose-reels
Pipeline pressure unit
Non-return valve
Drain valve
Duplicated pumps
Drain valve
Supply from main

Fig.8.26 Direct pumping from main

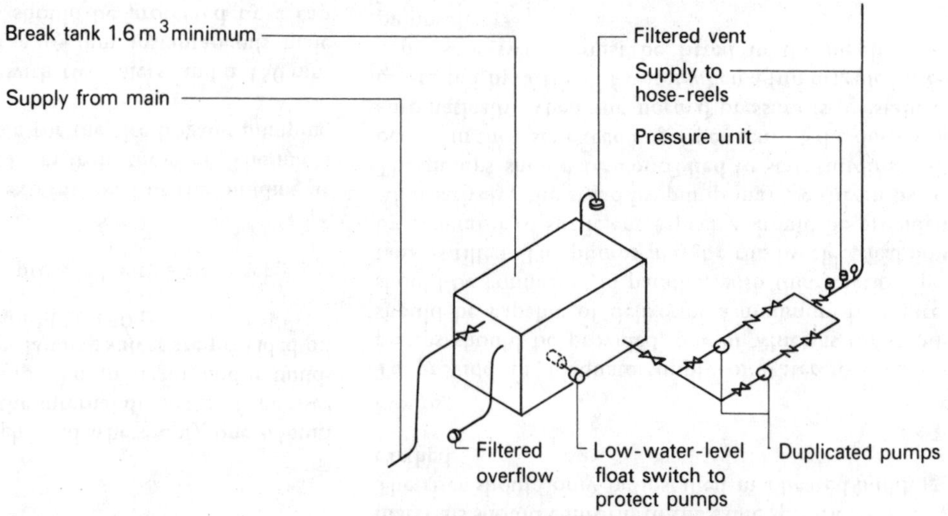
Break tank 1.6 m^3 minimum
Supply from main
Filtered vent
Supply to hose-reels
Pressure unit
Filtered overflow
Low-water-level float switch to protect pumps
Duplicated pumps

Fig.8.27 Pumping from low-level break tank

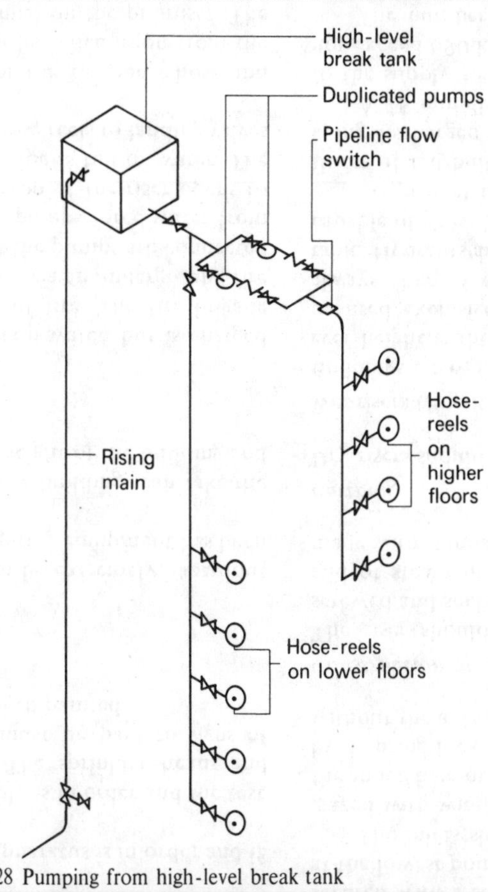
High-level break tank
Duplicated pumps
Pipeline flow switch
Rising main
Hose-reels on higher floors
Hose-reels on lower floors

Fig.8.28 Pumping from high-level break tank

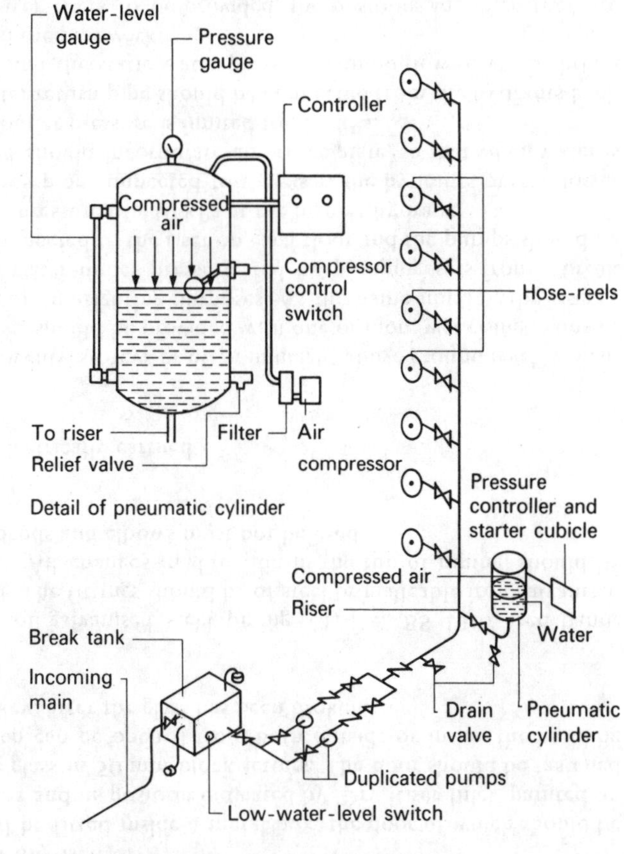
Water-level gauge
Pressure gauge
Controller
Compressed air
Compressor control switch
To riser
Relief valve
Filter
Air compressor
Detail of pneumatic cylinder
Hose-reels
Pressure controller and starter cubicle
Compressed air
Riser
Water
Break tank
Incoming main
Drain valve
Pneumatic cylinder
Duplicated pumps
Low-water-level switch

Fig.8.29 Auto-pneumatic system

Water supply

The water supply to the riser should be either from a break tank of not less than 11.5 m^3 in volume, supplied from a water main at the rate of not less than 27 litre/s, or from a break tank of not less than 45.5 m^3 in volume, supplied from a water main at the rate of not less than 8 litre/s. In addition to the normal supply through the ball valve, the tank should be furnished with a 150 mm diameter fire brigade breeching inlet at street level. This inlet should have four 64 mm internal diameter instantaneous male couplings for connection to the fire brigade pumps.

The supply pipe should not be connected directly to the break tank, but should deliver water through a bend above it. An overflow pipe capable of developing 375 litre/s to the open air should be connected to the break tank.

Note: During the construction of a building, fire risks can occur and wet risers should therefore be put into operation as construction work proceeds. To ensure a flow of water to the landing valves when the duty pump is in operation, all stop valves should be strapped and padlocked in an open position. The wet riser should be capable of withstanding a test pressure of twice the working pressure.

Where it is not possible to ensure that no part of the building is more than 61 m from the riser, an additional riser should be installed.

Hose-reel installations

These are for first-aid fire-fighting only, but it is often possible for a fire to be extinguished or contained by the occupants in its initial stage by the use of a jet of water from a hose-reel. It is not usual for portable fire extinguishers to be dispensed with when such a system is installed.

The hydraulic requirements for hose-reels is that they should be able to deliver 0.4 litre/s at a distance of 6 m from the nozzle, and that three should be capable of operating simultaneously. A pressure of 200 kPa is required at each nozzle, and if the water main cannot provide this at the highest reel pumping equipment must be installed.

Hose-reels should be provided at the rate of one reel for every 418 m^2 of floor area and the nozzle should reach to within 6 m of the furthest part of the building. They should be sited in an escape corridor, so that they may be used by people leaving the building, and if the hose has been used in a room full of smoke it can guide its user to safety.

Water supply

Some Water Companies will permit direct pumping from the main, providing that a reasonable supply of water is available at the highest reel without the use of pumps (see Fig. 8.26).

When the local Water Companies requires the use of a break tank, it should hold a minimum volume of water of 1.6 m^3. The break cistern may be sited at a low or a high level, as shown in Figs. 8.27 and 8.28, and duplicate pumps should be provided having a minimum discharging capacity of 2.3 litre/s. The stand-by pump may be required to be driven by either a petrol or a diesel engine.

Pipe sizes

For buildings up to 15 m in height the internal diameter of the supply pipe should be not less than 50 mm, and for buildings above 15 m in height not less

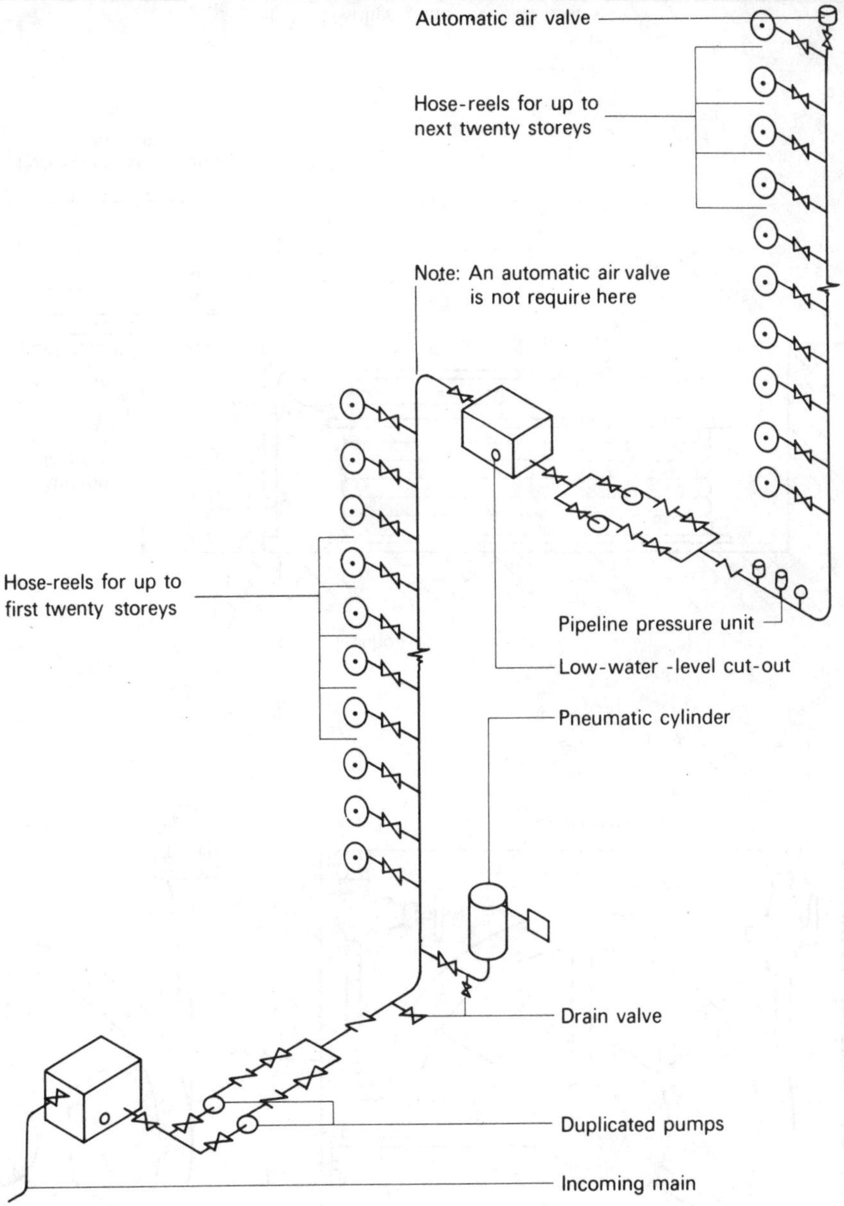
Automatic air valve
Hose-reels for up to
next twenty storeys
Note: An automatic air valve
is not require here
Hose-reels for up to
first twenty storeys
Pipeline pressure unit
Low-water -level cut-out
Pneumatic cylinder
Drain valve
Duplicated pumps
Incoming main

Fig.8.30 Pumping for buildings above twenty storeys

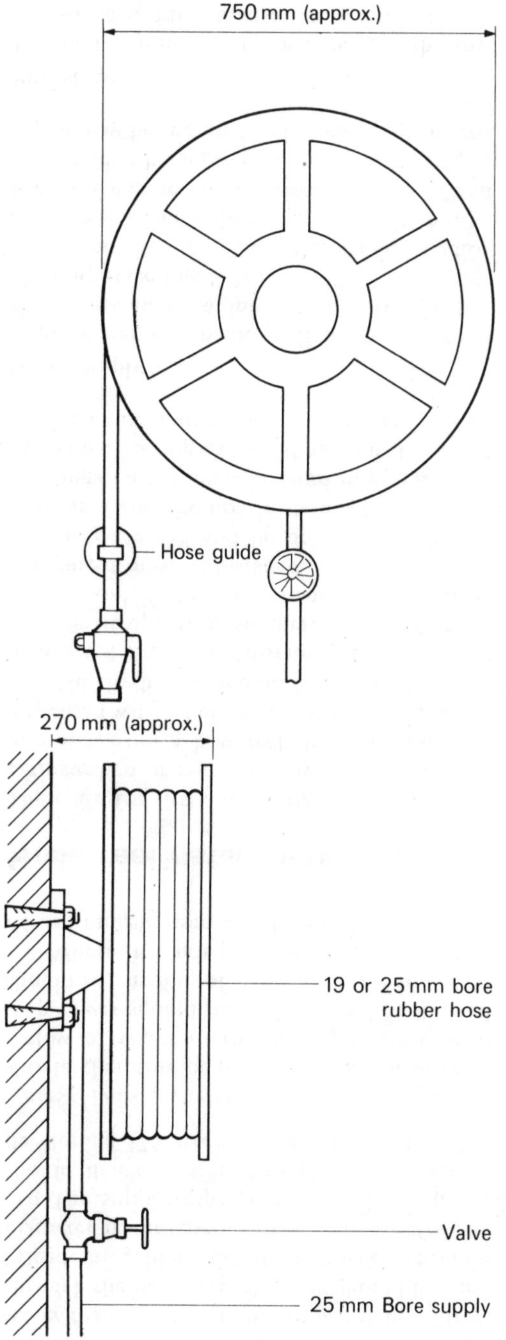
750 mm (approx.)
Hose guide
270 mm (approx.)
19 or 25 mm bore rubber hose
Valve
25 mm Bore supply

Fig. 8.31 Fixed hose-reel

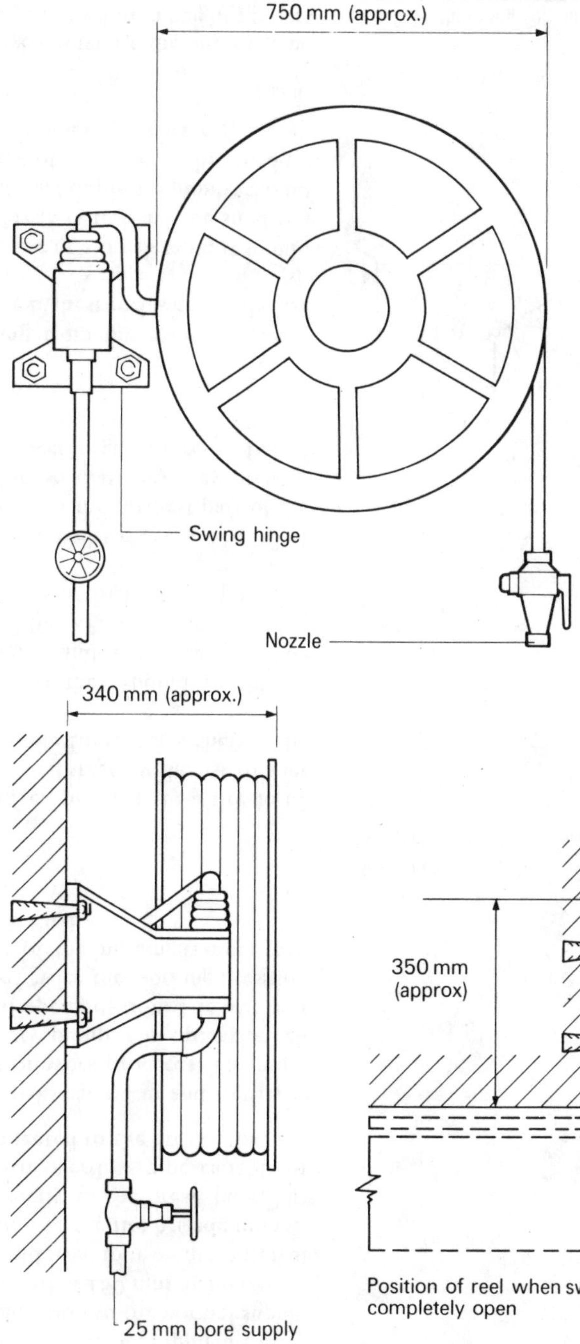
750 mm (approx.)
Swing hinge
Nozzle
340 mm (approx.)
25 mm bore supply

Fig. 8.32 Swinging-type hose-reel

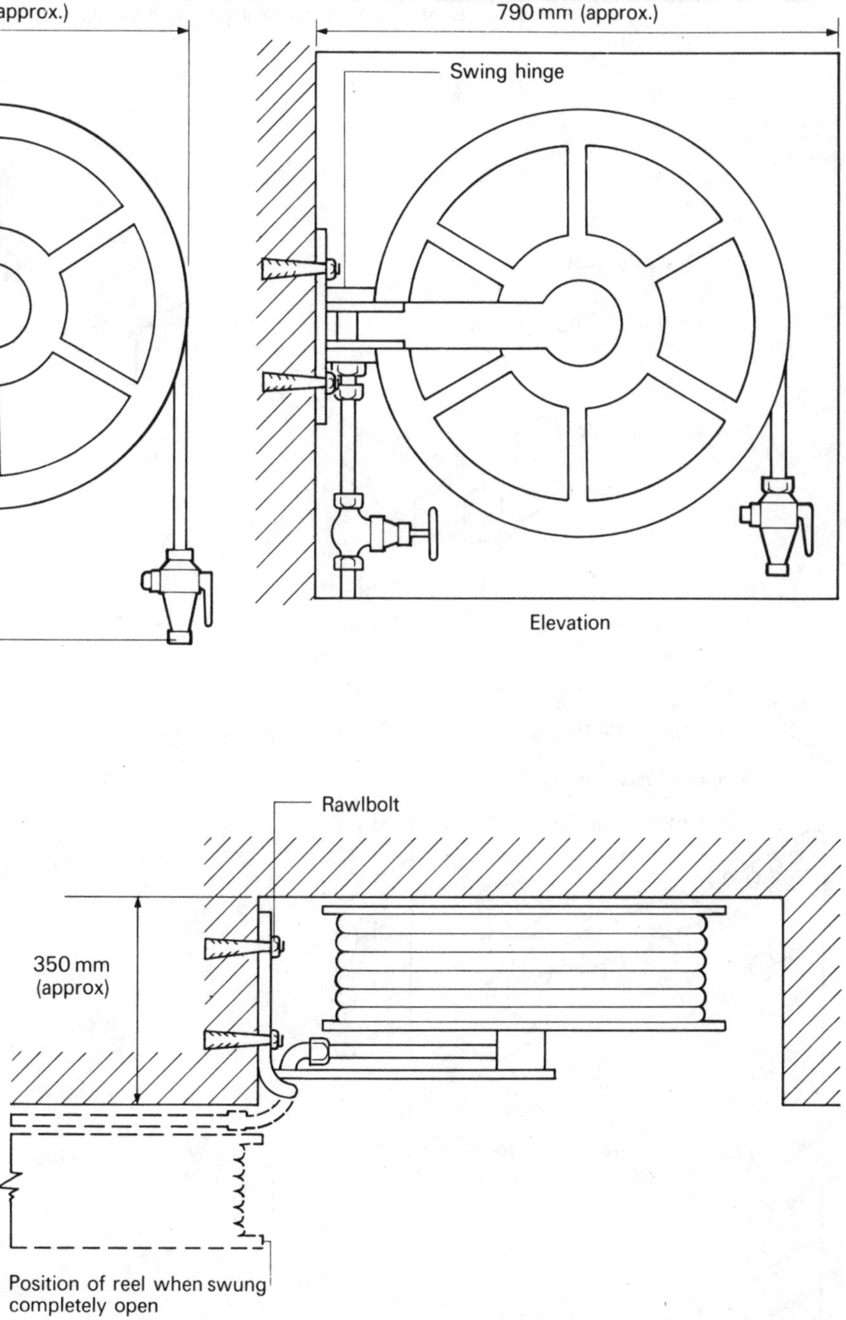
790 mm (approx.)
Swing hinge
Elevation
Rawlbolt
350 mm (approx)
Position of reel when swung completely open

Fig. 8.33 Swinging and recessed hose-reel

than 64 mm. The internal diameter of the pipe connection to each reel should be not less than 25 mm.

The riser may be used to supply the domestic cold-water storage cistern, but it is safer to provide an independent supply for hose-reels.

Automatic control of pumps

To switch on the duty pump when a hose-reel is used, a flow switch may be inserted in the pipeline: water passing through the switch causes the electrical circuit to be completed and the pump will start and run for as long as the water flows. When the flow of water stops, the flow switch opens the electrical circuit and stops the pump.

Alternatively, a pipeline pressure unit may be used which switches off the pump when the maximum pressure produced by the pump has been reached, this pressure being retained by a non-return valve. When a hose-reel is used, the pressure in the pipeline falls and the first pressure switch will start the duty pump. If the duty pump fails to operate within 10 s, the second pressure switch will start the stand-by pump and this will operate while any hose-reel is in use.

Automatic pneumatic system (Fig. 8.29)

Two sources of water supply may be provided for the hose-reels by the installation of an automatic pneumatic cylinder in addition to a break cistern. The system incorporates automatic water supply and pressure at all floor levels by the stored water in the cylinder pressurised by air. Immediately a hose-reel is used on any floor, the air in the cylinder forces water through the pipework to the nozzle and pressure or flow switches are therefore not required on the pipeline. In an emergency, the duty pump is automatically brought into operation and water is pumped from the break tank.

For buildings above twenty storeys in height it becomes necessary to pump in stages (see Fig. 8.30).

Hydraulic hose-reels

These consist of 19 mm or 25 mm i.d. (internal diameter), non-kinking, reinforced rubber hosepipe wound on a metal reel and fixed to the wall at a suitable height. The lengths of the hose are shown in Table 8.6.

Table 8.6 Lengths of rubber hose in metres

Internal diameter in 19 mm	Internal diameter in 25 mm
18	18
23	23
30	24
37	30
46	37

The outlet end of the hose is fitted with an approved nozzle, having an internal diameter of 5 mm or 6 mm, and with a pressure of 200 kPa at this point it will produce a horizontal throw of water of approximately 8 m, and about 5 m high. Some reels are designed to turn on the water automatically by drawing off the first few turns of the hose. When this type of reel is installed it is essential to place the isolating stop valve inside a service duct close to the reel to prevent the valve being shut off accidentally.

Hose-reels are made in three different types: fixed (Fig. 8.31), swinging (Fig. 8.32), and swinging-recessed (Fig. 8.33). The choice of reel is dependent upon its position relative to the fire risk. The fixed type is the cheapest but requires the hosepipe to be drawn off sideways, unless a special roller device is incorporated through which the hosepipe may be drawn off in various directions.

The swinging type permits the reel to turn through an angle of 180° and gives more flexibility for drawing off the hosepipe than the fixed type.

The recessed-swinging type is useful for fixing in corridors, as it prevents the reel from protruding from the wall and, if required, the reel may be concealed by a hinged or sliding door, providing its position is clearly indicated by suitable lettering.

Chapter 9

Portable and fixed extinguishers – detectors

Portable fire extinguishers

The term 'portable fire extinguishers' generally covers first-aid fire fighting appliances which can be carried by hand and from which the extinguishing agent can be expelled, usually under pressure. They are extremely valuable for extinguishing fires at the early stages, but they cannot be used successfully to deal with large fires. They can be included in buildings having sprinklers and hose-reels, but it is essential that staff are trained in their use.

Choice of extinguisher

There are several factors to consider in selecting the right type of appliance, but the most important are as follows:

1. It must contain the type of extinguishing agent suitable for the fire it may be required to extinguish.
2. It must not be dangerous to the user.
3. It must be simple to use.
4. It must be efficient and reliable.

Some extinguishers require more maintenance than others, and where there is a choice between extinguishers of similar type and equally suitable in all other respects, it is better to choose the type needing less maintenance.

To avoid confusion by the staff all extinguishers in the same building should, where possible, operate on the same principle, and where both water and foam extinguishers are installed they should all operate in the upright position.

If an extinguisher is installed to fight a possible outbreak of fire behind cupboards or under floorboards, it should be fitted with a short length of hose, and for rooms having unusually high ceilings the extinguisher chosen should be able to project the extinguishing agent on to the ceiling.

Some extinguishers may be too heavy or cumbersome for elderly people or women to use effectively, and therefore smaller extinguishers should be chosen for them.

High or low air temperatures (above 43 °C and below 4 °C) may have a detrimental effect on some types of extinguishers, and care must also be taken to choose one that will not prove dangerous or damaging in certain situations; for example, a water extinguisher should not be sited in a computer room.

Types of extinguishers

Portable fire extinguishers can be divided into the following five main groups, depending upon the extinguishing agent they contain.

Group 1: Water extinguishers:

(*a*) gas pressure,
(*b*) stored pressure,
(*c*) soda/acid.

Group 2: Dry powder extinguishers.

Group 3: Foam extinguishers:

(*a*) mechanical or gas pressure,
(*b*) chemical.

Group 4: Carbon dioxide extinguishers.

Group 5: Vaporising liquid extinguishers:

(*a*) Chlorobromomethane (CBM),
(*b*) Bromochlorodifluoromethane (BCF),
(*c*) Bromotrifluoromethane (BTM),
(*d*) Dibromotetrafluoroethane (DTE).

Water types

Water is used on Class A fires which involve fires in solid combustible materials such as wood, paper and textile fabrics. It has better cooling properties than other extinguishing agents and is preferable for use on fires which may re-ignite if not adequately cooled.

However, water is a conductor of electricity and must not be used on live electrical equipment.

Gas pressure type

The water contained in a steel cylinder is expelled by pressure from a carbon dioxide gas cartridge fitted inside the extinguishers (see Fig. 9.1). To operate the extinguisher the operating knob is struck, the plunger pierces a sealing disc and compressed carbon dioxide escapes from the cartridge into the space of the extinguisher body above the water level. The pressure exerted by the gas expels the water as a powerful jet through the nozzle out of the extinguisher.

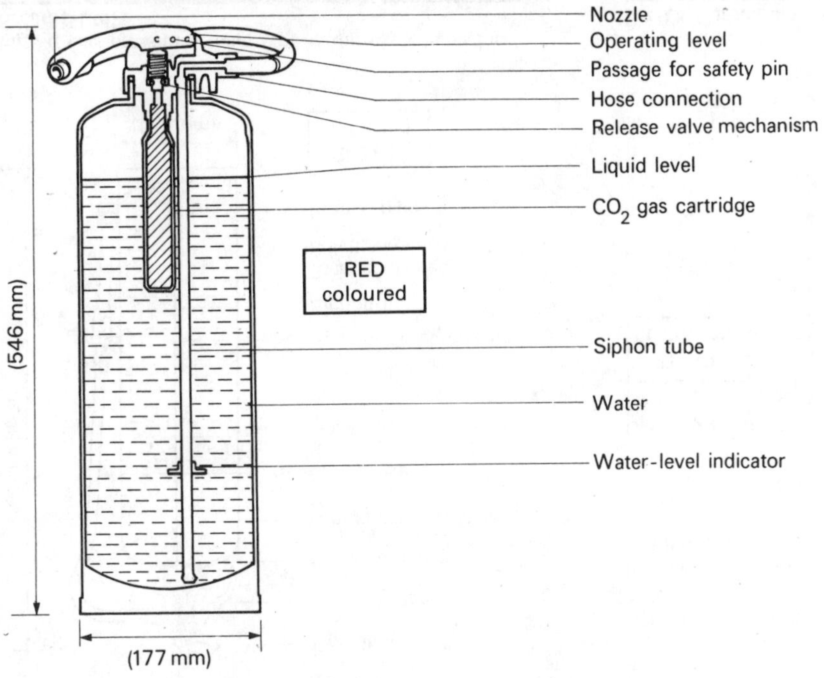
Nozzle
Operating level
Passage for safety pin
Hose connection
Release valve mechanism
Liquid level
CO_2 gas cartridge
RED coloured
Siphon tube
Water
Water-level indicator
(546 mm)
(177 mm)

Fig.9.1 Carbon dioxide - cartridge gas pressure type (Courtesy Chubb Fire Security Ltd)

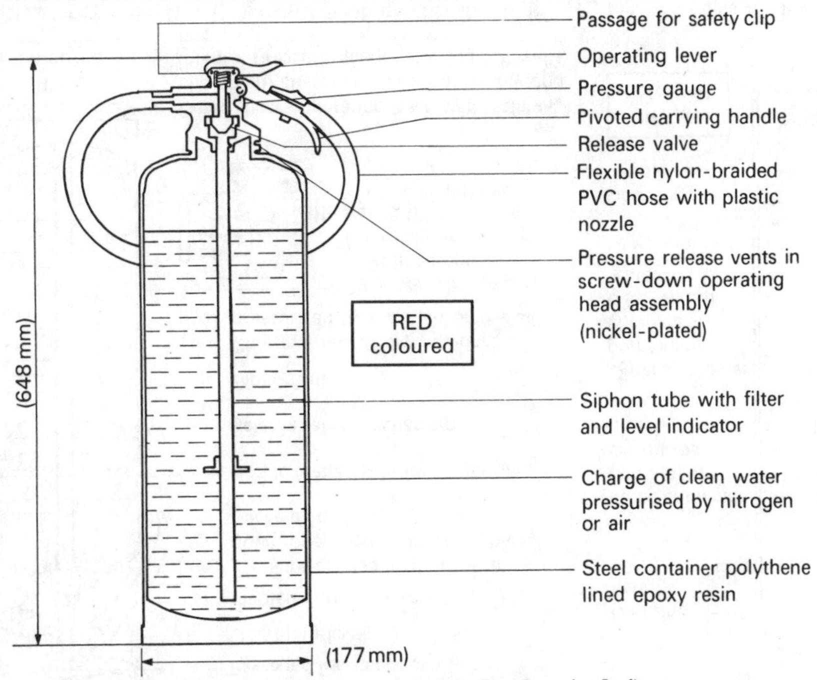
Passage for safety clip
Operating lever
Pressure gauge
Pivoted carrying handle
Release valve
Flexible nylon-braided PVC hose with plastic nozzle
Pressure release vents in screw-down operating head assembly (nickel-plated)
RED coloured
Siphon tube with filter and level indicator
Charge of clean water pressurised by nitrogen or air
Steel container polythene lined epoxy resin
(648 mm)
(177 mm)

Fig.9.2 Stored gas - pressure type (Courtesy Chubb Fire Security Ltd)

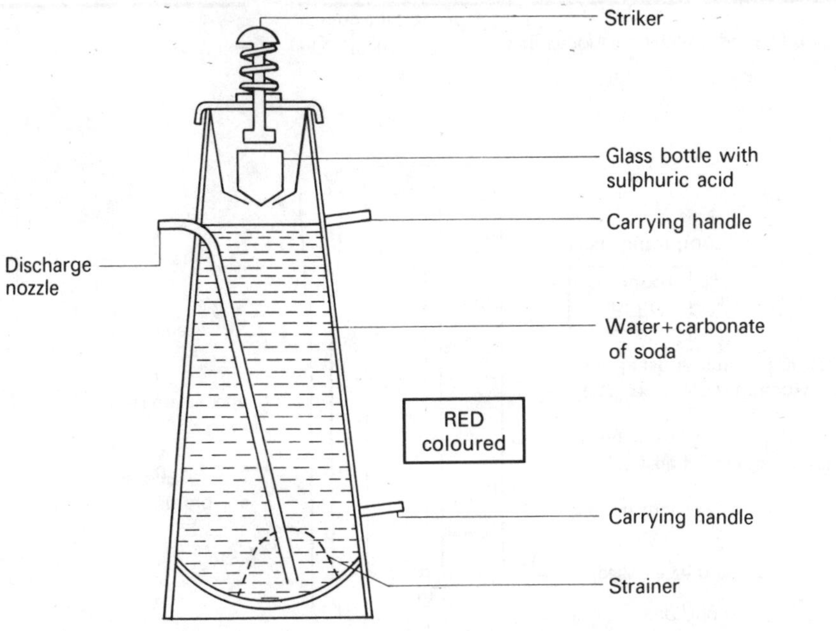
Striker
Glass bottle with sulphuric acid
Carrying handle
Discharge nozzle
Water+carbonate of soda
RED coloured
Carrying handle
Strainer

Fig.9.3 Striking type-soda acid water extinguisher

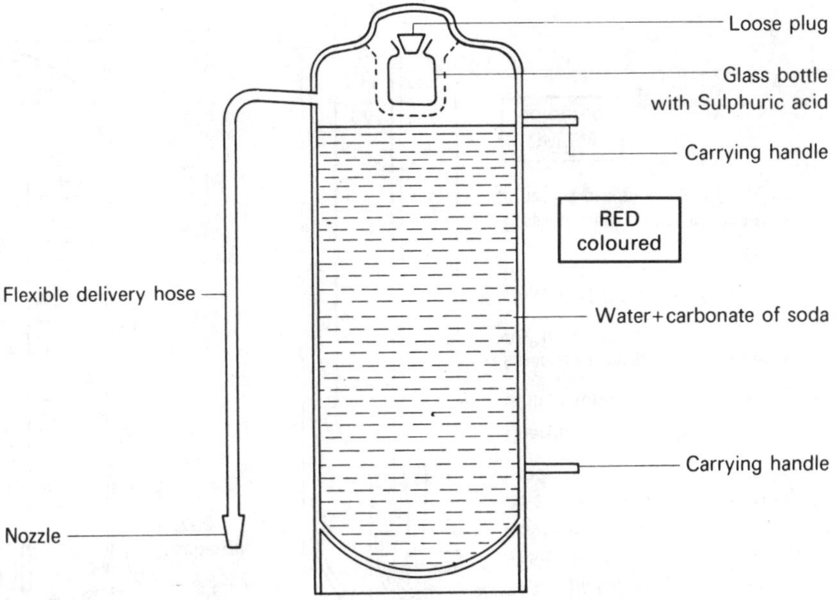
Loose plug
Glass bottle with Sulphuric acid
Carrying handle
RED coloured
Flexible delivery hose
Water+carbonate of soda
Carrying handle
Nozzle

Fig.9.4 Inversion type-soda acid water extinguisher

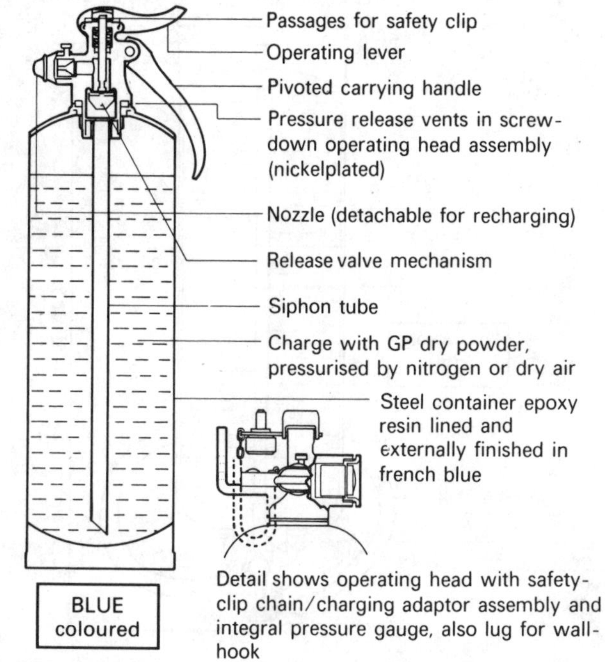
Passages for safety clip
Operating lever
Pivoted carrying handle
Pressure release vents in screw-down operating head assembly (nickelplated)
Nozzle (detachable for recharging)
Release valve mechanism
Siphon tube
Charge with GP dry powder, pressurised by nitrogen or dry air
Steel container epoxy resin lined and externally finished in french blue
Detail shows operating head with safety-clip chain/charging adaptor assembly and integral pressure gauge, also lug for wall-hook
BLUE coloured

Fig.9.5 Dry powder type (Courtesy Chubb Fire Security Ltd)

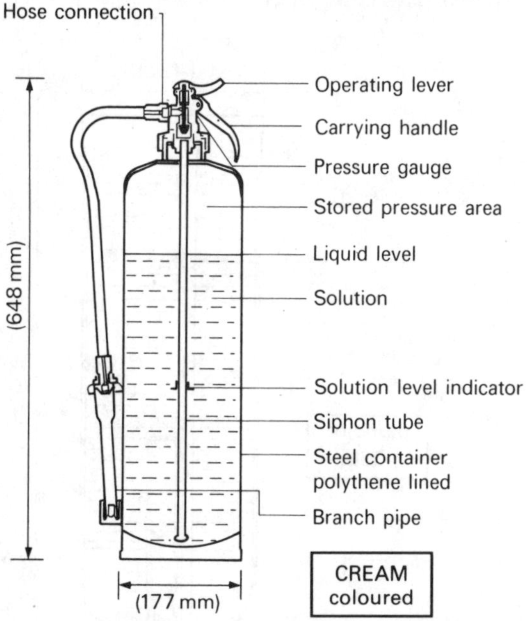
Hose connection
Operating lever
Carrying handle
Pressure gauge
Stored pressure area
Liquid level
Solution
Solution level indicator
Siphon tube
Steel container polythene lined
Branch pipe
(648 mm)
(177 mm)
CREAM coloured

Fig.9.6 Gas pressure - foam type (Courtesy Chubb Fire Security Ltd)

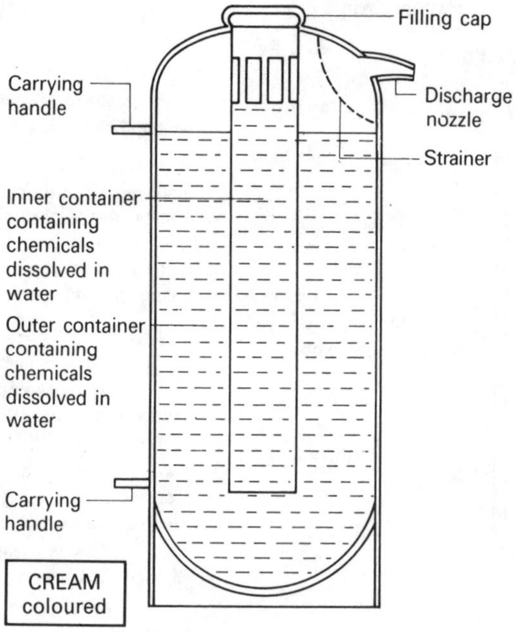
Filling cap
Carrying handle
Discharge nozzle
Strainer
Inner container containing chemicals dissolved in water
Outer container containing chemicals dissolved in water
Carrying handle
CREAM coloured

Fig.9.7 Chemical - foam type

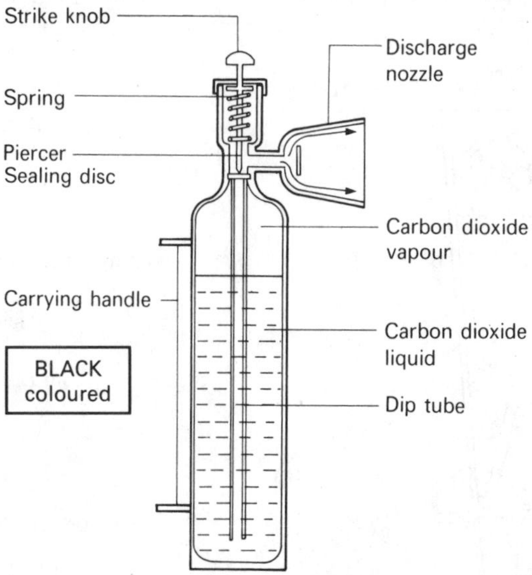
Strike knob
Discharge nozzle
Spring
Piercer
Sealing disc
Carbon dioxide vapour
Carrying handle
Carbon dioxide liquid
Dip tube
BLACK coloured

Fig.9.8 Carbon dioxide type

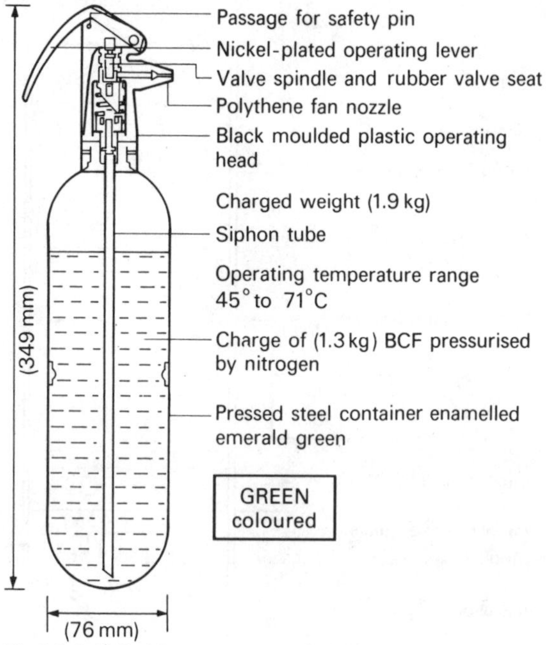
Passage for safety pin
Nickel-plated operating lever
Valve spindle and rubber valve seat
Polythene fan nozzle
Black moulded plastic operating head
Charged weight (1.9 kg)
Siphon tube
Operating temperature range 45° to 71°C
Charge of (1.3 kg) BCF pressurised by nitrogen
Pressed steel container enamelled emerald green
GREEN coloured
(349 mm)
(76 mm)

Fig 9.9 B.C.F. 1.3 kg vaporising liquid type (Courtesy Chubb Fire Security Ltd)

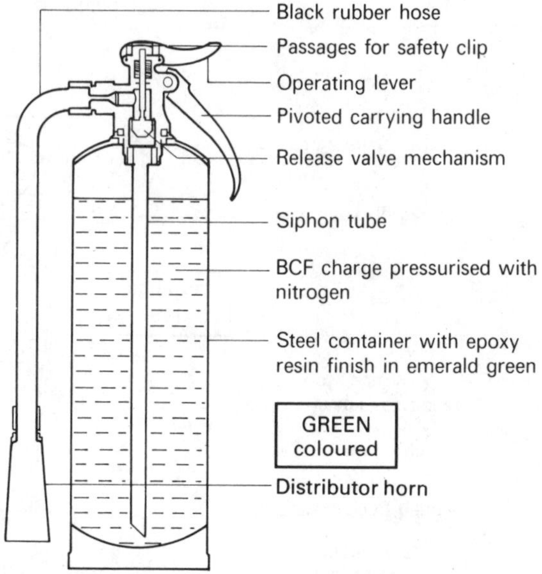
Black rubber hose
Passages for safety clip
Operating lever
Pivoted carrying handle
Release valve mechanism
Siphon tube
BCF charge pressurised with nitrogen
Steel container with epoxy resin finish in emerald green
GREEN coloured
Distributor horn

Fig.9.10 BCF 3.6 or 7.3 kg vaporising liquid type (Courtesy Chubb Fire Security Ltd)

Stored pressure type

The water inside the body of the extinguisher, is expelled by the pressure of either air or an inert gas such as nitrogen, which is compressed above the surface of the stored water. When the operating lever is depressed, the water is forced out of the extinguisher by the force of the compressed air or gas (see Fig. 9.2). The extinguisher is ready for instant use and does not require the release of pressurised carbon dioxide gas from a cartridge before water is expelled.

Soda/acid type

In this type the pressure necessary for discharge of the jet of water is dependent upon the chemical reaction between bicarbonate of soda dissolved in water (sodium sulphate) and sulphuric acid in a glass bottle. The sulphuric acid is released into the water when the bottle is shattered, either by a metal striker (Fig. 9.3) or by inverting the extinguisher (Fig. 9.4). Chemical reaction takes place between the sulphuric acid and the sodium sulphate solution which generates carbon dioxide gas and pressurises the container, thus forcing the contents of the extinguisher through the nozzle.

It is essential to keep the extinguishers the correct way up when being used, the striking type must be kept in the position shown in the diagram, while the other must be inverted. Using the extinguishers the wrong way up will allow carbon dioxide gas to be discharged instead of the liquid.

Dry powder extinguisher

General-purpose dry powder is suitable for all classes of fire risks and is particularly suitable for fires in inflammable liquids. The powder consists of a finely divided, non-conducting, non-toxic, water-repellent material which cools the flames, separates them from the burning material and excludes oxygen. It also acts as a form of screen, thus enabling the operator to approach close to the fire. The dry powder stored in the body of the extinguisher is pressurised by nitrogen or air, which expels the powder when the release valve is opened (see Fig. 9.5).

Alternatively, the dry powder may be expelled from the container by carbon dioxide gas after the breakage of a seal of a cartridge.

Foam extinguishers

Foam extinguishes fires by forming a blanket over the burning material, thus excluding oxygen, and stays in position long enough for the hot liquid to cool. It may be used on Class B fires involving burning liquids or liquefiable solids such as petrol, oil and grease.

Mechanical or gas pressure type

The foam compound is stored as a solution (Plus F fluoroprotein) in the body of the extinguisher which is lined with polythene to prevent corrosion. The foam is expelled by an inert compressed gas acting upon the surface of the foam (see Fig. 9.6).

Chemical type

This consists of an inner container, containing chemicals dissolved in water which mix with chemicals contained in the outer container when the extinguisher is inverted. Gas pressure and foam is thus generated and forced out of the nozzle by the pressure (see Fig. 9.7).

Carbon dioxide extinguisher

Carbon dioxide gas is suitable for use on Class B fires involving inflammable liquids and also on Class E fires which are those complicated by the presence of live electricity. It is particularly suitable for computer rooms and fires involving electronic equipment. The advantages of carbon dioxide as an extinguishing agent are as follows:

1. It is non-toxic and can therefore be only dangerous to life in a high concentration, when it may exclude oxygen from the atmosphere.
2. It is an inert gas and therefore does not damage materials or leave a contaminating agent.

The carbon dioxide extinguisher shown in Fig. 9.8 consists of a pressure cylinder containing liquid carbon dioxide under pressure. When the pressure is released the liquid changes to a gas, expands instantly and discharges through the nozzle as a cloud, thus extinguishing the fire by excluding oxygen.

Vaporising liquid extinguisher

Vaporising liquids act rapidly, but because they are small, they are useful only on very small fires and are widely used for motor vehicles and electrical apparatus. The liquid vaporises rapidly when in contact with a fire and acts by excluding oxygen. They should not be used or kept in a confined space, where there is a risk that a person may inhale the vapours or products formed when the vapours are heated by the fire.

Methyl bromide and carbon tetrachloride are highly toxic and are therefore not to be recommended. Figures 9.9 and 9.10 show diagrams of BCF vaporising liquid fire extinguishers.

Table 9.1 gives details of various types of extinguishers which is intended as a rough guide only. It should be noted that the area of fire in inflammable liquids which different types of appliances can extinguish is for a trained operator, and an untrained operator may be able to extinguish only about half the areas given.

The seals given relate to contained fires. Table 9.2 gives a summary of the type of fires extinguished by various agents.

Construction

Extinguisher bodies are usually made from mild steel, treated to avoid internal corrosion by a zinc coating followed by expoxy resin enamel or by lining with polythene. The small vaporising liquid types are made from either brass or copper.

While the body of any type of extinguisher may vary in construction material, the components such as control valves, hose couplings, piercing spindles and strike knobs are invariably made from brass.

Positioning

Extinguishers should be as near as possible to exits or on staircase landings. They should be in conspicuous positions and ready for immediate use. Where large individual floor areas necessitate positioning appliances away from exits or outer walls, they should be positioned on escape routes.

Table 9.1 Portable fire extinguishers

Type of extinguisher	British Standard No.	Typical capacities	Range (m)	Typical weights (kg)	Number required for protection: Per 200 m² of floor area	Number required for protection: Per floor	For flammable liquid fires (m²)
Water – gas pressure	1382	4.5–6.7 and 9 litre	9	9.5–11.8 15–18	2 1	4 2	
Water – stored pressure	3709	4.5–6.7 and 9 litre	9	9.5–11.8 15–18	2 1	4 2	
Water – soda/acid	138	4.5–6.7 and 9 litre	9	10.8 15	2 1	4 2	
Dry powder	3465	1.8–2.2 3 kg and 9 kg	3 4.6 6	4.5 8 18			0.9 1.4 3.7
Foam – mechanical	740, Part 2	4.5–6.7 and 9 litre	7	9 15–18			0.5 0.9
Foam – chemical	740, Part 1	4.5–6.7 and 9 litre	7	9 15–18			0.5 0.9
Carbon dioxide	3326	1.13 kg 3 kg 4.5 kg 6.8 kg	1.2 1.2 2 3	5.8 15 17–19 24.5–27			0.5 0.9
Vaporising liquid – CBM, BCF, BTM	1721	0.6 litre 2.4 litre	Up to a maximum of 6	Up to a maximum of 2.7			Up to a maximum of 0.3

Table 9.2 Extinguishing agent

Type of fire	Water	Dry powder	Foam	CO_2	BCF
Fires in paper, wood and cloth	Excellent. Water saturates the material and prevents re-ignition	Small surface fires only	Excellent. Has smothering and wetting action	Small surface fires only	Small surface fires only
Burning liquids such as petrol, oil paints and grease	Water will spread fire and should not be used	Excellent. Smothers the fire	Excellent. Smothers the fire	Excellent. Replaces oxygen	Excellent. Smothers the fire
Fires associated with live electrical equipment	Water conducts electricity and should not be used	Excellent. Non-conductor of electricity	Foam conducts electricity and should not be used	Excellent. Non-conductor of electricity	Excellent. Non-conductor of electricity

Note: The extinguishing agent should always be directed at the base of a fire.

Fire buckets

Although a water-type extinguisher has a better appearance and is more effective for extinguishing Class A fires than water used from a bucket, the installation of fire buckets is much cheaper and their use is generally well understood.

However, because of the difficulty in throwing water from a bucket to any point at high level, their use is limited to fires occurring at low level. A fire bucket should hold either 9 or 14 litres of water and be covered to reduce evaporation. A bucket may have a flat or rounded bottom – the latter type discourages its use for other purposes. It may be made from plastic or metal and painted red and hung so that it is not higher than 1 m above the floor or on a shelf 760 mm above the floor. As a guide, three buckets should be provided per 210 m^2 of floor area with a minimum of six buckets per floor.

Dry sand in fire buckets may also be installed where its non-conducting properties may be used to advantage, for example when the sand is used for extinguishing small fires associated with electrical equipment. Sand may also be used for extinguishing small liquid and metal fires.

Fixed extinguisher installations

For large quantities of inflammable liquids, industrial machinery, such as textile carding machines and large electrical fire risks that are beyond the scope of portable fire extinguishers, a fixed installation is required which is usually automatic in operation.

Fixed fire-extinguishing installations may be required for electrical substations, computer rooms, paint spraying booths, boiler rooms, textile machinery, kitchen equipment, oil and petroleum storerooms, printing machines, aircraft hangars and oil-tanker loading bays. The installation of a fixed extinguishing installation for these and many other fire risks may be required by an insurance company who will specify and approve the installation. An approved installation will almost certainly result in a reduction in fire premiums; generous Government grants are also available towards the cost of installation.

Foam systems

Foam is generally accepted as a highly effective agent for extinguishing fires arising from the use or storage of inflammable liquids and liquefiable substances. There are two main types of foam:

1. Chemical foam: this is formed by the chemical reaction between sodium bicarbonate and aluminium sulphate in aqueous solution in the presence of a foaming agent.
2. Mechanical foam: this is formed by the mixing of water, a foaming agent and air in a suitable generator or foam-making appliance. This type is often known as air foam.

Classification

There are three basic classifications for fire-fighting air foams:

1. *High-expansion foam:* This has an expansion ratio of up to 1200 : 1. When foam with an expansion ratio of 1000 : 1 is driven on to a fire, the one volume of liquid is flashed into steam and the resulting expansion of water into steam creates a mixture of 1700 volumes of steam and 1000 volumes of air. The oxygen content is less than 7.5 per cent which is well below that required to support combustion.

High-expansion foam has the ability to flow around obstacles and penetrate inaccessible areas, it is used for high rack warehouses, aircraft hangars, marine engine rooms and paint mix stores.

2. *Medium expansion foam:* This has an expansion ratio up to 300 : 1 and is produced by the mixture of foam concentrate, air and water. The foam is heavier than high-expansion foam and may be used for outdoor fire hazards such as oil-tanker loading bays, or indoor areas such as oil and petroleum stores.

3. *Low-expansion foam:* This has an expansion ratio of up to 20 : 1 and is used for such fire hazards as boiler rooms, transformer areas and inflammable liquid storage tanks. To reduce the risk of an explosion, low-expansion foam may be injected below the surface of a burning liquid at the base of the tank. At the same time, foam is conveyed to the liquid surface by an independent piping system.

Uses of foam systems

A pre-mix high-expansion foam installation may be used for a variety of fire risks and is entirely independent of outside water supply or any extraneous source of working pressure. The system will produce high-quality foam immediately without any preliminary discharge of water, which is a vital factor in the case of certain fire risks.

The pre-mix foam installation shown in Fig. 9.11 may be used for transformer areas or inflammable liquid stores and other isolated situations. The installation consists of a storage cylinder constructed to BSS 1500, designed for a maximum working pressure of 1034.22 kPa. The cylinder is filled with a solution of foam compound and water, and its capacity is determined by the quantity and depth of foam coverage required.

The cylinder is fitted with an inlet connection from a carbon dioxide gas cylinder (or cylinders) of appropriate capacity, with a disc-closure valve and a lever-operated piercing head. The rate of discharge of carbon dioxide gas in the event of fire is controlled so that a continuous pressure of 689.5–827.4 kPa will be maintained within the storage cylinder, ensuring a constant rate of flow for the outgoing solution. The other components consist of distribution pipework, incorporating foam-makers and spreaders, a fusible link fire-detecting system; an alarm bell or similar warning device is usually fitted. If required, a combined foam-maker and sprinkler may be used.

Operation:

1. When a fire occurs, the heat breaks the fusible link which allows the weight to fall and the lever to rise, thus piercing a seal contained in the piercing head.
2. Carbon dioxide gas is released from the cylinder or cylinders and passes to the foam-solution storage cylinder.
3. Foam solution is forced up the siphon tube and along the outlet pipe to the

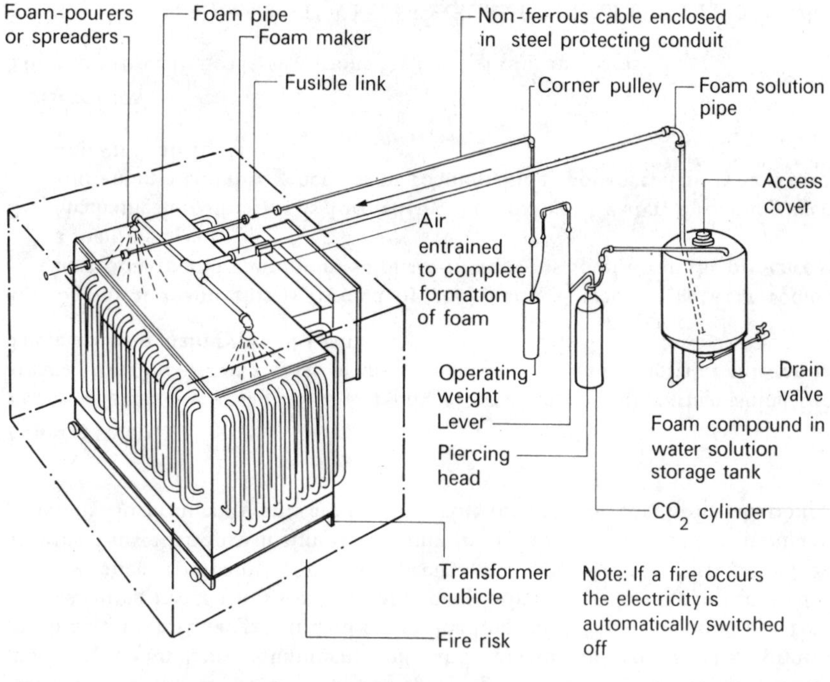
Foam-pourers or spreaders
Foam pipe
Foam maker
Fusible link
Non-ferrous cable enclosed in steel protecting conduit
Corner pulley
Foam solution pipe
Access cover
Air entrained to complete formation of foam
Operating weight
Lever
Piercing head
Drain valve
Foam compound in water solution storage tank
CO_2 cylinder
Transformer cubicle
Fire risk
Note: If a fire occurs the electricity is automatically switched off

Fig.9.11 Pre-mix foam fire extinguishing installation (Courtesy Chubb Fire Security Ltd)

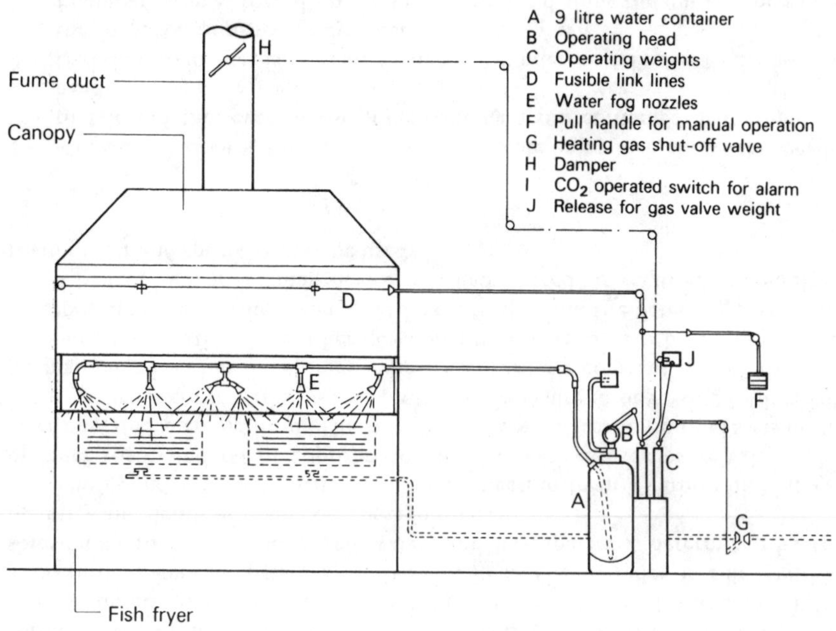
Fume duct
Canopy
Fish fryer
A 9 litre water container
B Operating head
C Operating weights
D Fusible link lines
E Water fog nozzles
F Pull handle for manual operation
G Heating gas shut-off valve
H Damper
I CO_2 operated switch for alarm
J Release for gas valve weight

Fig.9.13 Mechanically operated water-fog installation for two-pan fryer

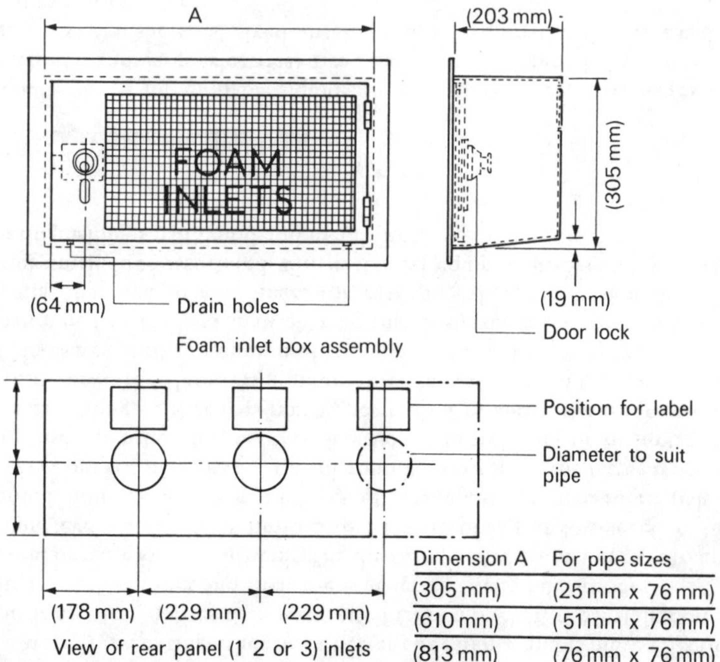
A
FOAM INLETS
(203 mm)
(305 mm)
(64 mm)
Drain holes
(19 mm)
Door lock
Foam inlet box assembly
Position for label
Diameter to suit pipe
(178 mm)
(229 mm)
(229 mm)
View of rear panel (1 2 or 3) inlets

Dimension A	For pipe sizes
(305 mm)	(25 mm x 76 mm)
(610 mm)	(51 mm x 76 mm)
(813 mm)	(76 mm x 76 mm)

Fig.9.12 Foam inlet box (Courtesy Chubb Fire Security Ltd)

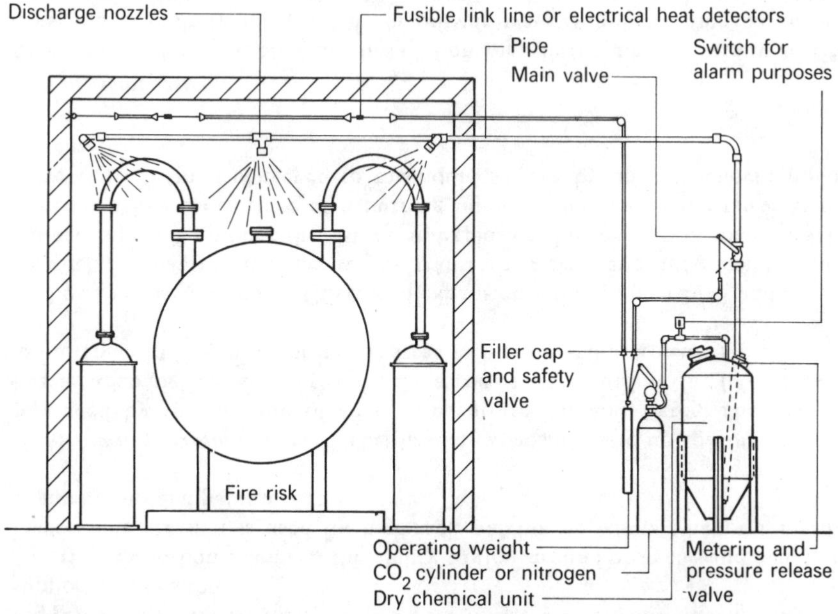
Discharge nozzles
Fusible link line or electrical heat detectors
Pipe
Main valve
Switch for alarm purposes
Filler cap and safety valve
Fire risk
Operating weight
CO_2 cylinder or nitrogen
Dry chemical unit
Metering and pressure release valve

Fig.9.14 Dry chemical fire-extinguishing installation

foam-maker device, where air is entrained to complete the formation of the foam.
4. The foam passes to the foam spreader and is sprayed over the fire.

Foam inlet system

The foam inlet system is frequently installed for the fire protection of oil-fired boilers and associated oil storage tanks. An inlet fitting is housed within a foam inlet box, which is built into the wall of the building at a point approved by the local Fire Authority. The box (see Fig. 9.12) is clearly labelled FOAM INLET and is usually placed about 600 mm above ground level; it should be clear of any openings through which flame, heat or smoke from the boiler room or oil-tank room could hinder firemen using the box.

The local Fire Authority should be consulted to ascertain which type of foam-making equipment will be used at the building being protected. If foam-making branch pipes are to be used, an inlet adapter sleeve to BSS 336: 1965, 64 mm or 76 mm diameter, will be required for each inlet pipe. If mechanical foam generators are to be used, 64 mm diameter male instantaneous inlet adapters to BSS 336: 1965 will be required. The piping into the fire-risk area should be 76 mm (i.d.), galvanised mild steel, screwed and socketed finished with a cone-type foam spreader or an open-ended pipe. The layout of the piping should be as direct as possible and any change of direction should be made by means of easy bends. Elbows should not be used, nor should the pipe contain a vertical riser. All pipework should slope slightly towards the outlet so that the pipework is self-draining. If two foam outlets are to be installed, the branch pipes to the outlets must be the same length.

The internal discharge point should be not less than 150 mm above the fire risk and placed centrally above it. When installed in transformer chambers the pipework should terminate at ceiling level and be electrically earthed. Each inlet should supply the piping to which it is connected into one fire-risk area only, for example an oil storage room or a boiler room.

Water-fog system (Fig. 9.13)

This is water in an atomised state which has been developed mainly for fire in frying-pans. The system is fixed, with atomising nozzles and a fusible link over the pans. The water fog smothers the fire, excludes oxygen and causes rapid cooling of the burning fat, making it almost impossible for re-ignition to occur. Very little water enters the fat as most of it forms steam on contacting the burning fat surface.

The installation container holds 9 litres of water which is sufficient for three pans. A small cylinder outside the kitchen contains liquefied carbon dioxide, which is released automatically and changes to a gas by a break in a fusible link, due to a fire or by manually breaking the glass in a pull-handle box and operating a handle downwards. A weight falls, opening a valve in an operating head at the top of the water container, which allows carbon dioxide gas to flow into the container and discharge water through water-fog nozzles. The same weight also closes a damper in the fume duct and prevents the possible spread of fire to other parts of the building. At the same time the carbon dioxide gas flows into a special switch which operates an alarm and another weight falls which closes a heating gas shut-off valve.

Dry chemical systems

Dry chemical extinguishing agents are powders such as sodium bicarbonate, potassium chloride or monammonium chloride, specially treated with metallic steorite as a waterproofing agent to prevent caking and provide flowability. The small particles extinguish fires by interrupting the chain action of oxygen uniting with the burning fuel.

Dry powder is non-toxic and non-conductive and may be used for fires involving live electrical equipment, inflammable liquid and ordinary carbonaceous materials such as wood, paper or cloth.

The disadvantages of dry powder, however, are as follows:

1. It has a relatively low cooling power which increases the possibility of re-ignition.
2. It leaves a residue which must be removed to prevent corrosion problems, especially if machinery and electrical equipment are involved.
3. Although the powders used are non-toxic, the presence of fine powder in the atmosphere may cause distress to some people in the vicinity of the fire.

Dry powder systems can be completely automatic and be designed to achieve any coverage required (see Figs. 9.14 and 9.15). A system includes a fusible link line or electrical heat detectors which sense overheated conditions and signal the control head on a pressurised dry powder cylinder or open the valve on a carbon dioxide or nitrogen cylinder. This causes the release of dry chemical, through piping network to nozzles at the fire hazard.

Dry-chemical cylinders can be assembled in manifold for large-capacity needs, and the detection methods used include: electric thermostatic, pneumatic rate of rise, fusible link, remote manual and local manual.

Carbon dioxide systems

Carbon dioxide gas is probably the most versatile and, for many operations, the ideal extinguishing agent. The gas covers the flames with a blanket of heavy gas that suffocates the fire by reducing the oxygen content of the surrounding atmosphere to a point where combustion is impossible. The gas is dry, odourless, non-corrosive, non-conductive and is heavier than air so that it flows around obstacles.

It cannot damage clothing or equipment, and when its work is completed it disappears without leaving any residue. Carbon dioxide systems can be installed for the protection of computer rooms, textile mills, power stations, transformer cubicles, fur vaults, museums and archives.

High-pressure systems

Liquid carbon dioxide is stored at about 5200 kPa at 20 °C ready for immediate use in steel cylinders made to BS 1288.

Low-pressure systems

Liquid carbon dioxide is stored in a refrigerated tank at about 2000 kPa and tanks are available from 3.4 t to 30 t. Joint protection systems can be designed so that one tank or battery of cylinders discharge the gas into any of several risk areas. Low-pressure systems are generally used for very large areas when it is often more economical in terms of cost and space to use a low-pressure bulk

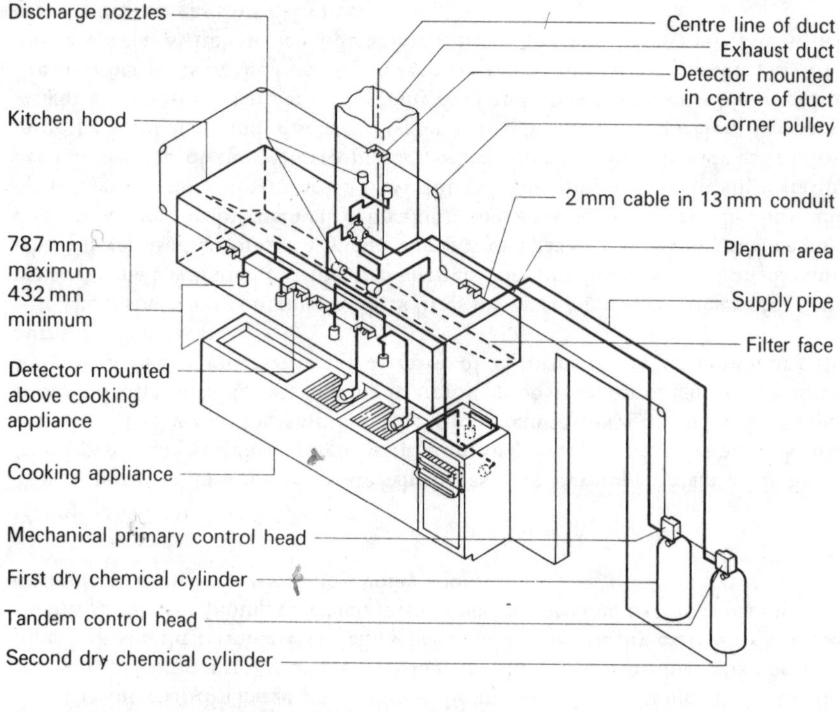
Discharge nozzles
Centre line of duct
Exhaust duct
Detector mounted in centre of duct
Kitchen hood
Corner pulley
2 mm cable in 13 mm conduit
787 mm maximum
432 mm minimum
Plenum area
Supply pipe
Filter face
Detector mounted above cooking appliance
Cooking appliance
Mechanical primary control head
First dry chemical cylinder
Tandem control head
Second dry chemical cylinder

Fig.9.15 A typical sentinel dry powder system protecting cooking range (Courtesy Chubb Fire Security Ltd)

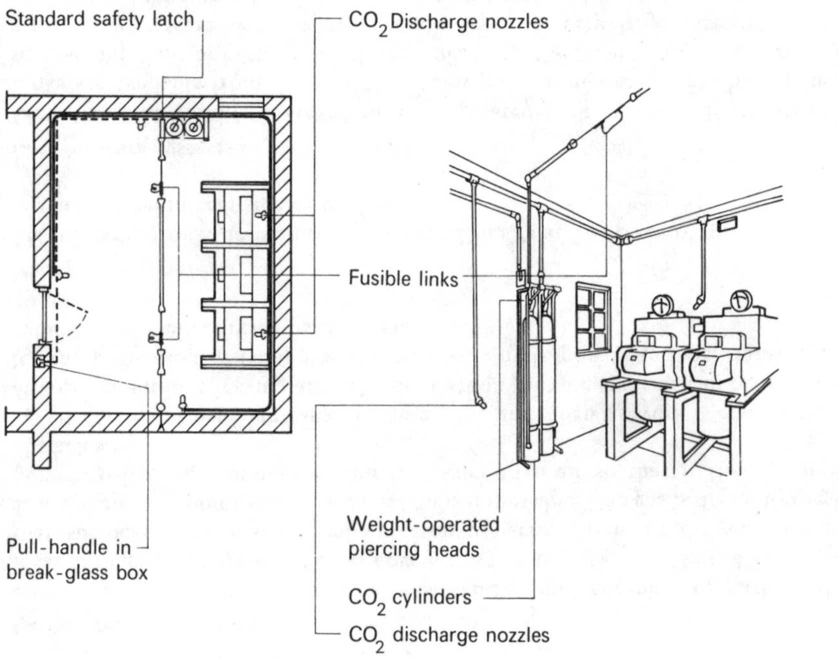
Standard safety latch
CO_2 Discharge nozzles
Fusible links
Pull-handle in break-glass box
Weight-operated piercing heads
CO_2 cylinders
CO_2 discharge nozzles

Fig.9.17 Typical automatic CO_2 fire-extinguishing installation for an electrical sub-station (Courtesy Chubb Fire Security Ltd)

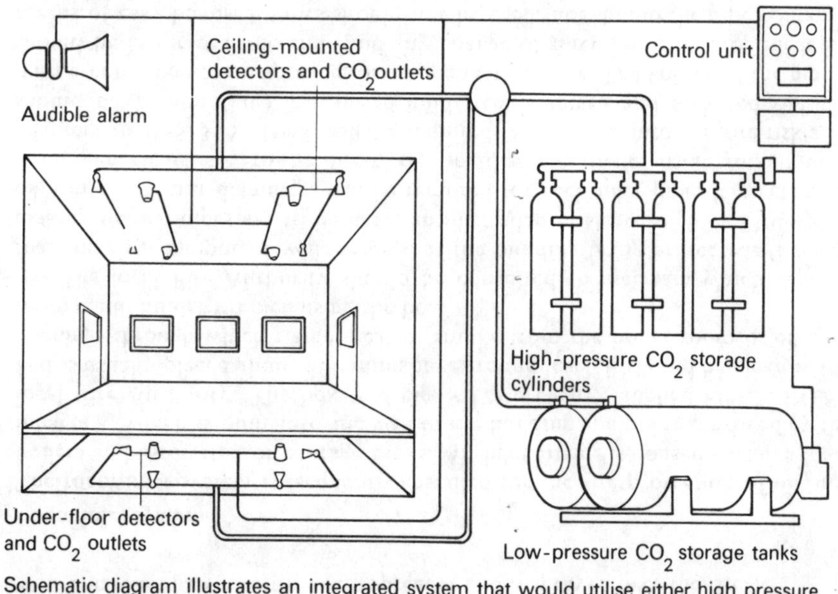
Ceiling-mounted detectors and CO_2 outlets
Control unit
Audible alarm
High-pressure CO_2 storage cylinders
Under-floor detectors and CO_2 outlets
Low-pressure CO_2 storage tanks

Schematic diagram illustrates an integrated system that would utilise either high pressure cylinders or low-pressure tanks

Fig.9.16 CO_2 fire-extinguishing installation for a computer room (Courtesy C F S Ltd)

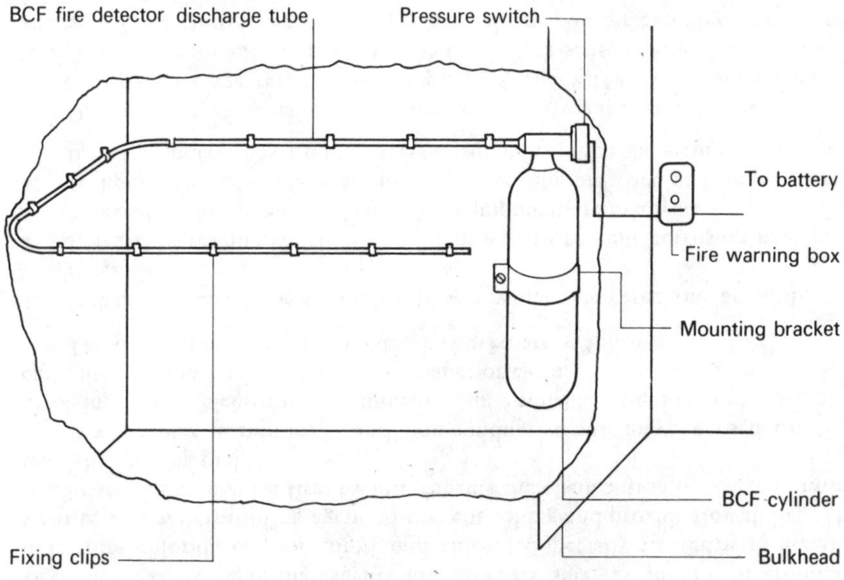
BCF fire detector discharge tube
Pressure switch
To battery
Fire warning box
Mounting bracket
BCF cylinder
Fixing clips
Bulkhead

Fig 9.18 Automatic BCF fire-fighting system (Courtesy Chubb Fire Security Ltd)

storage tank rather than a number of high-pressure cylinders. Figure 9.16 shows an integrated high- and low-pressure system for a computer or similar room.

The detectors located at the hazard, both on the ceiling and underfloor, sense overheat conditions and send a signal to the system control unit which causes the release of carbon dioxide gas from the cylinders and storage tanks.

Figure 9.17 shows a high-pressure system for an electrical sub-station. If a fire occurs the fusible link breaks and opens the cylinder valves, thus allowing the release of carbon dioxide gas. A pull-handle in a break-glass box is located outside the room, can be used on failure of the automatic system. The systems shown are designed for total flooding by carbon dioxide gas of the protected space, but systems are also designed for local application, for example risks involving textile machinery.

BCF system

The Pyrene Kill-Fire Automatic BCF system is shown in Fig. 9.18 and is suitable for the protection of small isolated risks which require a self-contained fire-fighting system. The system is suitable for vehicle and boat engine compartments, fuel stores, electrical power supply cabinets and small cable ducts.

Using BCF vaporising liquid the system instantly smothers the fire of inflammable liquids, and being non-conductive it is used for fires where live electricity is involved.

The system is pre-wired and therefore very simple to install. The tubing which is of nylon, usually 5 m long, is arranged around the area to be protected. This tubing, which acts as both detector and discharger, softens in the area of most intense heat, bursts and discharges the BCF on to the fire. At the same time, the reduction in pressure operates the pressure switch and activates an external visual fire-warning panel.

Lock-off devices

Every installation must be provided with means of immobilising the equipment, since it is important that automatic operation should not occur whilst people are in the protected zone. Provision is usually made to:

(*a*) lock-off the automatic equipment only (leaving over-riding manual control), and
(*b*) completely immobilise the installation.

These devices can usually be operated from a remote position outside the protected area.

Other automatic devices

By diverting a small amount of gas to pressure-operated switches and trip mechanisms it is possible to automatically:

(*a*) operate door-closing devices;
(*b*) switch off ventilating systems;
(*c*) operate fire curtains;
(*d*) close openings in ventilating ducts.

Automatic fire detectors

Unless a fire is detected at an early stage it will spread rapidly and get out of control. The purpose of an automatic fire detector, therefore, is to sense an outbreak or possible outbreak of fire at the earliest possible moment and give an alarm signal. The alarm signal may alert persons inside or near the building, and in addition the alarm may be connected to the local fire brigade or private fire station on the premises. The system should be completely reliable and regularly tested.

There are several groups of fire-detector systems briefly described as follows:

Group 1 (detectors for hazardous situations)

These are not fire detectors in the true sense of the word, but by providing early warning of a hazardous situation (e.g. a leak from a petrol tank, or overheating of a process plant) they can prevent fire. They can be linked to an extinguishing system as described under fixed extinguishers.

There are four main sub-groups:

1. *Flammable vapour detectors*

These operate on a catalytic principle and can be used for more than a thousand different types of flammable vapours including petrol and vinyl chloride.

2. *Butane and propane leakage detectors*

These operate on the principle that a heavy gas will diffuse more slowly through a membrane than a lighter gas. The detector operates over a smaller range of vapours than the flammable vapour detector.

3. *Overheat detector*

There are several types of these, the basis of each being a long, flexible, temperature-sensitive element that can be fitted to plant boiler rooms and kitchens, wherever excessive heating can result in damage unless heat is reduced within a certain time.

4. *Explosion detector*

This operates when either a predetermined rate of pressure rise, or static pressure setting is exceeded. Almost any material which will burn in bulk can explode if it is in a sufficiently divided state, and an explosion of this nature is in fact only an ultra-fast fire. However, there is a measurable interval of time between ignition of a combustible mixture and a build-up of pressure to destructive proportions. The detector is designed to activate a device that will suppress, vent or initiate other action to prevent the spread and effects of the explosion.

Group 2 (ionisation smoke detector)

An ion is an atom or group of atoms which have gained or lost one or more electrons and thus carries a positive or negative charge. Radiation from a radioactive source can cause ionisation of air. There are two distinct types of detectors, and one type balances an open ionisation chamber against a closed and sealed ionisation chamber (see Fig. 9.19). Another type uses an open ionisation sampling chamber, but the closed chamber is replaced by a transistorised

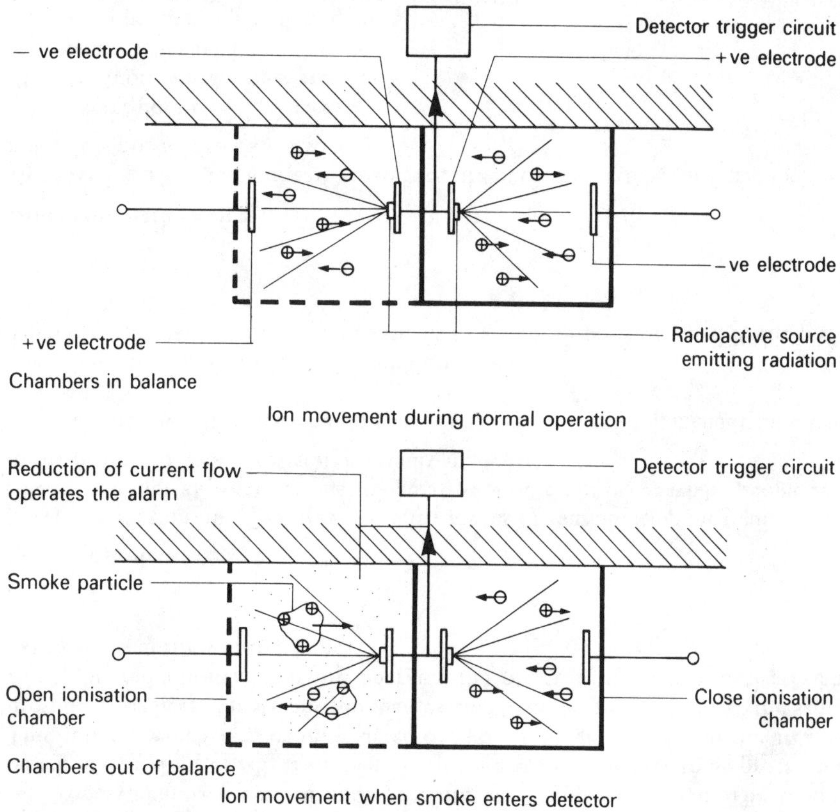
— ve electrode
Detector trigger circuit
+ve electrode
–ve electrode
+ve electrode
Radioactive source emitting radiation
Chambers in balance
Ion movement during normal operation
Reduction of current flow operates the alarm
Detector trigger circuit
Smoke particle
Open ionisation chamber
Close ionisation chamber
Chambers out of balance
Ion movement when smoke enters detector

Fig. 9.19 Ionisation smoke detector

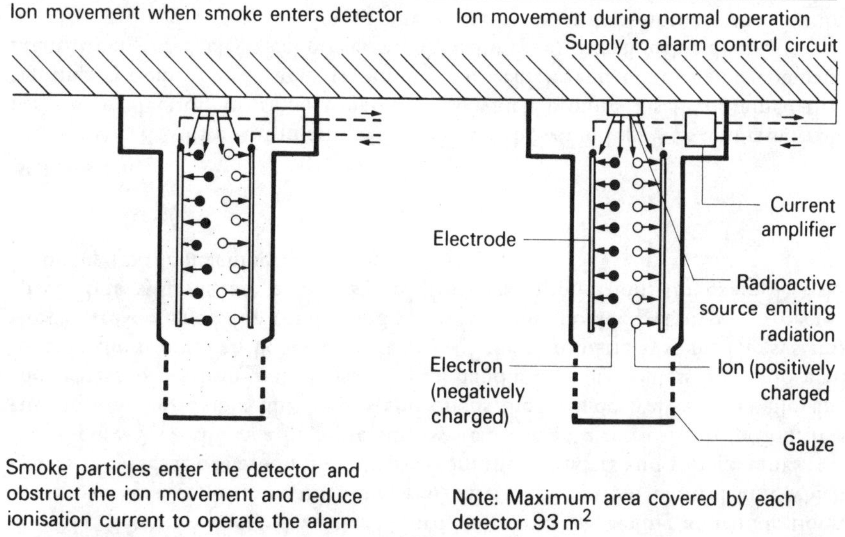
Ion movement when smoke enters detector
Ion movement during normal operation
Supply to alarm control circuit
Current amplifier
Electrode
Radioactive source emitting radiation
Electron (negatively charged)
Ion (positively charged
Gauze
Smoke particles enter the detector and obstruct the ion movement and reduce ionisation current to operate the alarm
Note: Maximum area covered by each detector 93 m²

Fig. 9.20 Ionisation smoke detector

circuit (see Fig. 9.20). The air in the open chamber is ionised from a radiation source, so that a minute electric current passes between two electrodes.

If smoke enters the chamber through the open wire gauze, ions are attracted to the particles, reducing the ionisation current flow, which then triggers an electronic relay circuit to operate an alarm.

The detectors usually react more quickly than heat detectors, but they are very sensitive and great care is required when installing them where products of combustion are generated by normal production processes. They are subject to control under various regulations, while being manufactured, repaired or stored. Storage includes transit time and period between delivery to site and fixing in position.

Any loss or damage to ionisation detectors at any time must be reported to the police and in writing to the Factory Inspectorate, in the case of industrial premises, or to the Department of the Environment for all other premises.

Air duct smoke detector

An ionisation smoke detector and a duct adapter fitted to a ventilating duct, will initiate an alarm signal when the air stream is polluted by smoke or products of combustion. The careful siting of these units will provide protection of the duct, rooms served by the ducts and air-conditioning or ventilating motors and filters.

The detectors are wired to an indicator panel which contain changeover alarm-signalling contacts to fire dampers, shutter and door release mechanisms and electric motors. It is often necessary to fit units on opposite sides of wide ducts to ensure positive sampling. Figure 9.21 shows a typical arrangement of duct-mounted smoke detectors. A fire in room A would initially be detected by detector A. Air drawn into ducts B, C and D would dilute the concentration of smoke in the main duct, preventing detectors at E from operating until the fire in the room A has sufficiently developed. Detectors at point E are installed to monitor the air flow in the ducts downstream of detectors A, B, C and D.

Group 3 (visible smoke detectors)

These are known as the true smoke detectors and are used where combustion particles have light-scattering and light-obscuring properties. There are three sub-groups:

1. Those using light-scattering techniques.
2. Those using light-obscuring techniques.
3. Sampling systems using either light-scattering or obscuring techniques.

Light-scattering detector (Fig. 9.22)

This utilises the Tyndell principle, which is related to the way particles of dust can be seen in a shaft of sunlight by reflection. Various systems can be built in to show that the detector is operational. The absence of light indicates a fault and that the detector is inoperative.

The operation of the detector depends upon detecting the light scattered by smoke particles. A detecting photocell is shielded from a light source so that no current flows. When smoke enters the detector the light is scattered by reflection from smoke particles on to the photocell, which then generates a small electric current. This small electric charge is amplified and actuates a relay system which automatically closes or opens an electrical circuit and activates the alarm.

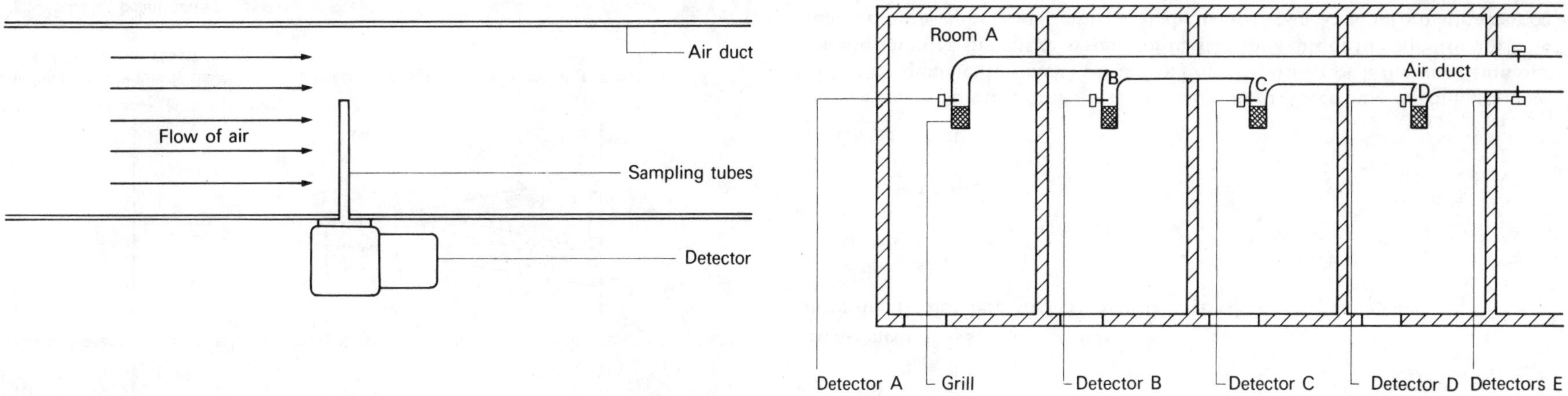
Air duct
Flow of air
Sampling tubes
Detector
Room A
Air duct
B
C
D
Detector A
Grill
Detector B
Detector C
Detector D
Detectors E

Fig 9.21 Air-duct smoke detector

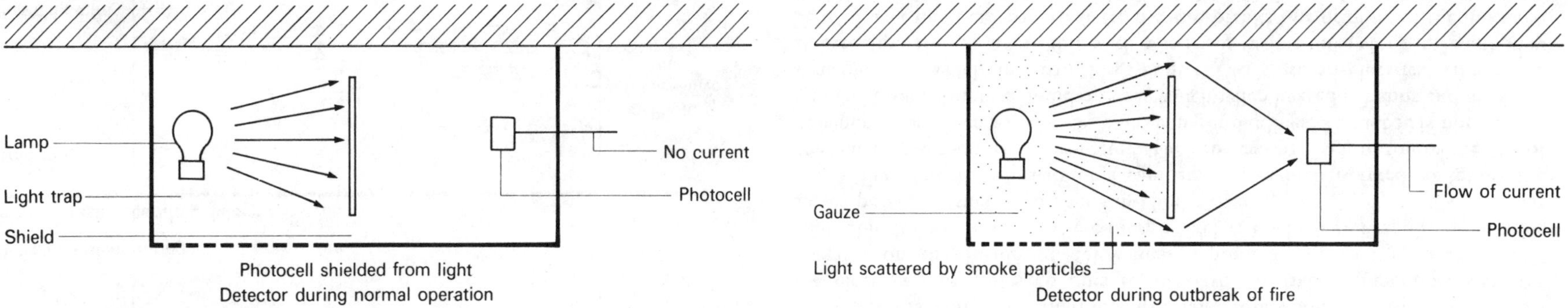
Lamp
Light trap
Shield
No current
Photocell
Photocell shielded from light
Detector during normal operation
Gauze
Flow of current
Photocell
Light scattered by smoke particles
Detector during outbreak of fire

Fig. 9.22 Light-scattering detector

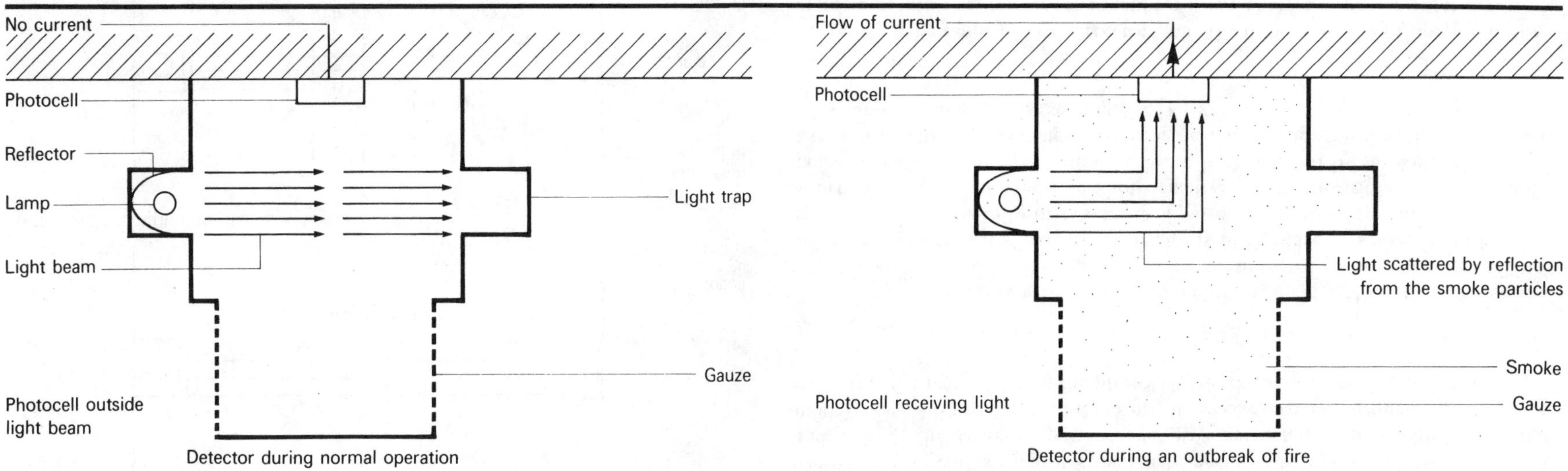
No current
Photocell
Reflector
Lamp
Light beam
Light trap
Gauze
Photocell outside
light beam
Detector during normal operation
Flow of current
Photocell
Light scattered by reflection
from the smoke particles
Smoke
Gauze
Photocell receiving light
Detector during an outbreak of fire

Fig. 9.23 Light-scattering detector

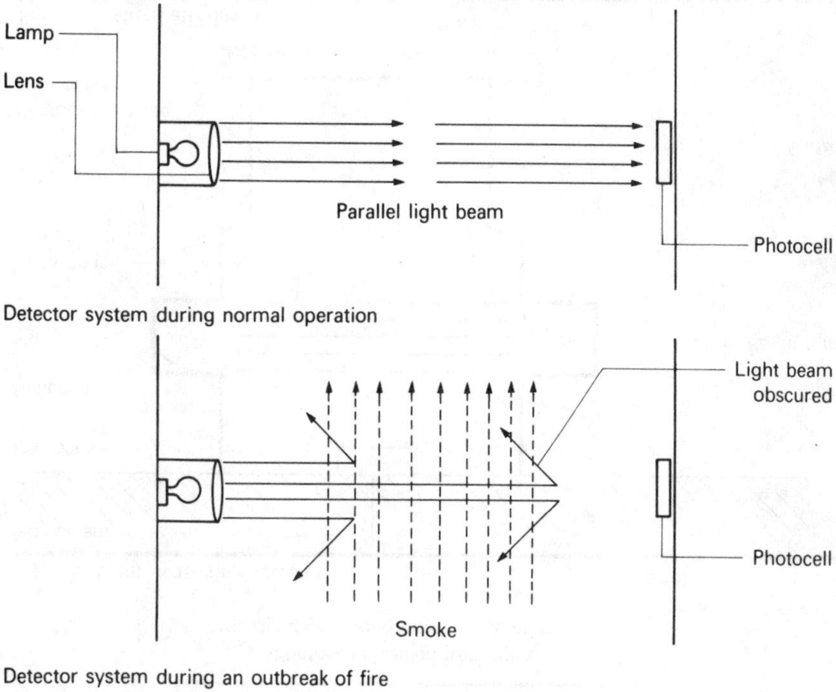
Lamp
Lens
Parallel light beam
Photocell
Detector system during normal operation
Light beam obscured
Photocell
Smoke
Detector system during an outbreak of fire

Fig.9.24 Light-obscuring detector system

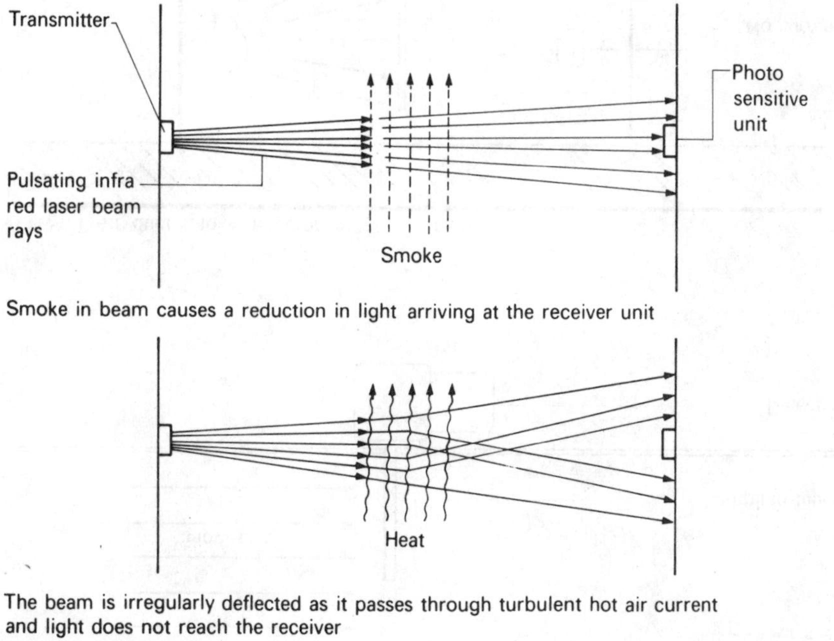
Transmitter
Photo sensitive unit
Pulsating infra red laser beam rays
Smoke
Smoke in beam causes a reduction in light arriving at the receiver unit
Heat
The beam is irregularly deflected as it passes through turbulent hot air current and light does not reach the receiver

Fig.9.25 Laser beam detector system

Light-obscuring detector

This system operates by projecting a light beam on to a photocell (see Figs. 9.23 and 9.24). The system operates over various distances depending upon the model, e.g. 4.5 m, 12 m and 15 m. If smoke passes through the beam it interrupts the light reaching the photocell and the alarm signal is activated.

Laser beam

The laser beam has been developed as a combined heat and smoke detector and will detect heat and smoke from any type of fire. It does not suffer the falling-off of sensitivity with increase of height as in the case of some detectors. The system consists basically of two units, a pulse transmitter emitting infrared rays, which is optically coupled to a photosensitive receiver positioned at a maximum range of 100 m. When the pulsating beam is attenuated by smoke or the refractive index of the air is changed by rising heat waves, appropriate circuits activate relays and warning signals (see Fig. 9.25).

Group 4 (heat detectors)

There is a wide range of heat detectors, but the basic principle of operation, that of temperature rise, remains the same for each. There are spot or point detectors using bimetal strips or coils (see Figs. 9.26 and 9.27).

Detectors which use thermocouples and solid-state detectors (thermistor) use electrical principles to operate an alarm. A fixed temperature detector operates on the fusion of a low-melting-point alloy (see Fig. 9.28) and is applicable for use in areas where there are relatively high and rapidly fluctuating air temperatures, such as kitchens and boiler rooms.

A line detector consists of a capillary tube which is fixed to follow the contour of the room to be protected. The tube can be fixed to follow the line of ornamentation, such as cornices and ceiling mouldings, so that it is not easy to see. The tube contains a liquid or a gas which when heated expands and displaces a diaphragm, which in turn closes contacts to close an electrical circuit and actuate an alarm. A steel cable with a series of fusible links along its length may be used as a line detector. Heat from a fire breaks one or more of the links and the release of a weight at the end of a cable operates an alarm. An air-type heat detector operates on the expansion, due to heating of air inside an air vessel (see Fig. 9.29).

Arrangement of systems (Fig. 9.30)

An automatic fire detector system usually consists of the following:

1. Detectors.
2. Alarms.
3. Control panel.
4. Mains-fed power unit.
5. Batteries.
6. Wiring circuit.

The main control unit may use advanced electronic techniques and computer equipment, which will not only give warning of a fire condition but will also monitor the entire system for faults. Any fault or fire condition will be indicated both audibly and visually at the control panel and information can be relayed to the local fire brigade or other remote point. The panel should be

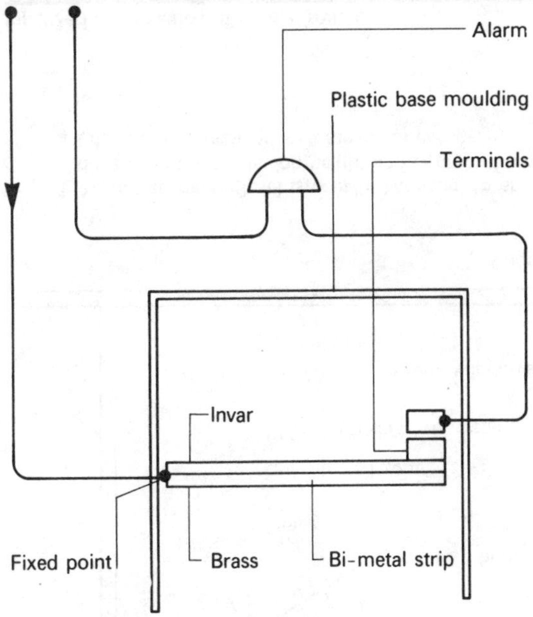
Alarm
Plastic base moulding
Terminals
Invar
Fixed point
Brass
Bi-metal strip

Heat from a fire causes the strip to bend and closes the circuit

Fig 9.26 Bimetal strip heat detector

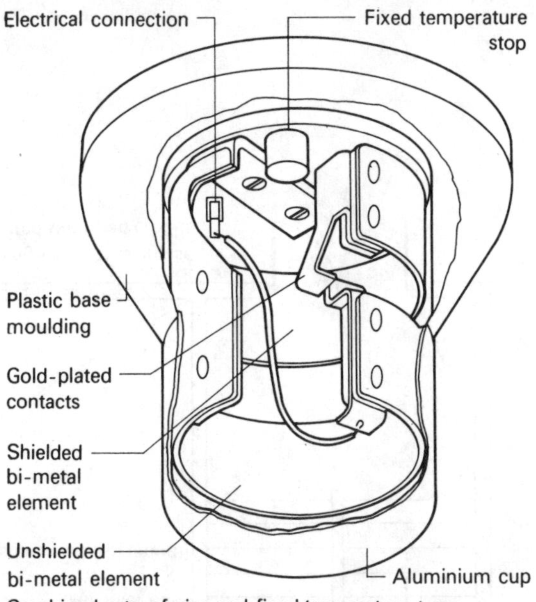
Electrical connection
Fixed temperature stop
Plastic base moulding
Gold-plated contacts
Shielded bi-metal element
Unshielded bi-metal element
Aluminium cup

Combined rate-of-rise and fixed temperature type: Maximum temperature of operation 57° or 100°C Area covered by each detector 46.5 m^2

Fig 9.27 Bimetal coil heat detector (Courtesy Chubb Fire Security Ltd.)

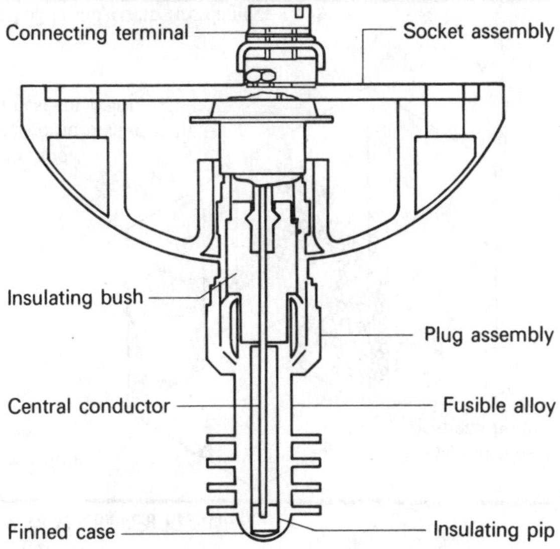
Connecting terminal
Socket assembly
Insulating bush
Plug assembly
Central conductor
Fusible alloy
Finned case
Insulating pip

Fixed temperature type: Maximum temperature setting 57° 68° and 102°C area covered by each detector 37 m^2

Fig 9.28 Fusible alloy heat detector (Courtesy Chubb Fire Security Ltd.)

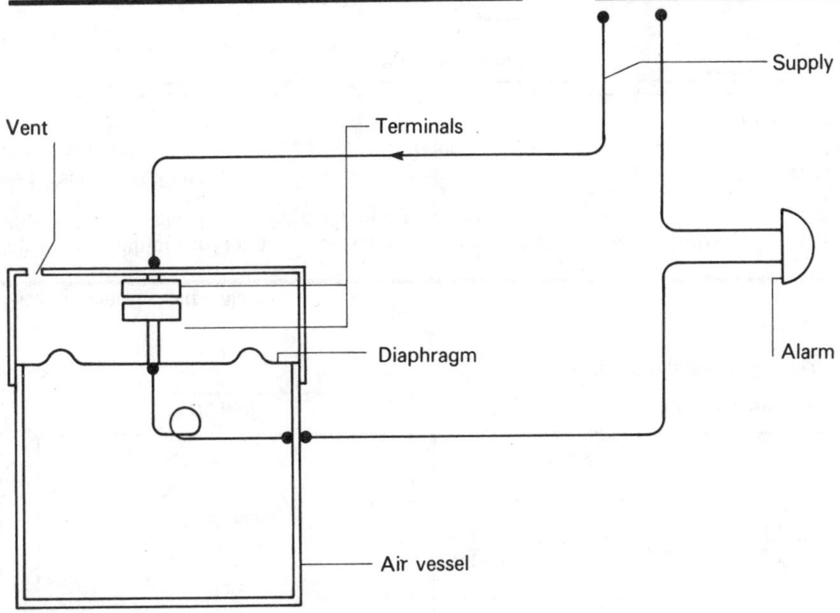
Supply
Vent
Terminals
Diaphragm
Alarm
Air vessel

Heat expands the air inside the detector and causes the diaphragm to rise, thus closing the circuit and sounding the alarm

Fig.9.29 Heat detector air type

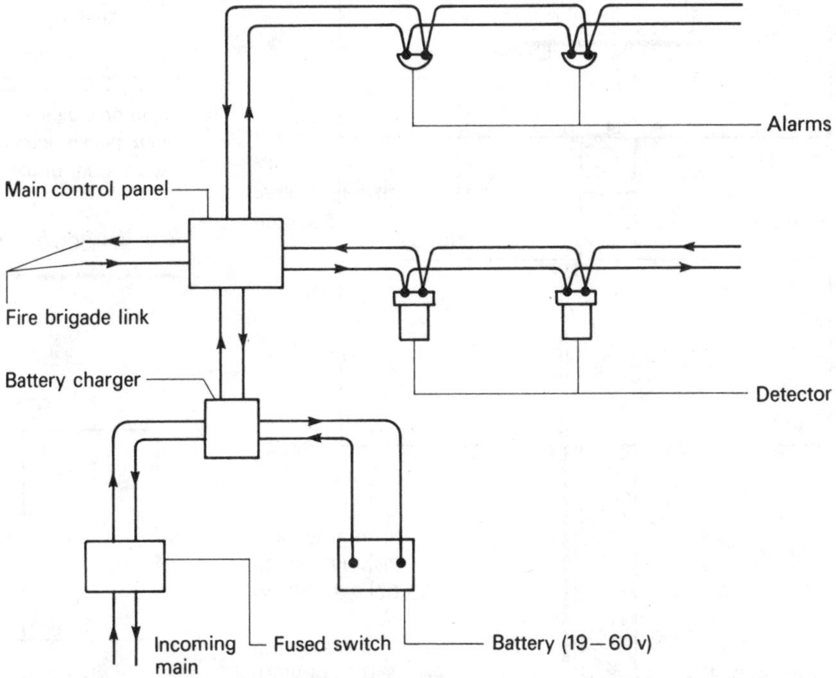
Alarms
Main control panel
Fire brigade link
Detector
Battery charger
Incoming main
Fused switch
Battery (19—60 v)

Fig 9.30 Diagrammatic arrangement of fire-detector system

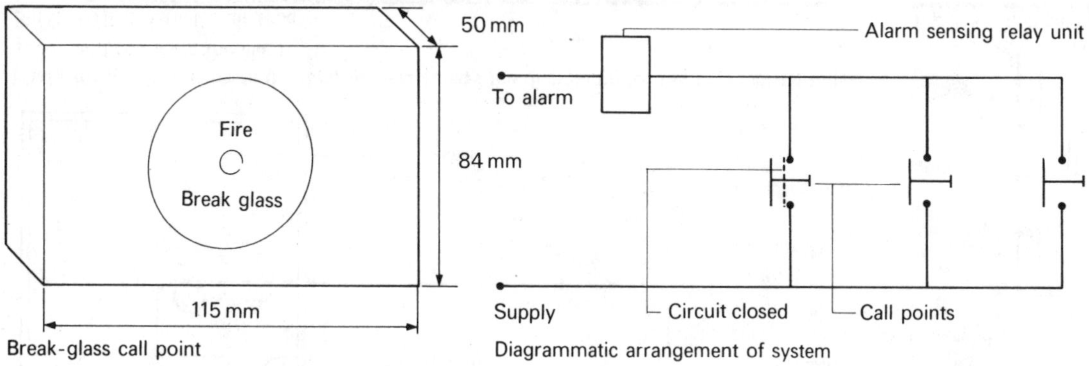
50 mm
84 mm
Fire
Break glass
115 mm
Break-glass call point
Alarm sensing relay unit
To alarm
Supply
Circuit closed
Call points
Diagrammatic arrangement of system

Fig. 9.31 Manual electric alarms

Minimum width of corridor and doors dependent upon gross floor areas, 630 mm for up to 232 m^2 and 930 mm from 232 to 1958 m^2 plus 75 mm for every 140 m^2 above 1958 m^2

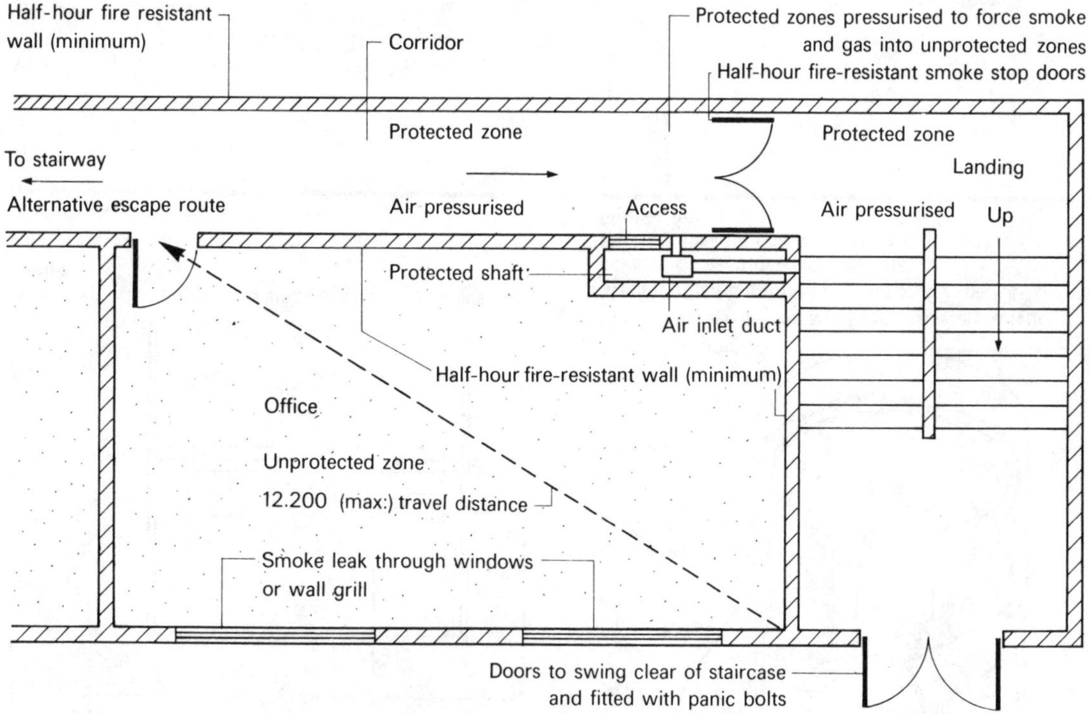
Half-hour fire resistant wall (minimum)
Corridor
Protected zones pressurised to force smoke and gas into unprotected zones
Half-hour fire-resistant smoke stop doors
Protected zone
Protected zone
Landing
To stairway
Alternative escape route
Air pressurised
Access
Air pressurised
Up
Protected shaft
Air inlet duct
Half-hour fire-resistant wall (minimum)
Office
Unprotected zone
12.200 (max:) travel distance
Smoke leak through windows or wall grill
Doors to swing clear of staircase and fitted with panic bolts

Note: 1 Air pressurisation level is usually between 25 and 50 Pa depending upon building height and degree of exposure.
2 Lifts are not classified as a means of escape.

Fig. 9.34 Pressurisation of escape routes

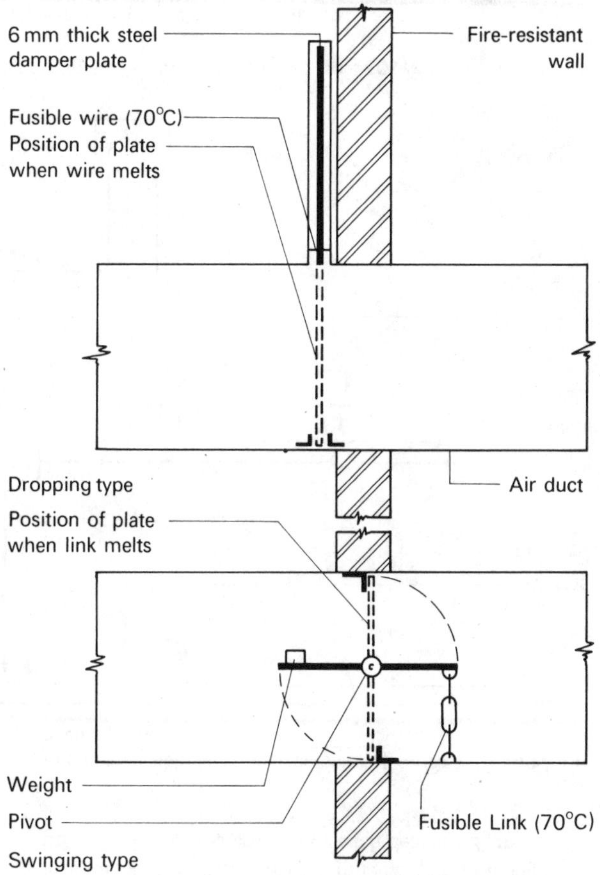
6 mm thick steel damper plate
Fire-resistant wall
Fusible wire (70°C)
Position of plate when wire melts
Air duct
Dropping type
Position of plate when link melts
Weight
Pivot
Fusible Link (70°C)
Swinging type

Fig. 9.32 Mechanical fire dampers

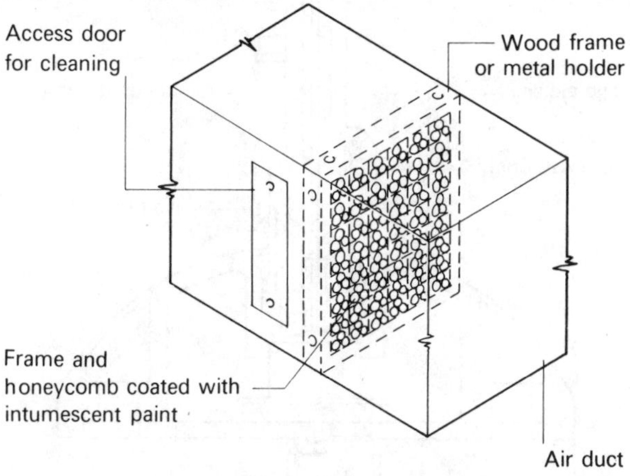
Access door for cleaning
Wood frame or metal holder
Frame and honeycomb coated with intumescent paint
Air duct

Fig. 9.33 Honeycomb fire damper

placed in a reception area or other prominent position. The mains power unit carries out all the system switching functions, such as automatic trickle and full charging of the batteries. It can be mounted either adjacent to the main control panel or in a remote position.

It is possible to connect up to fifty detectors in one circuit, which means that very large areas can be protected at comparatively low capital costs.

Alarms

Although bells are usually associated with fire alarms, sirens, klaxons, hooters and buzzers may also be used, providing the sound is distinct from any other sound normally heard on or near the premises.

Besides audible alarms, visual alarms should be included, especially for deaf persons and it is possible to obtain combined audible and visual alarms. Alarms should be sited so that they can be heard (and, if possible, seen) throughout the building.

Manual-electrical alarms (see Fig. 9.31)

Electrically operated alarms may be operated from break-glass call points and, once operated, the alarms will continue to sound automatically. The usual maximum travel distance to operate the alarm is 30 m and the call points should be fitted at a height of about 1.4 m above the floor, either on landings or corridors. The call point contains a depressed plunger pressing against a glass front. When the glass is broken the plunger is released and operates the alarm system.

Legislation

The Fire Precautions (Hotels and Boarding Houses) Act 1971 requires that in all premises means of giving warning of fire should be installed, which is:

1. Readily available at all times.
2. Capable of being operated without exposing any person to undue risk.
3. Perceptible throughout the premises and capable of waking staff and residents, and distinctive in that it will not be confused with any other signal in the premises.

In buildings of one or two floors manually operated sounders may be used, provided that any one sounder is audible throughout the premises, and at least two sounders are provided in such positions that a fire occurring anywhere on the premises will not render all sounders inaccessible due to heat or smoke.

Alternatively, self-contained electrical or mechanical sounders which, once activated, will continue to operate for a sufficient length of time, may be used providing at least two sounders are installed. Premises of more than two floors should have an electrical fire-warning system, and in most premises the operation of any one call point should cause the alarm signal to operate throughout the premises. In larger premises having an adequate degree of compartmentation and fire resistance, a two-stage alarm system may be installed.

The Factory Act 1961 requires that premises falling within the scope of the Act have effective means provided and maintained for giving warning in the case of fire, and capable of being operated without exposing any person to undue risk. The warning should be audible throughout the building and in every part of the building which is used for the purpose of the factory.

The Offices Shops and Railway Premises Act 1963 also mentions the installation of fire alarms.

Fire dampers

The Building Regulations 1976, require that where ventilation systems pass through fire-compartment walls or floors, or if they form part of a fire-compartment wall, they must have a fire resistance of not less than half the minimum fire resistance of the wall or floor (Regulations E9 (5) and (6)). If the ventilation duct lies within a protected shaft or if the shaft itself serves as a ventilating duct, then the duct must be fitted internally with automatic fire shutters to reduce the risk of fire spreading from one compartment to another (Regulations E10 (9a)).

Mechanical fire dampers (see Fig. 9.32)

These dampers that are actuated by a fusible link and which may be used to reduce the risk of fire spread through ventilating ducts; they usually operate at a temperature of 70 °C. However, any mechanical device, requires regular maintenance if it is to provide the standard of efficiency required for fire-protection devices and a honeycomb fire damper has been developed which is more reliable.

Honeycomb damper

This depends for its action on the property of intumescent paint which swells to many times its original volume when heated (see Fig. 9.33). When heated, the paint film froths into a dough-like mass which immediately seals and insulates the surface that is being protected from fire. The honeycomb damper made from metal or paper is completely coated with intumescent paint and under normal conditions permits the free flow of air through the duct. However, hot gases from a fire pass through the damper, the paint is heated and swells to completely fill the cells of the honeycomb and form a seal.

Resistance to air flow

The pressure drop across a honeycomb damper 50 mm thick and with a cell size of 10 mm covered by a wire grill, is 9 Pa at an air speed of 5 m/s. This low resistance is no doubt due to the fact that the honeycomb apertures are aligned to the direction of air flow and thus give a streamlining effect.

Effect of dirt and moisture

The intumescent paint is not affected by fluff or oil spray, or air containing 5 per cent of sulphur dioxide, irrespective of whether the air is dry or has a relative humidity of up to 95 per cent. However, dampers should not be used where water may condense on them, as this could lead to salt migration which would interfere with correct closure.

Pressurisation of escape routes (see Fig. 9.34)

Smoke and toxic gases produced by an outbreak of fire in a building can sometimes be a more serious hazard to the occupants than the heat produced. The filling of escape routes with smoke and toxic gases can account for a high proportion of lives lost in fires. The escape route normally takes the form of a corridor leading to a staircase. In some cases a lobby is interposed between the corridor and the staircase which helps to prevent the passage of smoke, and to achieve this the lobby is ventilated to the outside air.

However, where no lobby is provided, the staircase is ventilated to the outside air. These arrangements have the following disadvantages:

1. The lobby takes up valuable space.
2. The staircase and lobby have to be sited on an external wall which restricts the layout of the building.
3. The occupants complain of draughts and rain entering through the ventilating openings.

In order to overcome these disadvantages, the air in the corridor and staircase may be pressurised by means of a fan and ductwork which are more effective and cost less than the loss of valuable space taken up by a lobby. This pressurised air keeps the escape route clear of smoke and toxic gases.

Automatic fire ventilation

Principles (see Fig. 9.35)

Automatic fire ventilation of buildings is designed to facilitate fire-fighting and so reduce fire damage. Smoke, heat and unburnt gases are the products of combustion that make fire-fighting hazardous, and all three have one major common factor: they rise in accordance with the laws of convection. Therefore, if a means of direct access to the outside is provided above the area in which the fire occurs, then the smoke, heat and gases will flow directly upwards and out of the building.

Fire venting is primarily an aid to fire-fighting in that it controls conditions within a burning building, preventing smoke, heat and gases from hampering the firemen. However, fires require oxygen to support combustion and the fact that fire venting supplies oxygen is well understood. It is not possible to seal off an industrial building, and the volume of air inside even a small building is sufficient to support combustion for several hours. Furthermore, in its early stages a fire controls its own supply of oxygen so that the rate of combustion is little affected by fire venting. Fire venting is complementary to protection afforded by sprinklers, hose-reels, extinguishers, detectors and alarms.

Design

Fire vents should be weatherproof and designed to open automatically when the air temperature surrounding them reaches a predetermined limit. A fusible link is incorporated in the control assembly of a fire ventilator, and when the air surrounding the link reaches a temperature of usually 70 °C the link parts and allows the ventilator to spring fully open under the influence of strong springs.

Siting

Fire vents should be sited at the highest point within each control area and for a pitched roof they should be sited near to the ridge.

Advantages

If a fire occurs within a building, fire ventilation provides the following advantages:

1. Prevents smoke logging inside a building and thus enables firemen to see the fire and approach and tackle it without the use of breathing apparatus.
2. Removes partially burnt gases which might otherwise accumulate in the roof space, thus preventing the risk of an explosion.
3. Removes heat and thus lowers the air temperature in the roof space of a building to well below the softening temperature of the structural steelwork. This will prevent distortion of the roof, bulging of the walls and ultimate collapse of the building.
4. Removes heat and lowers the temperature of the room or rooms, thus reducing 'flashover' and lateral spread of fire.
5. Reduces water damage by enabling the jets to be applied directly on to the fire, instead of generally towards smoke-logged areas.
6. Reduces number of sprinkler heads by reducing the lateral spread of heat.

Control of smoke in covered shopping centres

Shopping centres with covered malls can present a hazard in the event of fire, because smoke and combustion products could spread throughout the centres. The rate of production of smoke depends mainly on the area of the fire and the height of the shop or mall, while the density of the smoke depends on the materials burning and the rate of burning of the fire. The smoke forms a layer beneath the ceiling which may move sufficiently fast to overtake people on escape routes.

Beneath the outward flowing layer of smoke and hot gases, air flows back towards the fire and smoke will mix with this return air, particularly when the layer of hot gases has reached an obstruction or the open end of the mall, so that the mall can become completely filled with smoke (see Fig. 9.36).

If shop fronts in the mall are fire resisting, hot smoky gases may be confined to the shops of origin. If shop fronts are not fire resisting, the size of the fire and hence the rate of production of hot smoky gases must be limited (preferably by sprinklers) and it is then possible to confine the hot smoky gases to a stratified layer beneath the ceiling, while the air beneath them is relatively cool and clear. The extent of the smoke layer may be limited by dividing the space beneath the ceiling into smoke reservoirs. This may be achieved by making the ceiling height in the mall greater than that in the shops, or by forming facias extending part of the way towards the floor (see Figs. 9.37 and 9.38).

Arrangements must be made to extract smoke from the reservoirs as fast as it flows into them, while fresh air must be introduced or allowed to flow into the building to replace the extracted hot gases. Such a system is satisfactory for single-storey malls. Natural or mechanical ventilating systems may be used. Natural ventilating systems are liable to be adversely affected by the wind, whereas mechanical systems can overcome wind effects, but it is sometimes difficult to obtain fans which will handle the large volumes of hot gases involved.

Spread of smoke along the mall

The rate of flow of smoke into the mall depends upon whether the shop front is open or closed. If the shop front is open but has a facia extending down below the ceiling level, the hot smoky gases will start to flow beneath the facia as soon as they reach it, unless there is a system for exhausting the hot smoky gases from the shop. The gases will flow out of the upper part of the open shop front giving the appearance of an inverted waterfall, while air to replace them will flow into the shop through the lower part of the shop front.

Once in the mall, the gases form a layer beneath the ceiling and the rate of flow of gases in this layer is greater than the rate of flow of the gases out of the shop, because mixing between gases and the air beneath occurs where they flow out of the shop and subsequently some distance along the mall. The average speed of advance of the smoke layer over a distance of 100 m has been found to be about 1 m/s if the fire is held in check by sprinklers, but if the fire is not held in check by sprinklers, the rate of spread of smoke and heat can be much faster.

A speed of 1 m/s is only slow walking pace if the movement is not impeded by the presence of other people. It is very likely that many people in the shops will use the mall as an escape route, even though alternative means of escape should be provided. These people could emerge from the shops at a time when there is a layer of smoke above their heads but the exits are clearly visible, only to find that the ends of the mall are filled with smoke before they can reach them. Even if sprinklers are used, it is evident that a smoke-control system must be installed to ensure safe evacuation of a covered shopping centre in the event of a fire.

Spread of smoke in multi-storey malls

Shopping malls on two or more levels are a feature of many shopping centres. The upper levels may be no more than balconies overlooking the lower levels. There may be malls at various levels leading to a central concourse with a large well extending through all levels, or there may be openings between levels for stairs or escalators.

With all these arrangements smoke from a fire on a lower level will flow into the upper levels. The hot smoky gases flowing over the edge of a balcony or through a well connecting two malls entrain further air as they do so, so the volume is increased and they become cooler and more dilute. It is seldom, if ever, that the dilution is sufficient to reduce the concentration of smoke to an acceptable level, but should a fire occur on the lower level, the depth of the layer of hot smoky gases beneath the ceiling of the upper mall is greater than if the same fire occurred on the upper level. This means that in many instances people on the upper level will rapidly be within the layer of smoke and hot gases produced by a fire on the lower floor level (see Fig. 9.39).

The control of smoke spread

The ideal system of smoke control would be to keep the mall free from smoke by confining the hot smoky gases to the shop of origin. However, this may frequently be impracticable and it may be necessary to allow the hot smoky gases to flow into the mall and to take measures to ensure that they do not hinder people escaping. Generally this is achieved by confining the smoke to a restricted length of the mall in a layer considerably above the heads of the people, while keeping the air beneath cool and relatively free from smoke.

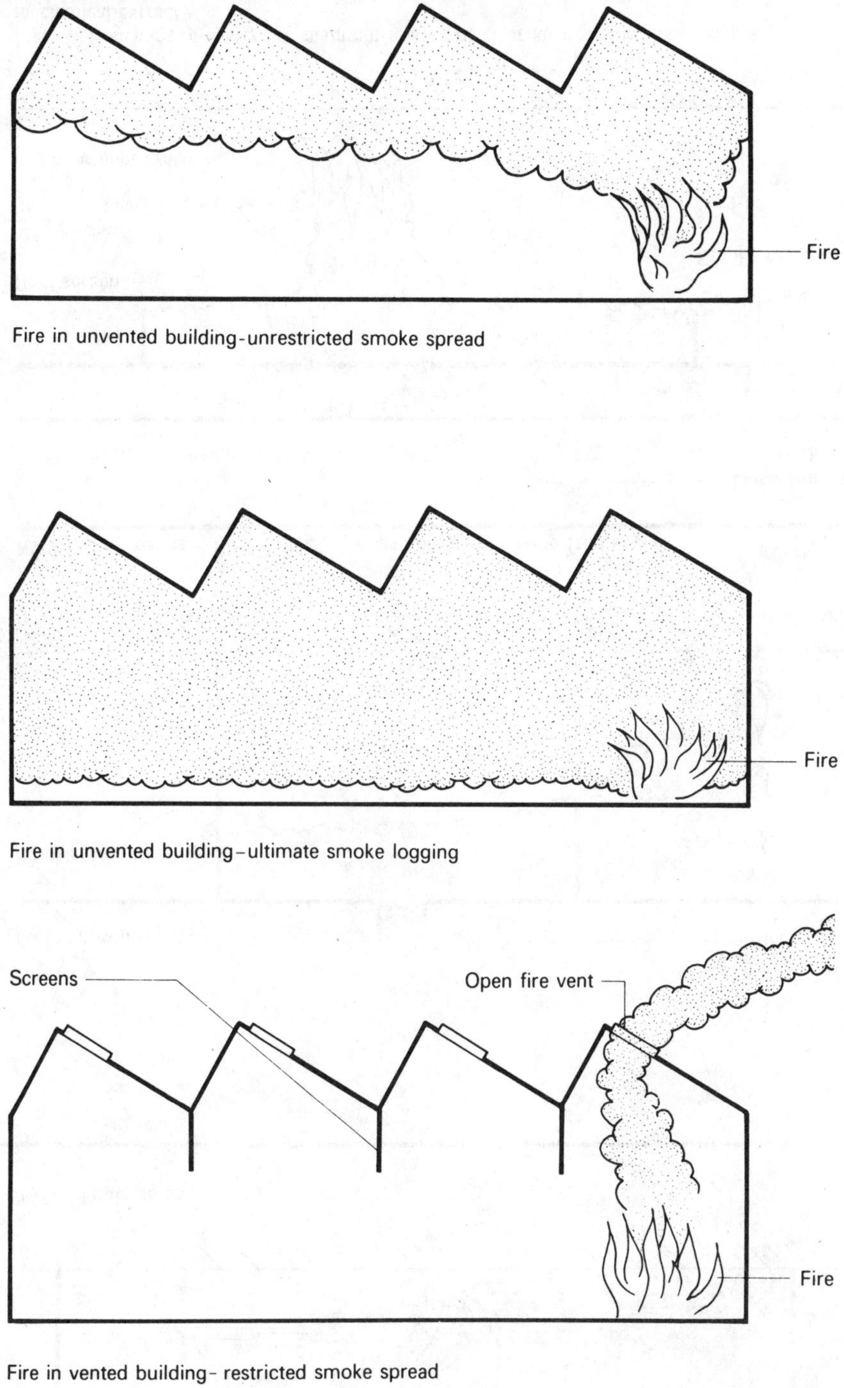
Fire
Fire in unvented building-unrestricted smoke spread
Fire
Fire in unvented building–ultimate smoke logging
Screens
Open fire vent
Fire
Fire in vented building- restricted smoke spread

Fig. 9.35 Automatic fire ventilation

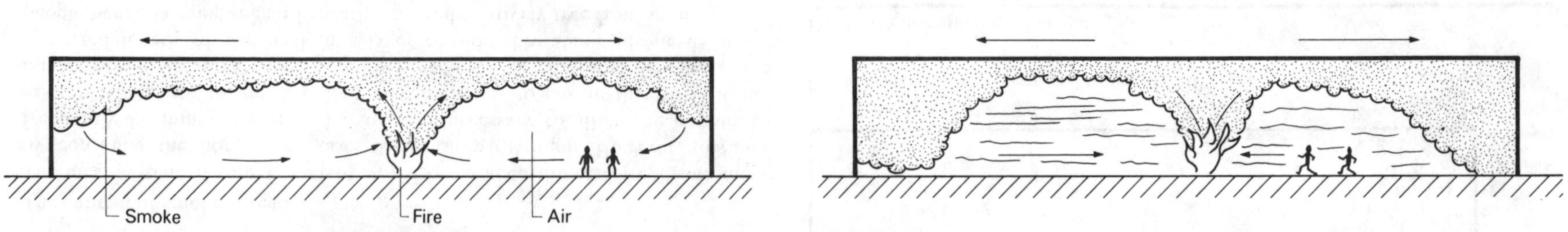
Smoke
Fire
Air

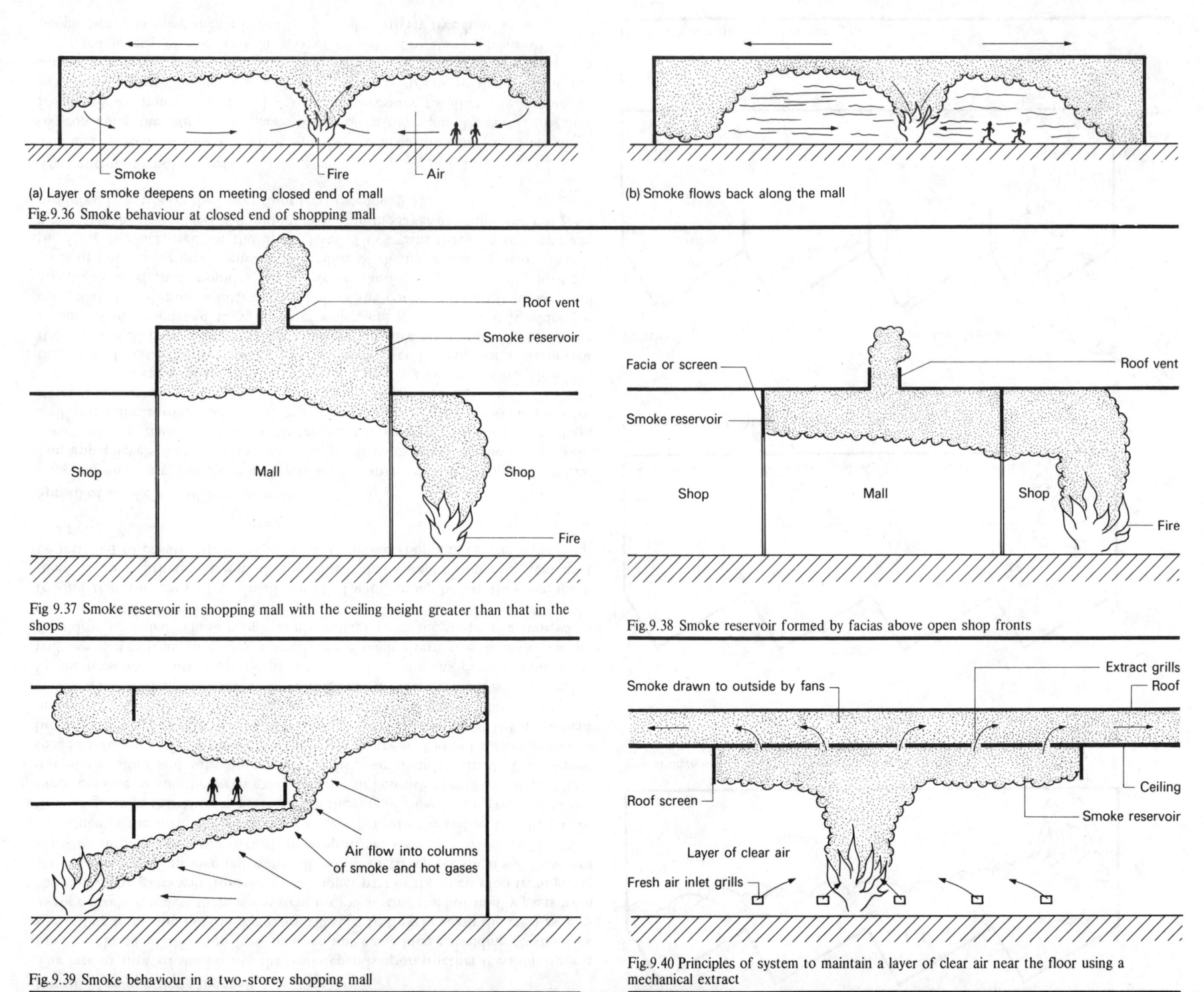

(a) Layer of smoke deepens on meeting closed end of mall

(b) Smoke flows back along the mall

Fig.9.36 Smoke behaviour at closed end of shopping mall

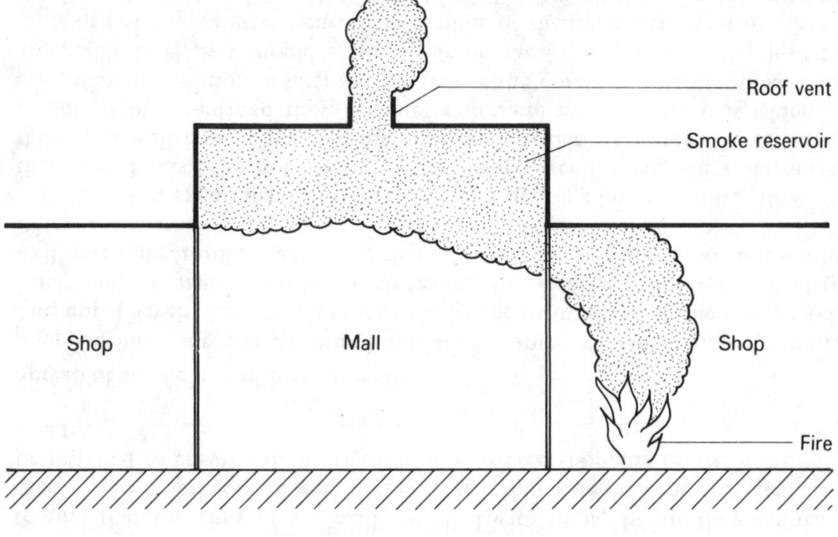
Roof vent
Smoke reservoir
Shop
Mall
Shop
Fire

Fig 9.37 Smoke reservoir in shopping mall with the ceiling height greater than that in the shops

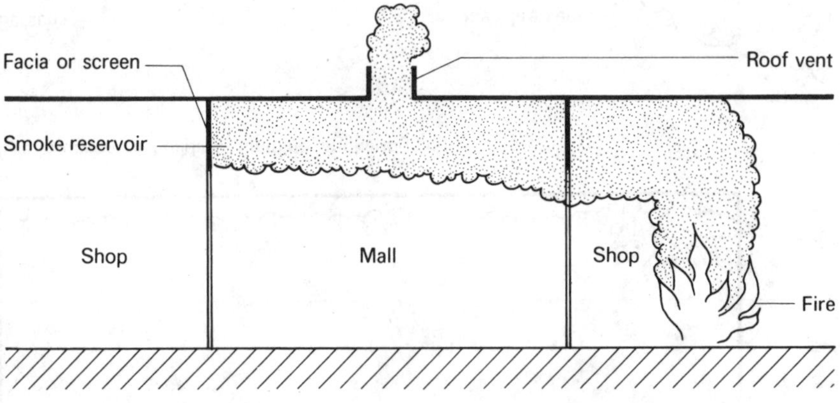
Facia or screen
Roof vent
Smoke reservoir
Shop
Mall
Shop
Fire

Fig.9.38 Smoke reservoir formed by facias above open shop fronts

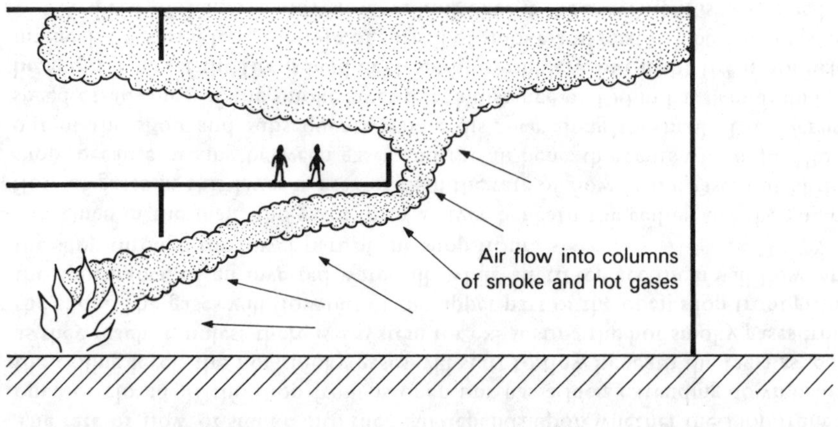
Air flow into columns
of smoke and hot gases

Fig.9.39 Smoke behaviour in a two-storey shopping mall

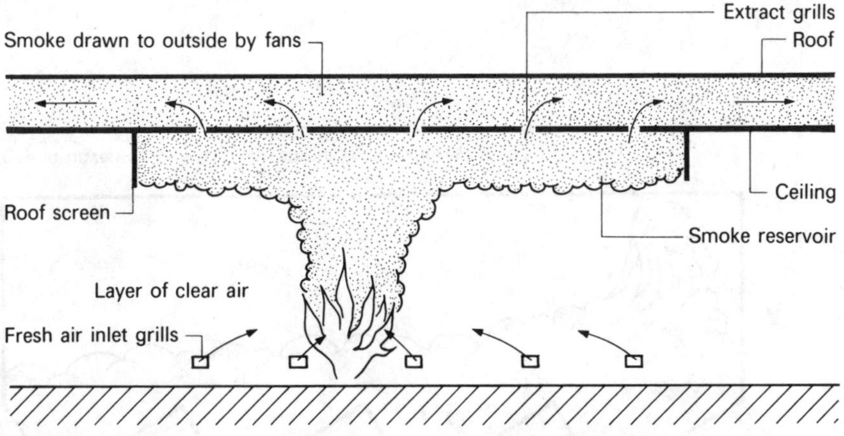
Extract grills
Smoke drawn to outside by fans
Roof
Ceiling
Roof screen
Smoke reservoir
Layer of clear air
Fresh air inlet grills

Fig.9.40 Principles of system to maintain a layer of clear air near the floor using a mechanical extract

Initiation of smoke-control devices

Smoke detectors are a suitable means of initiation of smoke-control devices used. When the smoke-control system is intended to restrict the hot smoky gases to the shop of origin, the detectors must be situated in the individual shops. Otherwise it may be satisfactory for them to be sited in the mall if the shops are only small, so that the hot smoky gases spill into the mall soon after the fire is started.

Extraction systems for limiting the travel of smoke

The formation of a layer of hot gases beneath the ceiling when there is a fire, can only be prevented by having a very large vent immediately above the fire, through which all the hot smoky gases can flow. It is possible to design a smoke-control system which will restrict the area of the hot gases and minimise mixing between it and the air beneath. Such a smoke-control system has four main elements:

1. The rate of production of hot smoky gases must be restricted by limiting the size which fire is likely to attain. In theory this could be done by strict control of the combustible contents of the shops, but in practice sprinklers are required to control the fire.
2. The space beneath the ceiling is divided into 'smoke reservoirs' either by screens extending from the ceiling part of the way down towards the floor, or by making use of suitable features in the building.
3. Means must be provided for extracting the smoke from the reservoirs as fast as it flows into them either by natural or mechanical ventilation.
4. Fresh air must be introduced or allowed to flow into the building to replace the extracted gases.

The system may be used to confine the smoke to the shop of origin, although it is more commonly used to limit the spread of smoke in the mall itself (see Fig. 9.40).

Table 9.3 gives the area of vents or vent stacks required for a 3 m x 3 m fire.

Table 9.3 Area of vents in (square metres) required in smoke reservoirs in shops

Height of vent outlet above edge of screen (metres)	Height of lower edge of screen above floor (metres)					
	2.5	3	3.5	4	5	6
1.5	5.0	6.6				
2.0	4.4	5.8	7.3	8.9		
3.0	3.6	4.7	5.9	7.2	10.1	13.2
5.0	2.8	3.6	4.6	5.6	7.8	10.3
8.0	2.2	2.9	3.6	4.4	6.2	8.1
10.0	1.9	2.6	3.2	3.9	5.5	7.3

Note: The areas required in smoke reservoirs in single-storey malls are twice those given in the above table.

Requirements of smoke reservoirs

The maximum area of a smoke reservoir must be a compromise between the requirements for limiting the extent of the spread of smoke layer and the design and economic consideration of the shopping mall. If the hot gases are to be confined to the shop of origin it may be necessary to provide a smoke reservoir having a maximum area of 1000 m^2. With shops having an area less than 1000 m^2 the smoke reservoir would be enclosed by walls of the shop with a deep facia over the shop front. With shops much larger than this, the ceiling would be sub-divided into smoke reservoirs by screens.

To limit the distance of smoke travel in malls, the maximum length of a smoke reservoir should not normally exceed 60 m. Suspended ceilings are often fitted in malls and shops and unless they are both fire-resisting and sufficiently air-tight to prevent substantial smoke leakage, it is necessary for ceiling screens to be continued up through the space in the suspended ceiling. If the suspended ceiling has an open type of construction it is possible to confine the smoke reservoir entirely to the space above the suspended ceiling (providing it is deep enough and not obstructed by beams, etc.) so the screens need not extend below the ceiling.

Note: Before commencing the design of a fire- or smoke-control system for a building, it is essential to consult the work by the Fire Research Station; Local Fire Prevention and Building Control Officer; Fire Insurance Company and the equipment manufacturer.

Chapter 10

Burglar alarms

Principles

A burglar alarm is a device intended to detect illegal entry or attempted entry of a building. It may be a simple alarm bell over a door or a highly sophisticated electronic system. The device gives no protection in itself but only warning of an attack, and since even the best system can sometimes be defeated by a criminal with sufficient time and equipment, the alarm itself must be protected.

Elements of installation

A burglar alarm installation consists of five elements: detection device or devices, battery, control box, signalling or alarm link and relay in series or in parallel.

In a closed circuit shown in Fig. 10.1, the battery and relay are in series and the detection devices are in series with one another and the relay and battery. Current flows until a contact is opened or the wiring is broken, this de-energises the relay and closes the alarm circuit, causing the alarm to operate. This arrangement is the basis of almost all modern alarm systems. A common variation uses a polar-sensitive relay in the control box connected to a battery via door or window contacts.

Single-core cable should be used – well concealed and unmarked to make it difficult to locate and identify and thus increase the chances of joining the wrong wires when attempting to bypass the detection device, thus setting off the alarm.

In an open circuit shown in Fig. 10.2, the battery and relay are in series, but the detection devices consisting of open-circuit contacts, are connected in parallel. Current can only flow in the circuit when one or more contacts are closed, thus energising the relay which closes the alarm circuit and operates a warning device. However, the use of an open circuit is not as efficient as a closed circuit, since a break in the wiring renders the system useless and would not be detected under normal conditions.

The most common open-circuit devices are pressure mats which can be sealed under floor coverings, in addition they can be fitted to window-sills, on top of high fixtures, underneath roof lights or in any position likely to be used as a pressure point for a burglar.

Detection devices

A number of detection devices are in common use, but they may be divided into two main types:

1. Linear – which give protection only in a single plane.
2. Volumetric – which give three-dimensional protection.

Linear detection devices include magnetic reed contacts, recessed micro-switches, non-recessed plungers and knockout bars – all of which are operated by the opening of a door or window. They also include wiring systems.

Magnet reed contacts

These consist of a hermetically sealed glass tube in each end of which is inserted a flat strip of metal (see Fig. 10.3). The ends of the strips overlap each other at the centre and become polarised by the influence of a permanent magnet. The metal strips are polarised so that dissimilar poles are produced at the overlap, and since unlike poles attract each other and make contact, they immediately separate when the magnet is removed by the opening of a door or window. They are connected to a closed circuit, therefore the opening of the circuit sets off an alarm. The system may also be used for roller shutters and sliding doors, providing that care is taken to prevent damage to the glass tube by shock.

Recessed micro-switch

A recessed micro-switch is fitted to a door- or window-frame, as shown in Fig. 10.4. A spring-loaded plunger fitted to the door or window is arranged so that when they are closed the plunger passes through the hole in the plate and depresses the button of the micro-switch, thus opening a closed circuit and setting off the alarm. The system is suitable for open- or closed-circuit operation. The spring-loaded plunger permits some degree of movement between the door and the frame, but it can be seen when the door is open.

A non-recessed plunger consists of a plunger-operated switch mounted in the door- or window-frame (see Fig. 10.5). The plunger projects through the frame so that the door depresses it when closed. When the door or the window-frame is opened, the plunger is released and the switch closes a circuit to operate an alarm.

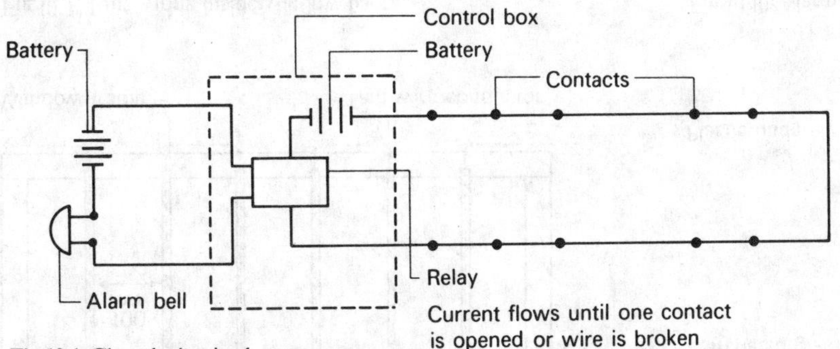
Control box
Battery
Battery
Contacts
Relay
Alarm bell
Current flows until one contact
is opened or wire is broken

Fig.10.1 Closed-circuit alarm system

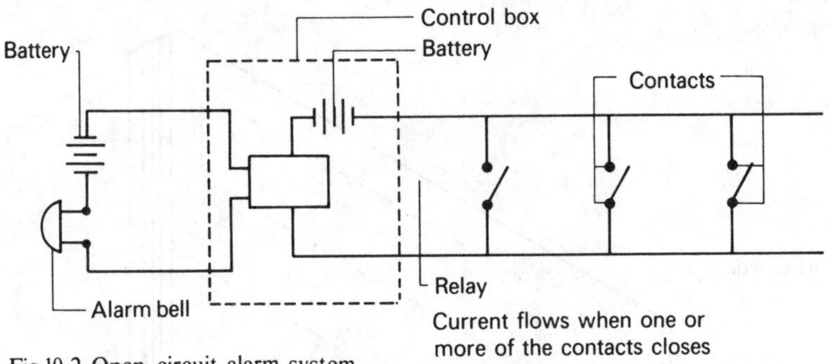
Control box
Battery
Battery
Contacts
Relay
Alarm bell
Current flows when one or
more of the contacts closes

Fig.10.2 Open-circuit alarm system

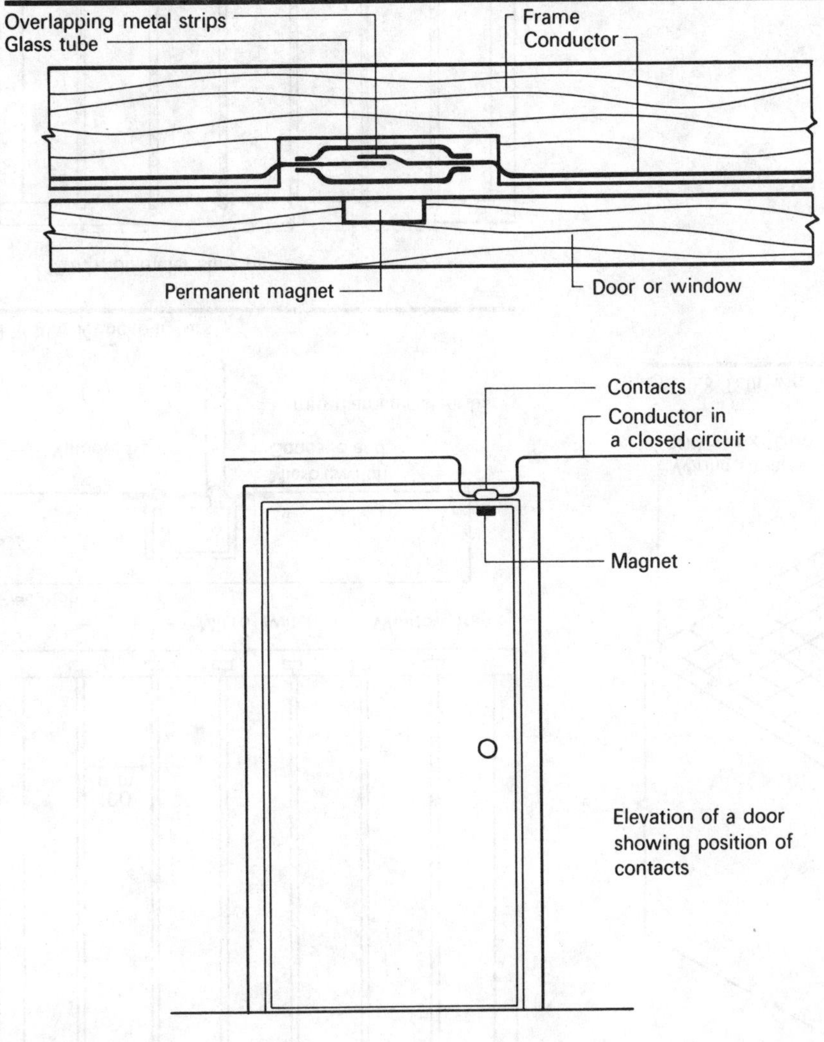
Overlapping metal strips
Glass tube
Frame
Conductor
Permanent magnet
Door or window
Contacts
Conductor in
a closed circuit
Magnet
Elevation of a door
showing position of
contacts

Fig.10.3 Magnetic reed contacts

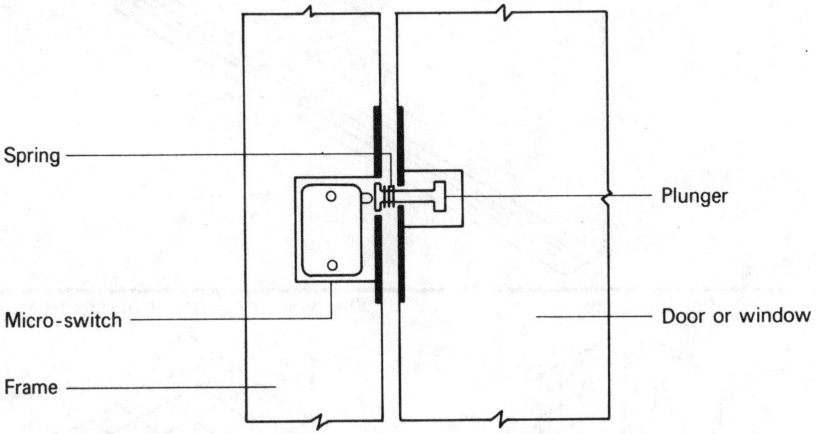
Spring
Plunger
Micro-switch
Door or window
Frame

Fig.10.4 Recessed micro-switch with separate plunger

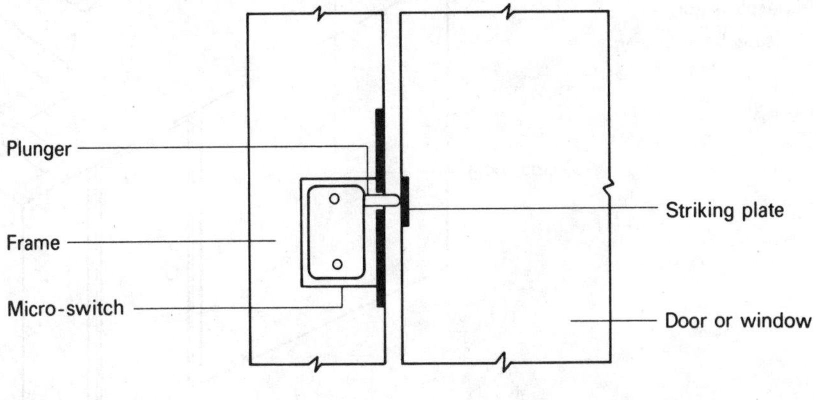
Plunger
Striking plate
Frame
Micro-switch
Door or window

Fig.10.5 Recessed micro-switch with integral plunger

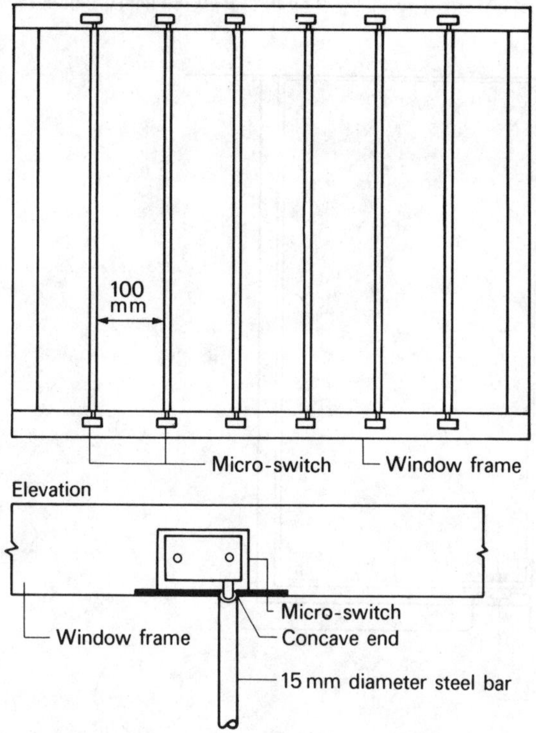
100
mm
Micro-switch
Window frame
Elevation
Micro-switch
Window frame
Concave end
15 mm diameter steel bar

Fig.10.6 Knockout bars

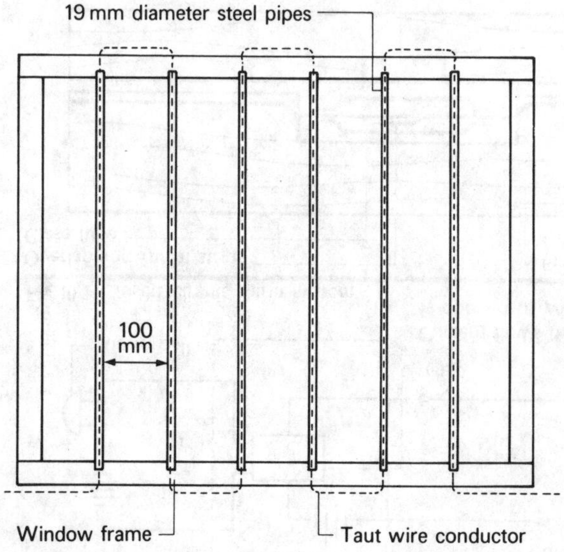
19 mm diameter steel pipes
100
mm
Window frame
Taut wire conductor

Fig.10.7 Taut wiring inside window bars

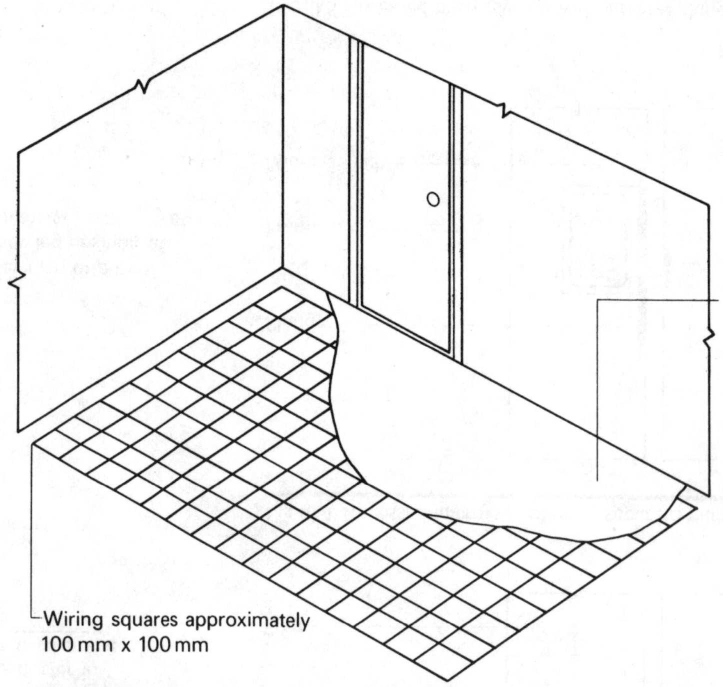
Floor covering
Wiring squares approximately
100 mm x 100 mm

Fig.10.8 Taut wiring to protect a room from entry from below floor

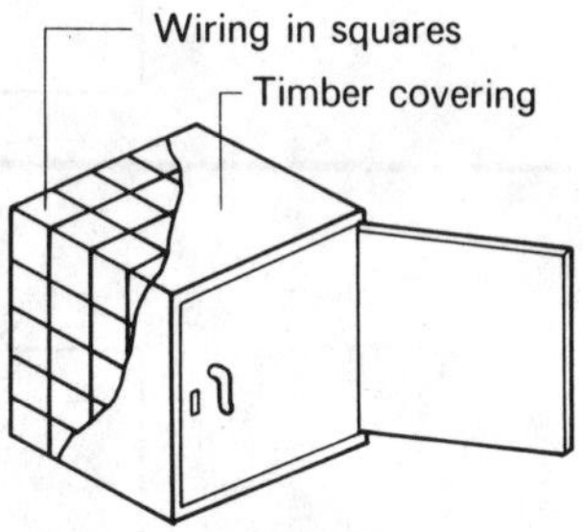
Wiring in squares
Timber covering

Fig.10.9 Taut wiring to protect a safe

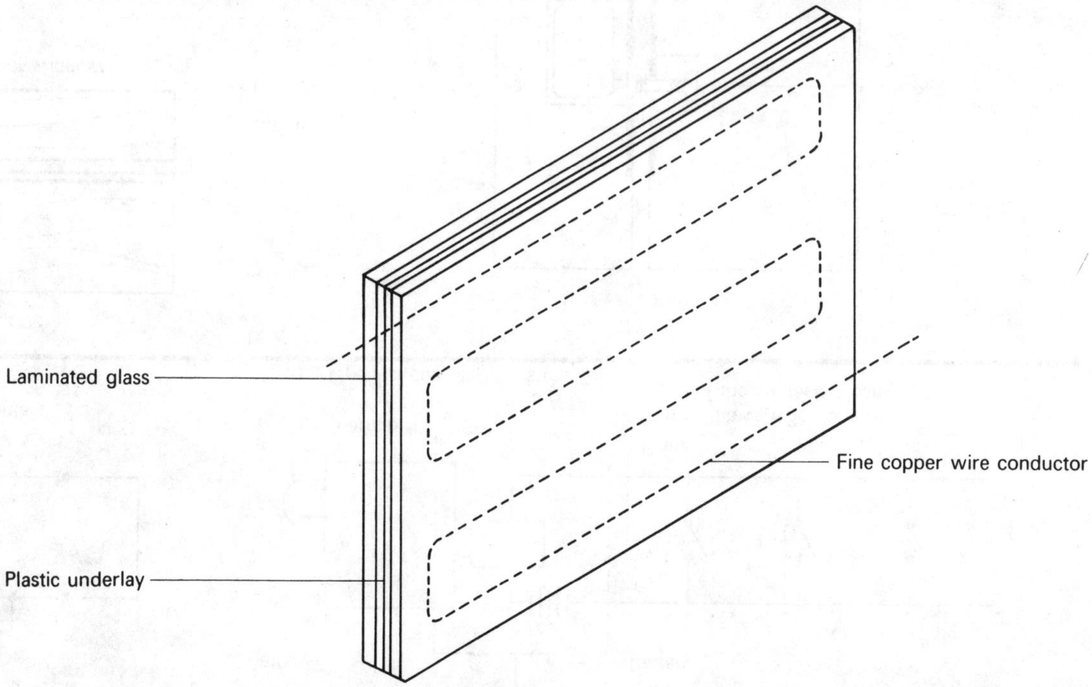
Laminated glass
Fine copper wire conductor
Plastic underlay

Fig.10.10 Alarm glass

Knockout bars

In this system, micro-switches are mounted in both the window-sill and the window-head, with plungers facing each other. A bar is mounted between each switch so that it can operate both switches (see Fig. 10.6). Each end of the bar is concave-shaped so that it engages the plunger of each switch. If an intruder moves a bar the plunger is released and the switch closes a circuit to operate an alarm.

Taut wiring systems

These consist of taut conductors arranged so that an intruder attempting to break through the protected area breaks one or more of the conductors and opens a closed circuit to operate an alarm. The conductors – which are usually single hard-drawn copper wires with plastic insulation – may be inserted inside tubes to protect a window, as shown in Fig. 10.7. If an intruder spreads the tubes the conductor is broken and this operates an alarm.

The conductors may also be used to protect a building from entry by breaking through a structure. They are fixed behind a wall or ceiling lining or below the floor covering, as shown in Fig. 10.8.

The system may also be used to protect a safe by fixing the conductors behind a wooden covering around the safe. The covering is fixed to the floor so that lifting the safe will break the conductors (see Fig. 10.9). More sophisticated linear devices include vacuum glass which consists of two pieces of glass kept apart by distance pieces. A partial vacuum is formed between the glass, and a diaphragm switch is fitted in the space so that the pressure is only slightly above that of atmosphere. If the glass is broken, the equalising air pressure acting upon the diaphragm operates a switch and triggers an alarm.

Alarm glass may be used as an alternative, consisting of laminated glass with a fine copper wire inserted in a plastic underlay (see Fig. 10.10). Breaking the glass also breaks the conductors and sets off an alarm. An alternative, cheaper method than alarm glass consists of a number of strands of thin copper wire sandwiched between two panes of ordinary glass. The wire is sometimes glued to the glass to keep it in position. This method may be used for both windows and doors.

Optical systems (see Fig. 10.11)

These make use of the invisible or infrared parts of the electromagnet spectrum. They generate a current when the beam falls on a photocell. If the beam is to be interrupted by an intruder, the current ceases to flow from the photocell and gives an alarm. For efficiency the source is modulated mechanically or electronically and the photocell is designed to respond only to the frequency of the modulated beam.

Electronic modulation is to be preferred since mechanical devices are more expensive to maintain, are more likely to fail and cause a greater incidence of false alarms.

Radiation in the near-infrared region of the spectrum lies outside the visible range and an intruder cannot therefore detect it. It also penetrates fog or mist better than visible light. The beam may be made to cover a large area by use of mirrors, but the strength is reduced by 20–30 per cent each time it is reflected by a mirror.

The system should be mounted about 60 mm above the floor level and the transmitters and receivers must be well guarded to prevent damage. The system is not recommended for external use, due to the risk of the beam being interrupted by dense fog or heavy precipitation, giving rise to false alarms.

Volumetric detection devices

These work on a number of different principles. A pressure-differential system operates by maintaining different air pressures inside and outside the protected zone and the opening of a door or window equalises the air pressure and triggers an alarm circuit. The equipment includes a fan, a pressure-sensing diaphragm, which senses the difference in pressure between the inside and outside of the zone, and a control box (see Fig. 10.12).

The system can protect spaces having an air volume of 850 m^3 and over, depending upon the degree of sealing that can be affected. The fan is usually operated by mains electricity, and in the event of a failure the system goes into an alarm state. Small zones may be protected by a battery-operated fan.

The case of the diaphragm unit should be placed slightly away from the wall, so that if the vent hole through the wall to the outside is blocked the pressure on both sides of the diaphragm will be equalised and trigger the alarm circuit. The fan and diaphragm unit must be fitted on interior walls, so that changing outside atmospheric conditions do not cause a false alarm.

Proximity systems

These employ the principle of electrical capacitance. The value of a capacitor is determined by the area of two conducting plates, the distance between them and the insulating medium. The variation of any of these parameters will change the value of the capacitance. The insulation used for an alarm capacitor is air, and if an object is introduced near the plates the electrostatic field between them is increased, causing the capacitor to discharge and triggering the alarm (see Fig. 10.13).

Electronic detectors

These are designed to operate on the vibrations an intruder makes and are designed to detect airborne noise and structural vibrations.

The airborne noise-detector system consists of a microphone in the protected area with an amplifier to operate the alarm. The system is designed to discriminate between normal ambient noise such as traffic, having a frequency of between 1 Hz and 400 Hz, and anything above 400 Hz of sufficient amplitude triggers the alarm.

The structural vibrations detector consists of a microphone fixed to the target, for example a safe. The flexing caused by a thin piece of quartz crystal inside the microphone, when the safe is tampered with, produces a small current which is used to operate the alarm.

Mechanical vibration detectors

These operate on the very small movement caused when a structure is vibrated by an intruder. Energy in the form of compression waves transmitted through the structure causes it to be momentarily displaced, and the detector amplifies the movement by means of a spring leaf, or a pendulum (see Fig. 10.14).

The most commonly used is the spring-leaf detector which is mounted

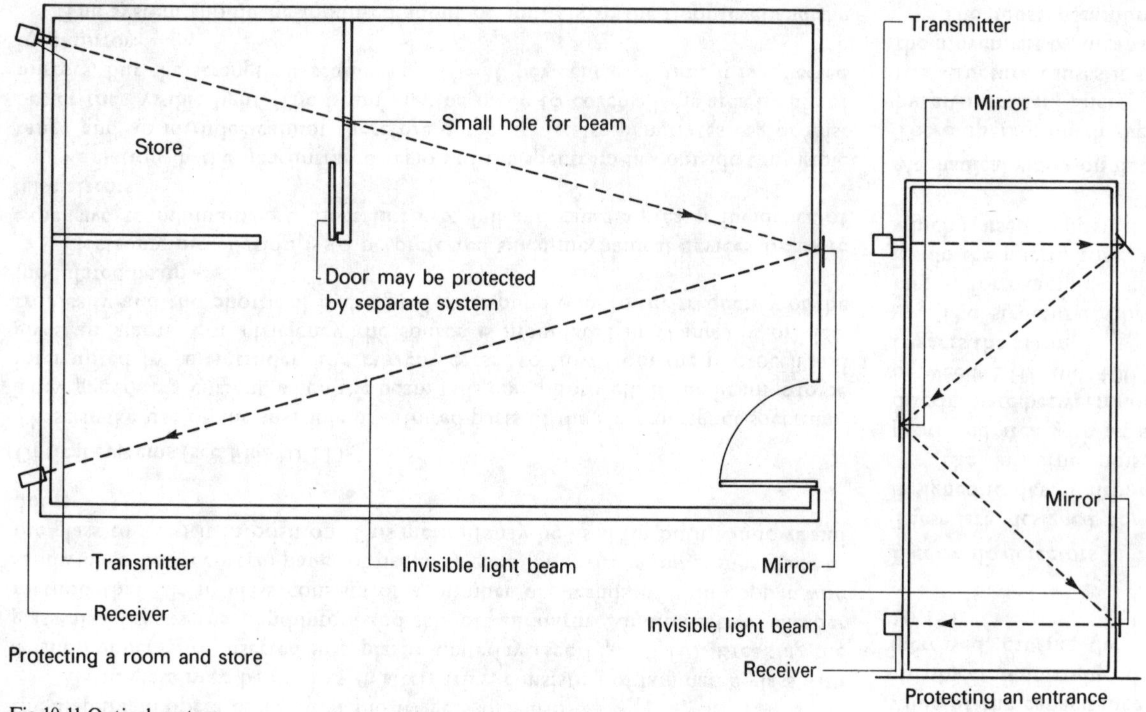
Store
Small hole for beam
Door may be protected
by separate system
Transmitter
Invisible light beam
Mirror
Receiver
Protecting a room and store
Transmitter
Mirror
Mirror
Invisible light beam
Receiver
Protecting an entrance

Fig.10.11 Optical system

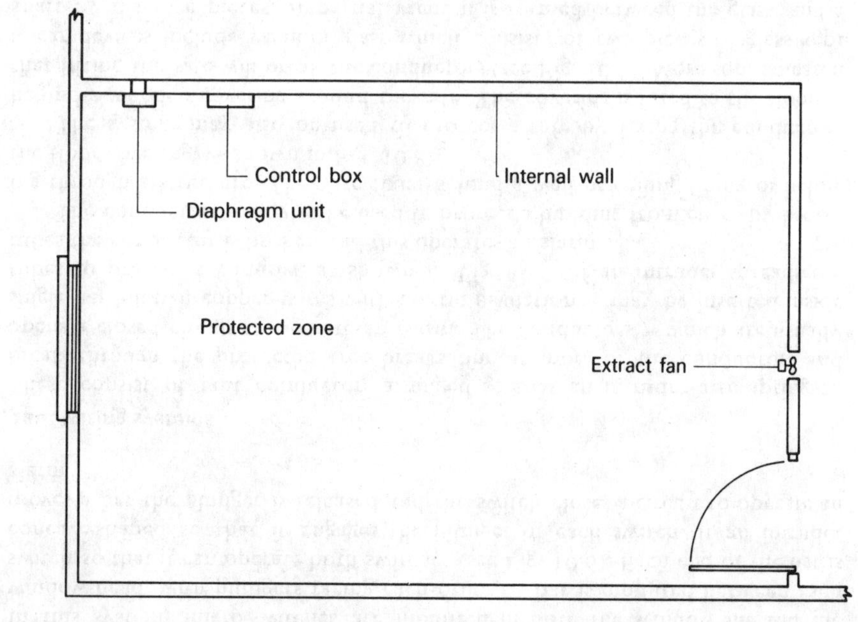
Control box
Internal wall
Diaphragm unit
Protected zone
Extract fan

Fig.10.12 Pressure-differential system

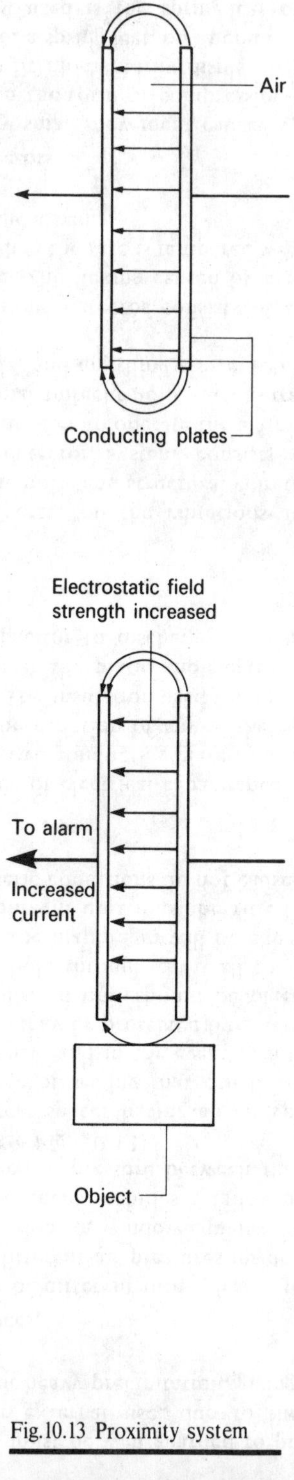
Air
Conducting plates
Electrostatic field
strength increased
To alarm
Increased
current
Object

Fig.10.13 Proximity system

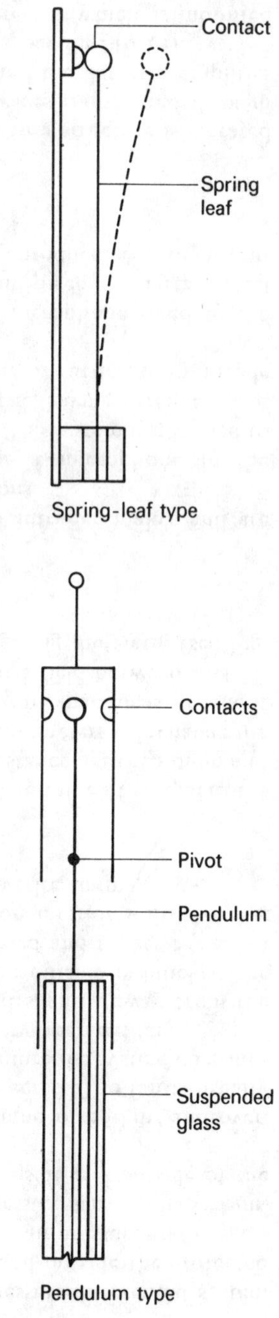
Contact
Spring
leaf
Spring-leaf type
Contacts
Pivot
Pendulum
Suspended
glass
Pendulum type

Fig.10.14 Mechanical vibration detectors

directly on the structure to be protected. Vibrations in the structure will cause the spring leaf to jump, break contact and trigger the alarm.

Ultrasonic detectors

These employ high-frequency sound waves (20–70 kHz) generated inside the protected zone. Movement of an intruder within the zone causes a change in the frequency of the signal, this reaches the receiver and thus triggers the alarm.

Signalling devices

The detection device must be arranged to trigger an alarm – either by signalling to the police or a commercial security organisation, or by sounding an alarm. The alarm system may be terminated at the commercial security organisation's central station which is manned throughout the day and night.

A range of services can be provided which include alarm and line-fault signals and the monitoring of the closing and opening of the building. The most serious drawback of private line connection is the high cost of Post Office lines, but a number of techniques have been devised where unmanned satellite stations are used to transmit signals to a central station.

An automatic 999 dialling machine can be installed and used on the Post Office telephone network. The machine consists of a tape-recorder which is triggered by the action of a detection device. The machine then connects the 999 line and transmits a recorded message to the police. At a prescribed time interval after the transmission of the message, an audible alarm is operated on the premises. The system can be arranged so that if the 999 line is faulty or engaged the audible alarm is sounded immediately.

Alarm bells

Self-powered bell units are generally used to give the audible alarm signal. The units are fed from the mains electrical supply, but if the connecting cable is cut or interfered with, the bell – which has a detector unit – rings from its own power supply. The bell will also operate if it is interfered with.

Closed-circuit television systems

These cannot be used successfully on their own as an intruder alarm, but they can assist security guards in large factories or very high-risk premises. They are successful in identifying an intruder or pilferer and act as an effective deterrent in shops and departmental stores.

Lighting

This can be used to deter intruders and is a useful back-up aid to other security methods. A good standard is required so that an intruder can be easily seen and the protected zone should be provided with light during all hours of darkness.

Chapter 11

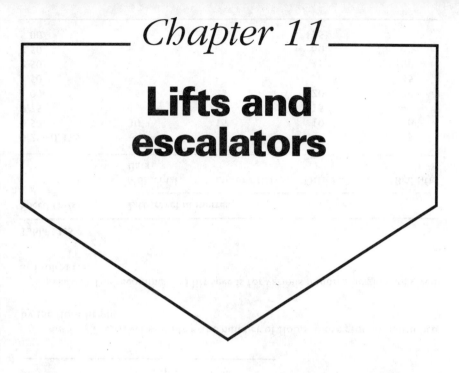

Lifts and escalators

Lifts

A lift installation has an important bearing on the efficient functioning of the building it serves, and to obtain an efficient service the number and type of lifts must take into account several factors including the type of building and nature of its occupancy.

Location

Lifts should be sited in the central area and take into account the proximity of entrances to the building and staircases. If the entrances to a building are not in a central position, there is still a strong case for centralising the lifts, since their use during the day may outweigh the inconvenience of reaching the lifts at morning arrival and evening departure.

When a building has to have a number of passenger lifts it is usually preferable to group them together rather than spread them throughout the building. Although passenger walking time is saved by spreading the lifts, this is more than offset by the increase in average waiting time for the lift service and passengers tend to be more impatient standing waiting for a lift than they are by walking to it. Grouping of lifts also reduces the cost of installation. If passengers have to pass a staircase on the way to a lift, the demand on the lift tends to be reduced. If they pass a lift before reaching a staircase, the demand on the lift tends to be increased.

In departmental stores shoppers must be encouraged to visit the upper sales floors and,therefore lifts in these buildings should be easily seen and accessible.

In hospitals, a bed lift will be required close to the operating theatre in addition to other lifts. In all types of buildings a lift lobby is desirable and should be large enough to allow traffic to move in both directions on the landing without being obstructed by people waiting for the lift. Lift lobbies should be visible from entrance halls, but intending passengers should not be able to see the entrance hall from the lift, as they may hold the lift for late arrivals, cause disturbances and also wear of the control system. Figure 11.1 shows the method of grouping lifts.

Number of lifts

The number and size of lifts must be related to the following:

1. Population of the building.
2. Type of building occupancy.
3. The starting and finishing times of the population, whether staggered or unified.
4. Number of floors and heights.
5. Position of building in relation to public transport services. A building near a traffic terminal generally has high passenger peaks during arrival hours.

Note: The choice of the number of lifts and their sizes usually lies between the convenience of the user and the overall building loading times, and a compromise is usually required to achieve a satisfactory balance between these two factors.

Several smaller lifts will provide a better service than fewer larger lifts, but the installation cost of the latter is lower.

Population

If a definite population figure is unobtainable, an estimate can be made from the net floor area and the probable population density per square metre. The average population density can vary between one person per 4 m^2 and one person per 20 m^2, but the building owner should be able to give a reasonable figure.

For general office buildings a population density of one person per 10 m^2 of net floor area may be assumed and for these buildings a guide to the minimum number of lifts required is given in Table 11.1.

Table 11.1 Minimum number of lifts for offices

Installation	Quality of service
One lift for every three floors	Excellent
One lift for every four floors	Average
One lift for every five floors	Below average

Note: A lower standard than the above would be acceptable for hotels and blocks of flats. Where large numbers of people have to be moved, cars smaller than twelve-person capacity are not satisfactory.

Round-trip time

The time in seconds taken by a single lift to travel from the ground floor to the top floor, including the probable number of stops, and return to the ground floor.

Flow rate

This is usually expressed as a percentage of the total population requiring lift service during a 5 min peak demand period. Surveys have shown that between 10 per cent and 25 per cent of the total population will require transportation during a 5 min peak demand period. If no information is available on the expected flow rate, 12 per cent may be assumed for speculative buildings or where staggered starting times will be practised, and 17 per cent for buildings which will have unified starting times.

Interval for lifts

The interval is expressed in seconds, and represents the round-trip time of one car divided by the number of cars in a common group system; it provides a criterion for measuring the quality of service. The average waiting time may therefore be expressed theoretically as half this interval, but in practice it is probably nearer three-quarters of the interval.

Intervals may be considered from Table 11.2.

Table 11.2

Interval (s)	Quality of service
25–35	Excellent
35–45	Acceptable for offices
60	Acceptable for hotels
90	Acceptable for flats

Lift travel: The travel of a lift is the number of floors above ground multiplied by the floor height.

Lift speeds: The recommended lift speeds for various building heights are given in Table 11.3.

Table 11.3

Speed (m/s)	Lift travel in metres			
	Municipal flats	Luxury flats	Offices	Bed lifts
0.25–0.375	–	–	–	5
0.50	30	15	10	10
0.75	45	20	15	–
1.00	55	25	20	20
1.50	–	–	30	45
2.50	–	–	45	100
3.50	–	–	60	–
5.00	–	–	125	–

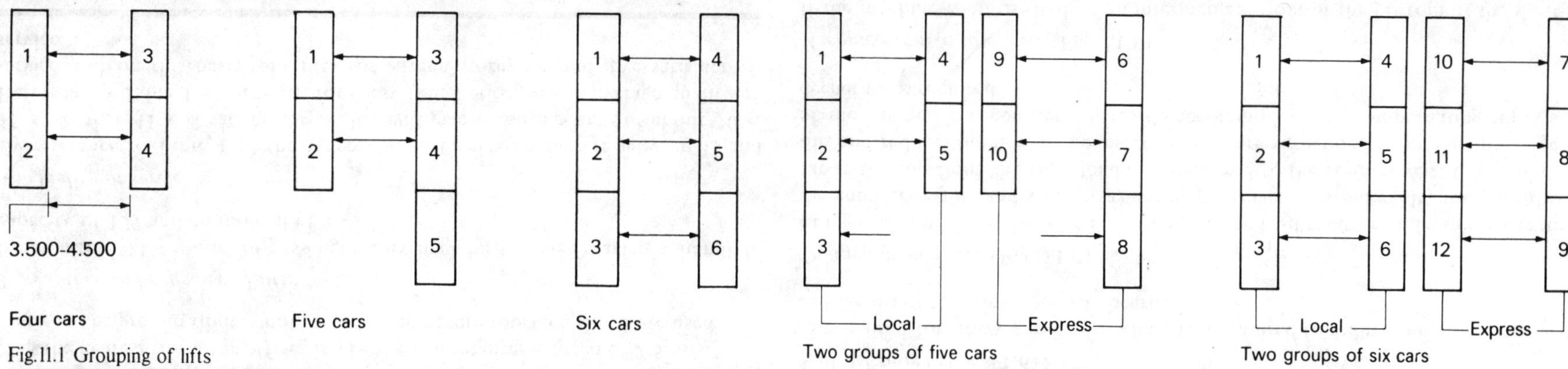
1
2
3
4
3.500-4.500
Four cars
1
2
3
4
5
Five cars
1
2
3
4
5
6
Six cars
1
2
3
4
5
9
10
6
7
8
Local
Express
Two groups of five cars
1
2
3
4
5
6
10
11
12
7
8
9
Local
Express
Two groups of six cars

Fig.11.1 Grouping of lifts

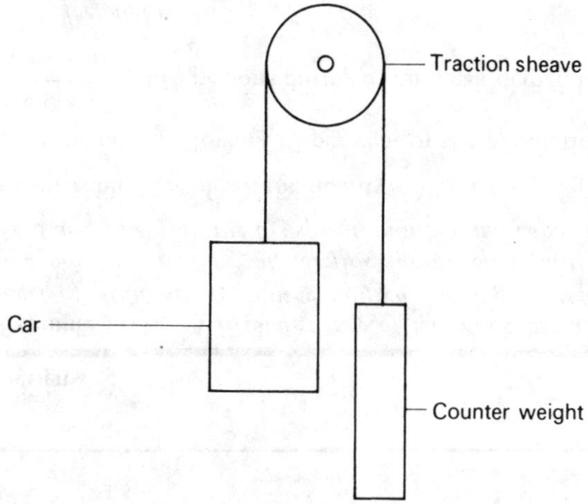
Traction sheave
Car
Counter weight

Fig.11.2 Single-wrap for small car

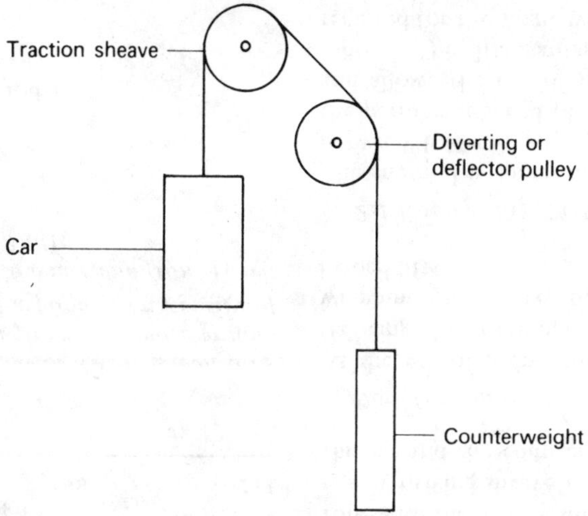
Traction sheave
Diverting or
deflector pulley
Car
Counterweight

Fig.11.3 Single-wrap with deflector pulley

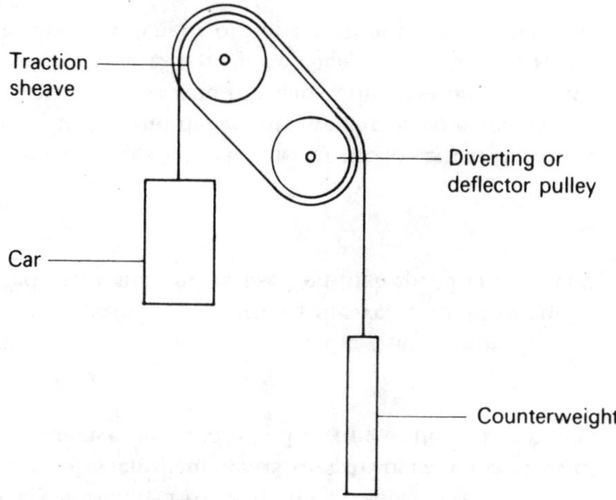
Traction
sheave
Diverting or
deflector pulley
Car
Counterweight

Fig.11.4 Double-wrap traction

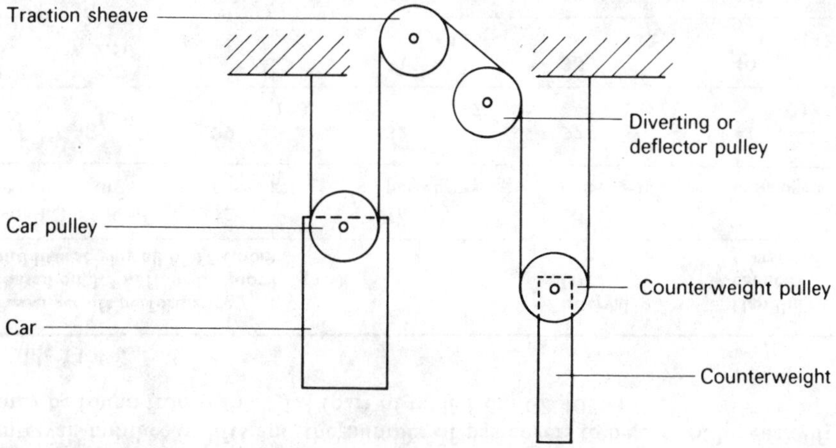
Traction sheave
Diverting or
deflector pulley
Car pulley
Counterweight pulley
Car
Counterweight

Fig.11.5 2:1 roping

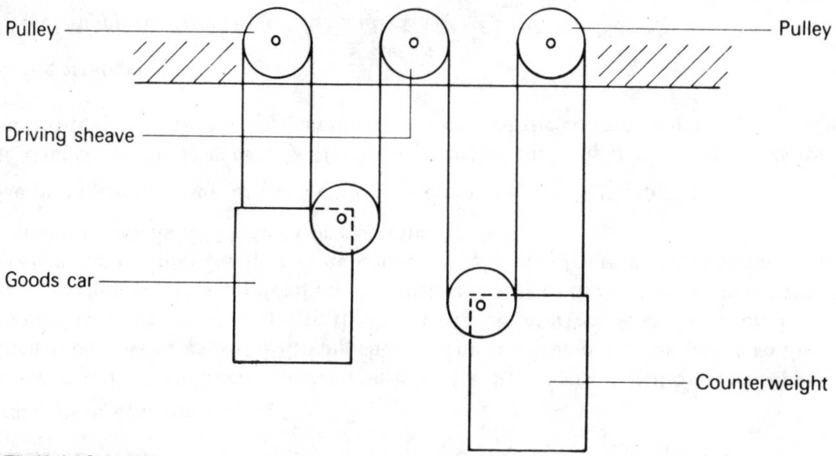
Pulley
Pulley
Driving sheave
Goods car
Counterweight

Fig.11.6 3:1 roping

Lift performances: If the travel, speed and building population are known, the interval, number of lifts and the number of passengers to be carried by each lift may be found from Table 11.4 (part of table from CP 407: 1972).

Table 11.4

Passenger lift performance (based on 3.3 m floor-to-floor heights) and lifts serving all of 15 floors

Number of cars	Speed (m/s)	12 passengers: Interval (s)	12 passengers: Handling capacity (persons)	16 passengers: Interval (s)	16 passengers: Handling capacity (persons)	20 passengers: Interval (s)	20 passengers: Handling capacity (persons)	24 passengers: Interval (s)	24 passengers: Handling capacity (persons)
4	2.50	29	103	32	112	37	127	41	137
4	3.50			31	116	36	132	40	142
5	3.50			25	146	29	165	32	178
6	3.50					24	198	27	213

Design

Example 11.1. *A fifteen-storey office block has a net floor area above ground level of 8000 m². Assuming unified starting times and a population density of one person per 10 m² of net floor area, calculate the flow rate and from Tables 11.3 and 11.4 find the lift speed, number and capacity of lifts.*

From Table 11.2 check the quality of service.

1. *Flow rate* – allowing 17 per cent of the population

$$\frac{8000 \times 17}{10 \times 100} = 136 \text{ persons during 5 min peak demand period}$$

2. *Travel and speed*

Assuming a floor height of 3.3 m, the lift travel = 14 × 3.3 = 46.2 m. From Table 11.3 the nearest travel for offices is 45 m which requires a speed of 2.5 m/s.
Note: The ground floor is not included, therefore fourteen floors are used.

3. *Number and capacity of lifts*

From Table 11.4, four 24-passenger cars may be installed having a handling capacity of 137 and an interval of 41 s.

4. *Quality of service*

By reference to Table 11.2 the acceptable interval for offices is between 35 and 45 s, so that 41 s is satisfactory. This will give a waiting time of about 30 s. Referring to Table 11.1, one lift for every four floors gives an average quality of service and on this basis four lifts for fifteen storeys would give satisfactory service.

Electric lifts

Principles of operation

An electric lift with traction drive consists of a lift car suspended by steel ropes which travel over a grooved driving sheave. The steel ropes are connected to the top of the car at one end and to the frame of a counterweight at the other. The counterweight reduces the load on the electric motor to the difference in weight between the car plus load, and the counterweight plus friction the difference is termed the 'unbalanced load'. For example

load on motor = weight of car + its load – counterweight + friction.

The counterweight is generally 40–50 per cent of the weight of the car plus its load and friction. Friction is generally 20 per cent of the counterweight.

Roping arrangements

1. *Single-wrap traction* (Figs. 11.2 and 11.3)

This arrangement is normally used with geared machines, but may be used for gearless machines for the lower speeds of 1.75–2.5 m/s. The angles of contact of rope with the driving sheave are normally 140° and 180°, respectively.

A driving sheave is seldom of such diameter as to span between centres of the car and the counterweight, hence the need of a diverting or deflector pulley.

2. *Double-wrap traction* (Fig. 11.4)

As the use of a diverting or deflector pulley increases the risk of rope slip, by reducing the frictional area of rope with the driving sheave, a double-wrap or wrapping pulley may be used. This method is used on high-speed and heavily loaded lifts.

3. *2:1 Roping* (Fig. 11.5)

This method is sometimes used with geared machines at the lower car speeds of between 1.75 m/s and 3 m/s. The car and the counterweight speed equal half of the peripheral speed of the driving sheave and this halves the load on the sheave and allows the use of high-speed motors which are cheaper than slower speed motors. The disadvantage is that the length of rope is about three times that required for the single-wrap system.

4. *3:1 Roping* (Fig. 11.6)

This is used for heavy goods lifts where it is required to reduce the motor power and the pressure acting upon the bearings.

Compensating ropes (Fig. 11.7)

In high-rise buildings above ten storeys the rope load transferred from the car to the counterweight (and vice versa) during car travel is considerable, and with the car at the top floor the rope load is transferred to the counterweight. To offset this and reduce 'bounce' compensating ropes are suspended from the underside of the car and the counterweight. To accommodate the compensating ropes a deeper pit is required.

Machine room at low level (Fig. 11.8)

If the machine room is sited on an intermediate floor or the bottom of the shaft

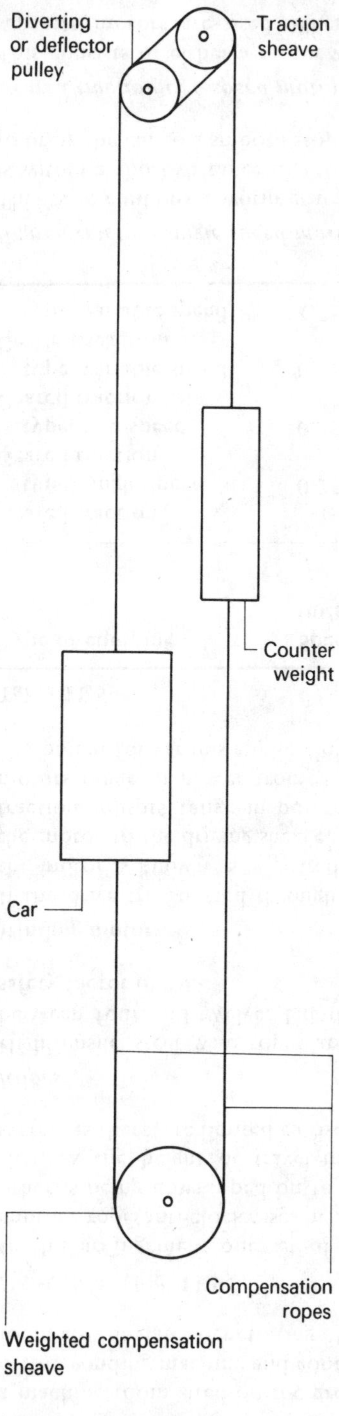
Diverting
or deflector
pulley
Traction
sheave
Counter
weight
Car
Compensation
ropes
Weighted compensation
sheave
Fig 11.7 Compensating ropes

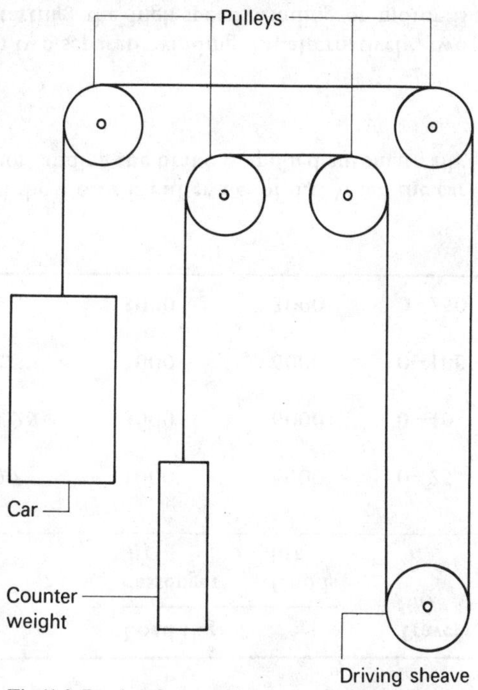
Pulleys
Car
Counter
weight
Driving sheave
Fig.11.8 Roping for machine room at intermediate floor or bottom of shaft

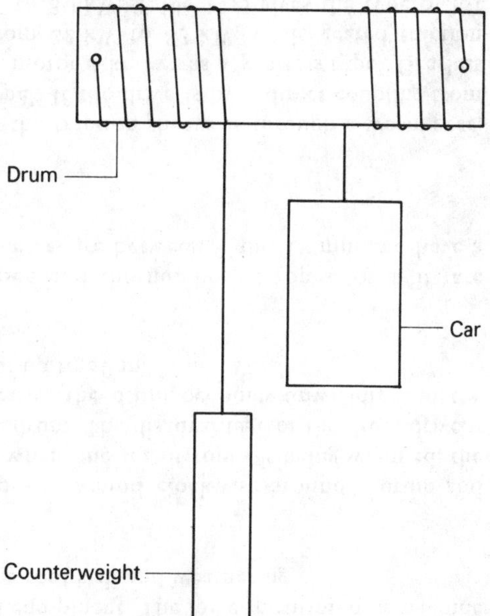
Drum
Car
Counterweight
Fig.11.9 Drum drive

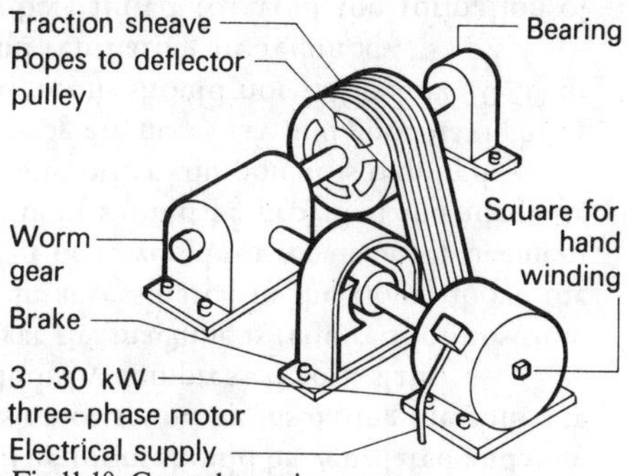
Traction sheave
Ropes to deflector
pulley
Bearing
Square for
hand
winding
Worm
gear
Brake
3–30 kW
three-phase motor
Electrical supply
Fig.11.10 Geared traction motor

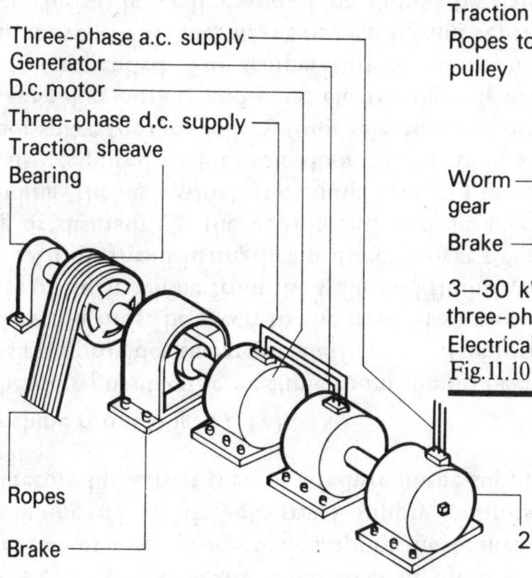
Three-phase a.c. supply
Generator
D.c.motor
Three-phase d.c. supply
Traction sheave
Bearing
Ropes
Brake
22—84kW three-phase induction motor
Fig.11.11 Gearless traction variable-voltage motor

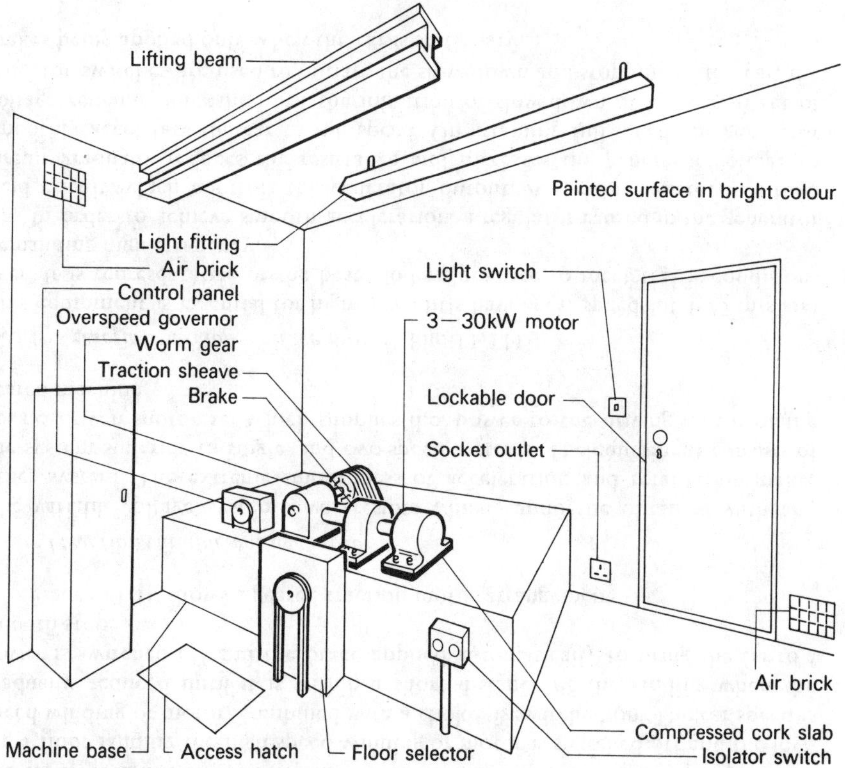
Lifting beam
Painted surface in bright colour
Light fitting
Air brick
Control panel
Overspeed governor
Worm gear
Traction sheave
Brake
Light switch
3—30kW motor
Lockable door
Socket outlet
Air brick
Compressed cork slab
Isolator switch
Machine base
Access hatch
Floor selector
Fig.11.12 View of lift machine room

a longer rope is needed, it travels round more pulleys, resulting in higher frictional resistance and therefore more maintenance is required. However, with a machine room sited on the ground floor the lift shaft is relieved of the weight of the winding machine and control equipment. The lower position of a machine room also obviates penetrating of the roof slab and weathering.

Drum drive (Fig. 11.9)

In this arrangement one set of ropes is wound clockwise around a drum and another rope anti-clockwise, hence when one set of ropes is being wrapped, the other is being unwrapped on to the drum. The disadvantage of the drum drive is that, as the height of travel increases, the drum becomes unwieldly and the system is therefore limited to rises of up to 30 m.

Ropes

High-tensile steel wire ropes are used and the number of ropes for a lift are between four and twelve. The diameters are between 9 and 19 mm and have a safety factor of 10.

Winding motors

If the drive transmitted through to the traction sheave is through a worm gear, the motor is known as a 'geared type'. If the drive is by a direct coupling from the motor to the driving sheave, the motor is known as a 'gearless type'. Gearless traction motors range in power from 22 kW to 83 kW, while geared traction motors range in power from 3 kW to 30 kW. Table 11.5 gives the type of lift equipment for various applications.

Table 11.5

Type of equipment	Speed (m/s)	Load (kg)		Travel (m)
		Passenger lift	Goods lift	
Geared traction type, single-speed	0.25–0.8	1000	6000	0–25
Geared traction type, two-speed	0.25–1.25	3000	6000	0–40
Geared traction type, variable-speed	0.75–2.5	3000	6000	0–100
Gearless traction type, variable-speed	1.75–8	3000	3000	0–250

Geared traction single-speed motor

This type contains a worm gear and the motor is either a.c. or d.c. When the car is within a short distance of the floor landing the brake is applied automatically to bring the car to a smooth stop.

Geared traction two-speed motor

This consists of either a motor with two separate windings or, alternatively, two separate motors are used. When starting, the high-speed winding or motor is switched on in series with a resistor to limit the current. Smooth acceleration of the car is obtained as the resistance field is progressively lowered. On approaching a floor landing the high-speed winding or motor is switched off and the low-speed winding or motor, combined with a choke, is switched on. The car speed is gradually reduced until it is within a short distance of the landing when the power is switched off and the brake applied automatically to bring the car to a smooth stop.

Figure 11.10 shows a geared traction motor arrangement.

Geared traction variable-voltage motor

The variable-voltage system gives results which cannot be obtained with any other system. The extreme smoothness of acceleration and retardation makes the system superior to single- and two-speed systems. The equipment consists of an a.c.-driven motor set which supplies d.c. power to the driving motor of the geared machine.

Gearless traction variable-voltage motor (Fig. 11.11)

This equipment is essential for high-speed lifts having car speeds of 1.75 m/s and over. It is representative of the best modern practice to meet traffic conditions demanding high efficiency.

In order to achieve smooth acceleration, a regulator is used in the generator field circuit which controls the generator output. A variable resistor in a field circuit gradually reduces the resistance and increases the generator voltage to smoothly accelerate the car to full speed. On attaining full speed, the generator voltage remains constant until the initiation of slow-down of the car. A set of inductor switches are used to initiate the slow-down and stopping of the car, the brakes being applied only when the car is stationary.

Brakes

For all types of lift machine equipment, an electrical-mechanical brake is required which is designed to fail safe. When the lift is running, the brake shoes are electro-mechanically lifted clear of the brake drum, overcoming the force of the coil or disc springs which apply the brakes when the car is stationary. The switching-off of the electrical supply permits the brake to be applied and therefore fail safe if there is a failure in the supply.

Machine room (Fig. 11.12)

Wherever possible the machine room should be at the top of the lift shaft, as this position provides the greatest efficiency. The room should be ventilated and consideration must be given to the transmission of sound by insulating the concrete base of the machine from the walls and floor by compressed cork slabs.

An overhead lifting beam directly over the machine is required for positioning or dismantling the equipment, and an access hatch in the floor, above the landing, through which the equipment can be lowered for repair or replacement is also required. A lockable door to the room should be provided and adequate floor space for controllers, floor selectors and other equipment is required.

Socket outlets and good electric lighting are necessary and good daylighting is recommended. The temperature of the room should not fall below 10 °C or rise above 40 °C, and means of heating and ventilating are required.

The walls, ceiling and floor should be painted to avoid the formation of

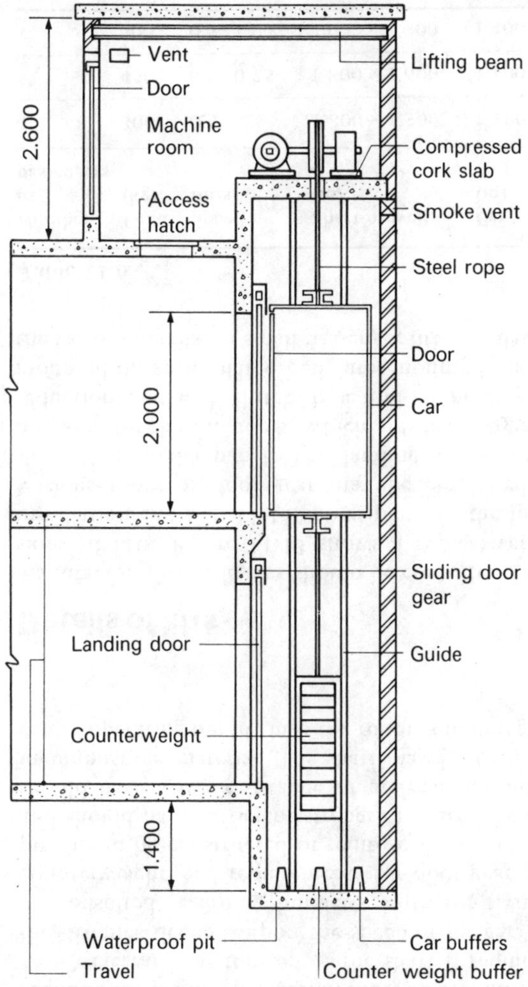
Vent
Door
Machine room
2.600
Lifting beam
Compressed cork slab
Access hatch
Smoke vent
Steel rope
Door
Car
2.000
Sliding door gear
Landing door
Guide
Counterweight
1.400
Waterproof pit
Car buffers
Travel
Counter weight buffer

Fig.11.13 Vertical section of electric lift

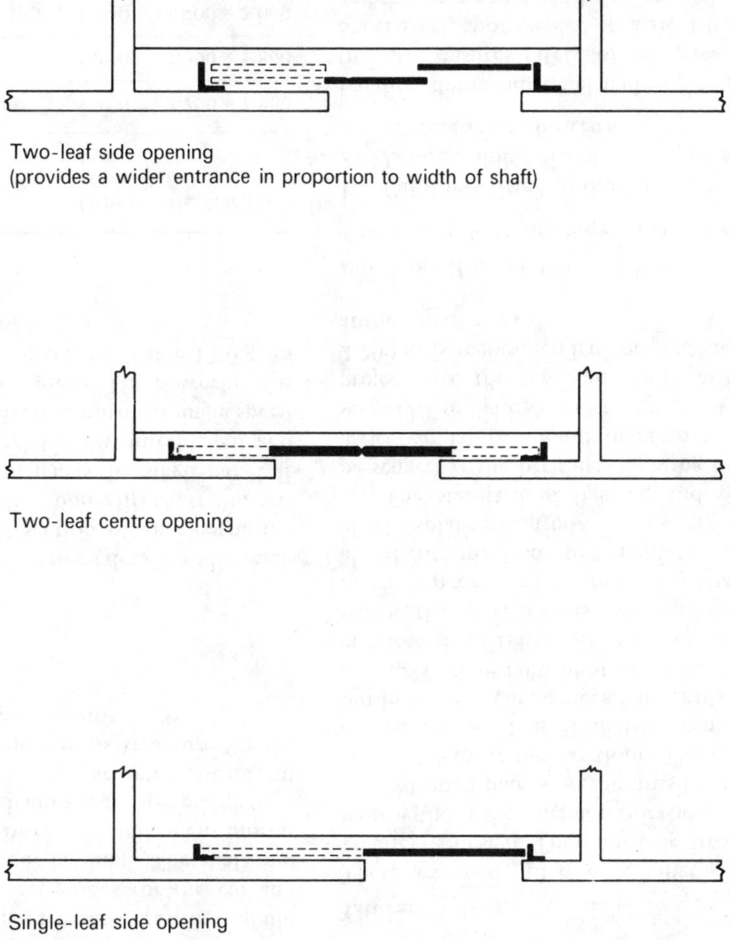
Two-leaf side opening
(provides a wider entrance in proportion to width of shaft)
Two-leaf centre opening
Single-leaf side opening

Fig.11.15 Lift doors

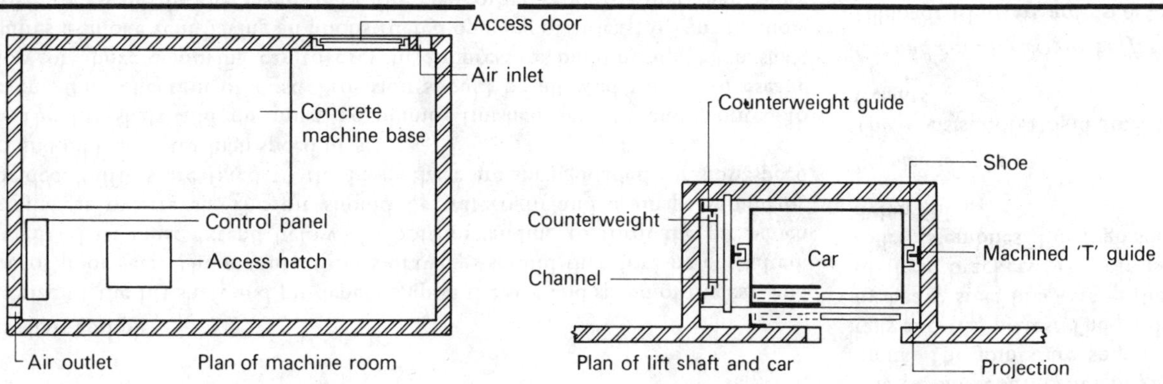
Access door
Air inlet
Concrete machine base
Control panel
Access hatch
Air outlet
Plan of machine room
Counterweight guide
Shoe
Counterweight
Car
Machined 'T' guide
Channel
Plan of lift shaft and car
Projection

Fig.11.14 Plans of machine room lift shaft and car

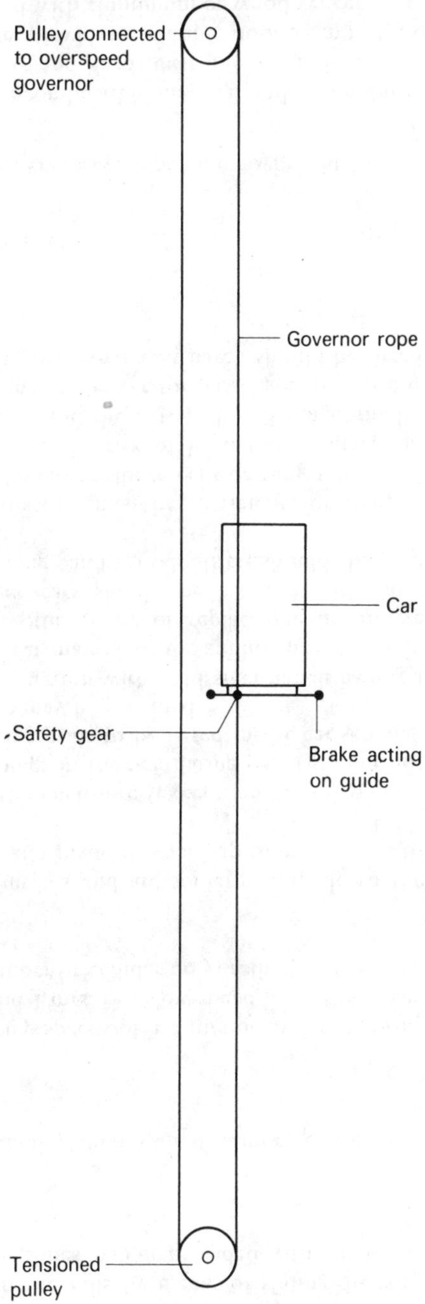
Pulley connected to overspeed governor
Governor rope
Car
Safety gear
Brake acting on guide
Tensioned pulley

Fig.11.16 Overspeed governor and equipment

dust, which can damage equipment and cause a breakdown of the electrical circuit due to poor contacts.

Lift shaft and pit

The size of the lift shaft and pit depends upon the size and speed of the car, and type of door gear. The manufacturer's drawings should therefore be consulted. The lift shaft must extend below the bottom landing to form the pit which permits car overtravel. The pit should be watertight and drainage should be provided. Buffers are fixed to the base: these are spring-loaded for slow-speed lifts and oil-loaded for high-speed lifts.

The lift shaft and pit must be plumb, finished smooth and painted to prevent the collection of dust. Provision should be allowed for air to escape below or above a moving car to prevent air pressure building up. Each shaft requires a smoke vent having an unobstructed opening of at least 0.1 m^2 to allow smoke to escape in the event of a fire. No other services other than those required for the lift installation must be accommodated in the shaft.

A clearance at the top of the shaft is required for overtravel of the car and the distance depends upon the speed of the car. Manufacturer's drawings should be consulted regarding builder's work for fixing steel guides for the car and counterweight, and requirements for door gear at floor landings. The shaft and pit should be constructed of reinforced concrete or brickwork in cement mortar and should have sufficient strength to carry the dead and superimposed loads. It should have a fire resistance of at least one hour and constructed entirely of incombustible material. The shaft may have an opening in its structure for the cables operating the lift into the room containing the lift motor.

Details of lifts

Manufacturer's catalogues should be consulted for the various details and dimensions of lifts. Figure 11.13 shows a typical vertical section of an electric lift installation and Fig. 11.14 shows plans of the machine room, lift shaft and car. Various types of door gear may be employed which may be single-panel as shown, or double-panel. The landing doors are mechanically interlocked and operated by the car doors, which are driven by an electric motor having a speed reduction unit with clutch drive and connecting mechanism. The power unit is mounted on an insulated bedplate, mounted on top of the car. Table 11.6 gives the various details for small passenger lift installations.

Table 11.6

Number of passengers	Load (kg)	Speed (m/s)	Shaft or well (m)	Car (m)	Depth of pit (m)	Machine room (m)
4	300	0.5	1.800 x 1.300	1.100 x 800	1.400	3.700 x 1.800
6	450	0.5–0.75	1.800 x 1.600	1.100 x 1.100	1.400	4.000 x 1.800
8	600	0.5–1.00	2.000 x 1.900	1.400 x 1.100	1.500	4.400 x 2.000

Guides

The car and counterweight guides are machined tee sections finished to very fine limits. The joints are spigoted and fish-plated with machined surfaces on both rails and fish-plates. The guides are erected plumb and fastened to the lift shaft by heavy steel brackets – the builder must leave holes in the shaft for the fixing of these brackets. The car is guided on the rails by means of sliding shoes or roller assemblies. For high-speed lifts the shoes may be provided with renewable nylon linings.

Balance weight

This consists of a rigid steel frame containing the required number of cast-iron weights.

Car and counterweight buffers

Oil-loaded buffers are used for lifts having speeds of 1.5 m/s or over and spring buffers used for speeds of between 0 and 1 m/s. For slow-speed lifts of between 0 and 0.25 m/s it is permissible to use timber or cellular polyurethane buffers.

Lift cars

These are made to a wide range of finishes and are sometimes made to the architect's design. They must be strong and made to stand up to many years of wear without showing deterioration.

No open panels are permissible in the enclosure (except ventilating panels) within 1.8 m of the car floor. The openings in the ventilating panel should not exceed 13 mm. It is normal for ventilating grills to be fitted in the car skirting, but in some circumstances fan ventilation may be provided.

Cars are usually of steel frame construction with mild steel panels having a cellulose paint finish. Alternatively, the car may be of the applied décor type in which the steel panels are finished by the fixing of replaceable decorative interior panels. The floor of passenger cars should be covered with non-absorbent material to facilitate cleaning and goods lifts should have a checker-plate steel floor.

The standard of lighting and ventilation must be sufficient for trapped passengers if the lift fails. Lighting can be either fluorescent or tungsten and at least two lamps should be used to reduce the risk of having an unlit car in service. The light should be on at all times and the fitting should not normally project into the car. The minimum height of the car for passengers is 2 m but 2.200 m is recommended for greater comfort. An access hatch should be fitted at the top of the car.

Lift doors (Fig. 11.15)

Two sets of doors are required at lift entrances:

1. Car doors fitted to the lift car.
2. Landing doors fitted to the lift shaft enclosure (open metalwork enclosures are no longer allowed).

Landing doors must be made of solid incombustible material and this reduces fire risk, ensures safety of the passengers and keeps dust out of the lift shaft. Sheet steel, spot welded to a steel frame, provides a suitable door 32 mm thick and the door panel can be painted or faced with aluminium or wood veneer.

Sliding doors should have robust steel angle frames to which top and bottom tracks may be fitted.

Landing doors must have no tangible means by which an unauthorised person can open them from a landing and each door is locked from within the lift shaft. If a maintenance engineer unlocks the landing door, a contact is broken which switches off the electrical supply and renders the lift inoperative.

Doors may be of the following types:

1. Two-leaf side opening.
2. Two-leaf centre opening.
3. Single-leaf side opening.

Single-hinged doors are suitable for residences and small hotels, they should be self-closing and provided with a locking mechanism.

Overspeed governor and equipment (Fig. 11.16)

To prevent the car from overspeeding due to the ropes breaking or stretching, or by some electrical fault, an overspeed governor is fitted. The governor is normally mounted in the motor room directly above the lift shaft and is fitted with a pulley at least 300 mm in diameter which is driven by a governor rope. One end of the rope is attached to a braking system underneath the car, while the other end extends upwards from the car, wraps around the governor pulley and extends down the shaft to a tension pulley at the bottom and then returns to the car. As the lift car travels so the governor rope drives the pulley.

Fly weights with spring control respond to overspeed and their movement is used to break electrical contacts and trigger a mechanical device which arrests the governor rope to operate the brake under the car.

Safety gears

Safety gears which are actuated when the governor arrests the governor rope in the event of overspeed in a downward direction, are fitted below the car. When operated they engage on a guide rail. The safety-gear mechanisms are of two types:

1. Instantaneous action for slow or medium speeds.
2. Gradual action for high speeds.

The instantaneous type consists of cams with machine-cut gripping surfaces of hardened steel fixed to the underside of the car. There are two cams on either side of the car and a mechanism is used to operate all four cams instantaneously. The cams make contact with a guide rail and the car is brought to rest abruptly. A switch is provided to cut off the current should the safety gear come into action.

The gradual-action type consists of hardened steel wedges which, when brought into contact with the guide rail, slide into position and exert a steadily increasing pressure on the surfaces of the rail, thus bringing the car to rest gradually without undue jar, within a predetermined distance. A switch is again provided to cut off the current should the safety gear come into action.

Lift controls

(a) Automatic push-button

This is the simplest and cheapest form of control, but is only capable of accepting a single call at a time. When a car is standing unoccupied at a floor and a passenger enters, a short delay in the landing call push-button circuit, after the doors have been closed, will allow time to press the car push-button corresponding to the floor the passenger requires, and the car will instantly respond and stop automatically at that floor.

Once the car commences to travel, all push-buttons are rendered inoperative until a few seconds after it has stopped and this enables the passenger to alight before the car can be called to another floor. The car, when stationary and unoccupied, can also be called to any floor in response to a landing push-button, but it will not answer to any landing push-button when travelling.

(b) Collective control

This system is the basis upon which modern supervisory controls operate, whether for a single or a group of lifts and overcomes the disadvantages of automatic push-button control. There are two types, namely down collective and directional collective.

The down-collective system operates as an automatic push-button control when the car is travelling upwards, but when the car is travelling in a downward direction it will stop for calls in order of floor landings and not in the order they are received. The system is suitable for office blocks with a variety of tenants and little interfloor travel.

In the directional-collective system UP and DOWN push-buttons are fitted on each landing and the car also has a set of buttons for each landing served. The landing push-buttons register the direction the passenger wishes to travel, and when all car and landing doors are closed the car will respond to all car and landing calls in floor sequence. When the car is standing at the ground floor, any call from the car or landing will cause the car to travel with the control system set in the upwards direction.

During its upward journey, the car will not answer DOWN calls, but these will be registered, the car will however, answer all UP calls within the car or from landings, in floor order. After dealing with the highest call, the car reverses and will answer all registered DOWN calls in floor sequence. Although the car is normally free to reverse direction at the highest floor, it may also reverse at an intermediate floor. A short delay time is provided before reversal takes place and if no further calls are registered the car is 'free' – passengers on entering can register a call and travel in either direction, or the car can be called from any landing.

(c) Group supervisory control

A group of intensive passenger lifts require a supervisory system to co-ordinate the operation of individual lifts, which are all on collective control and are interconnected. The system regulates the dispatching of individual cars and provides service to all floors as different traffic conditions arise, minimising uneven service, excessive waiting time and the presence of idle cars. It also determines which calls have waited the longest and which should be answered first.

A computer system dispatches the cars to answer calls according to the demand at any particular moment. Instead of having to make the full round trip as normally happens in the collective control system, the supervisory control can reverse at any floor if there is no demand beyond that floor. Therefore, the car can travel down to answer an UP call and travel up to answer a DOWN call. The system gives quick efficient service at both peak demand periods and throughout the day. Each individual person is given attention based on his priority and not when it suits the group system to answer the call.

(d) Attendant control

The movement of the car is by means of a handle-operated switch which has UP, DOWN and STOP positions. The car moves up by the attendant moving the handle in the UP position and down by moving the handle in the DOWN position. The car is brought to rest by moving the handle to the OFF position when approaching a landing, the car stopping automatically at the landing.

Lift cars have either drop flag or illuminated indicators, which are operated by push-buttons on the landings. Illuminated types have two push-buttons labelled UP and DOWN (except for top and bottom landings) so that the attendant knows both the landing at which the car is required and the direction the intending passenger wishes to travel.

The cancellation of the call, registered in the car, is done automatically when the car answers the call. The attendant control system is now less common and the group supervisory system for a group of passenger-controlled lifts gives a more efficient service.

(e) Dual control

This is essentially a directional-collective control system so arranged that the lift may be operated by either passengers or an attendant. When attendant operated, a NON-STOP or PASS button in the car becomes operative by use of a key-operated switch. Continuous pressure on this button permits the car to bypass landing calls and so give priority service to floors requiring special service.

Fireman's lift

Many local authorities, including the Greater London Council, require that buildings which extend out of reach of conventional fire-fighting equipment, i.e. above 24 m in height, must be provided with a special lift for firemen. The lift should have direct access from the street and the electricity supply should be separate to that provided for other lifts. A special switch on the ground floor, close to the entrance of the lift, can be used by the firemen to cancel all calls and bring the car down to the ground, after which the lift is under manual control. The lift car must be capable of reaching the top floor in 1 min and have a minimum loading of 544 kg and a floor area of 1.44 m^2.

Goods lifts and service lifts

The design of goods lifts is similar to that of passenger lifts, but they are usually larger and the car is less decorative. The car speed rarely exceeds 1 m/s and accurate levelling is usually essential to facilitate loading and unloading wheeled trolleys. Heavy goods require strong cars and efficient brakes, and the roping arrangement should be designed to prevent slipping of the rope.

Service lifts are designed and constructed so that only goods can be carried. The floor area does not usually exceed 1 m^2 and for small goods will carry 50 kg at a speed of 0.25 m/s or 0.50 m/s. The principles of design are similar to those for goods lifts, but the machine is much smaller and often safety gear is not fitted. They are used mainly by hotels and restaurants to provide a service from the kitchens to the dining rooms. The car may be constructed of stainless steel which can be polished and easily cleaned. A stainless steel removable shelf may be included and the car can be opened at the front only, or opened at the front and back. The doors to the hatches can be hinged – upward or sideways sliding – or a roller shutter may be provided.

The common classifications and speeds of goods and service lifts are given in Table 11.7.

Table 11.17

Type of lift	Speed (m/s)	Load (kg)
Document	0.4	10
Ledger	0.4	35
Food service	0.4	50
Small goods	0.25–0.5	100
Large goods	0.25–0.5	500–2000
Canteen service	0.25–0.5	110–150

Hospital bed lifts

Accurate floor levelling is essential to allow a bed to be wheeled into and out of the car. The car should be large enough to take a bed with room to spare for passengers, and the speed should be between 0.25 m/s and 1 m/s. A minimum depth of car of 2.4 m is required with a minimum width of 1.4 m. The minimum height should be 2.2 m.

Hand-powered lifts

There is a wide range of lifts available to meet the needs of food service and other goods. The load that can be carried by the car varies, but if it is above 25 kg the number of operations per day should be limited and the maximum load for the car should be 100 kg. The hatch doors may be hinged timber or multi-leaf metal sliding types. The usual method of operation of the lift is by an endless hauling rope external to the landing doors.

Paternosters

A paternoster consists of a series of two-person doorless lift cars, suspended from hoisting chains which run over sprocket wheels at the top and bottom of the shaft. The system provides both up and down movement of passengers in one shaft, and when the car reaches the limit of travel in one direction it moves across to the other set of chains and engages with the guides to travel in the other direction. A passenger enters and leaves a car while it is moving and therefore the speed must not exceed 0.4 m/s. The installation is suitable for use in universities, colleges of technology and other types of buildings where the passengers would normally be agile enough to use the lift, but it is not suitable for the elderly or infirm.

Figures 11.17 and 11.18 show the principles of operation of a paternoster. It will be seen from the diagrams that the supports for the cars are at the top and therefore the cars remain vertical at all times. If a passenger fails to alight at the top floor, no accident can occur because the car moves over from the left to the right and re-engages with the down guides for the downward journey.

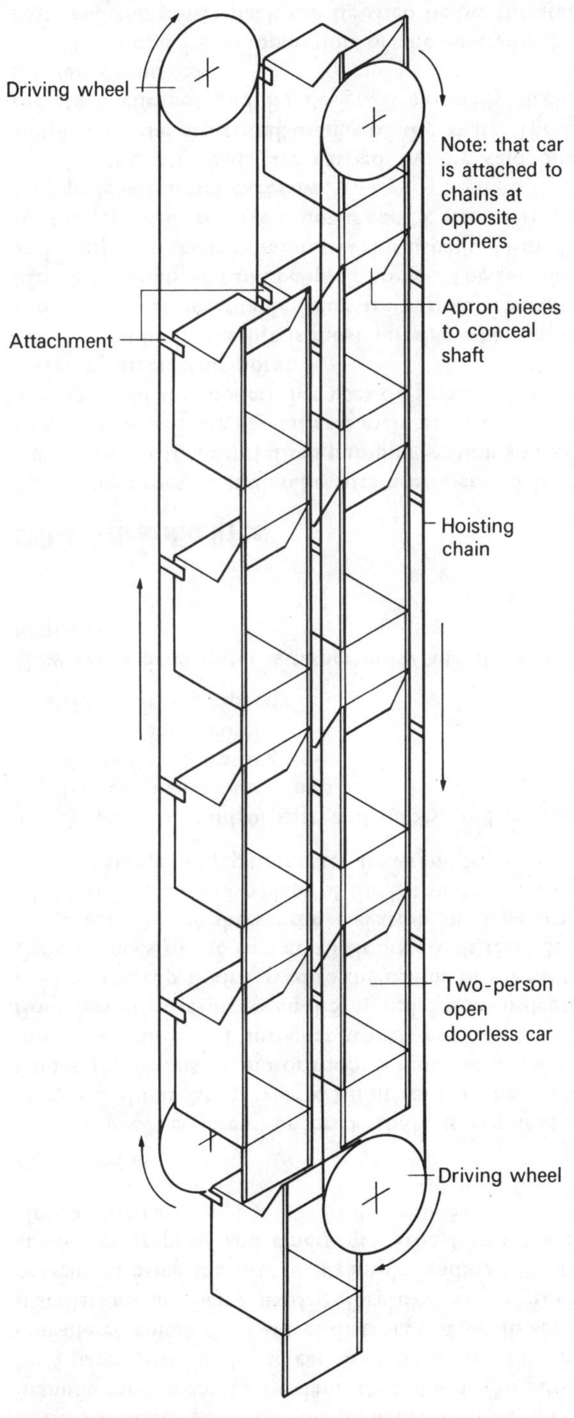
Driving wheel
Note: that car is attached to chains at opposite corners
Attachment
Apron pieces to conceal shaft
Hoisting chain
Two-person open doorless car
Driving wheel
Fig.11.17 Principle of operation of paternoster

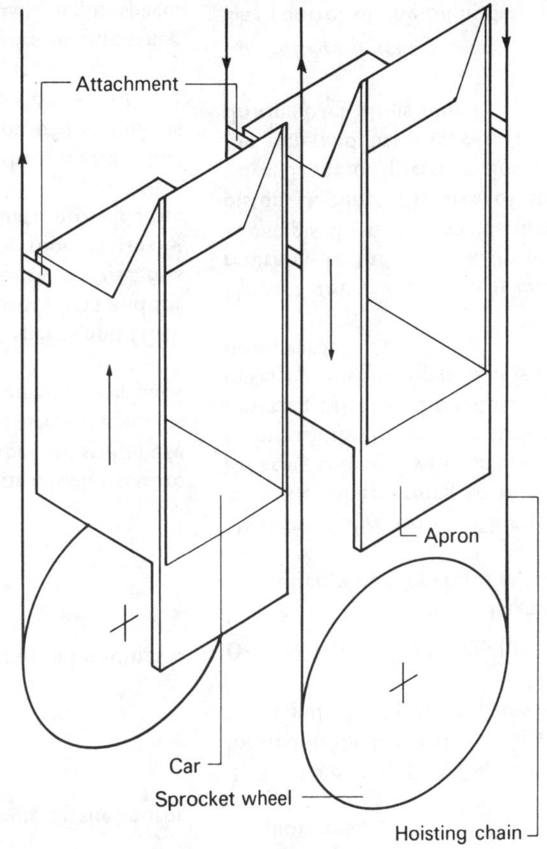
Attachment
Apron
Car
Sprocket wheel
Hoisting chain
Fig.11.18 Principle of operation (enlarged detail)

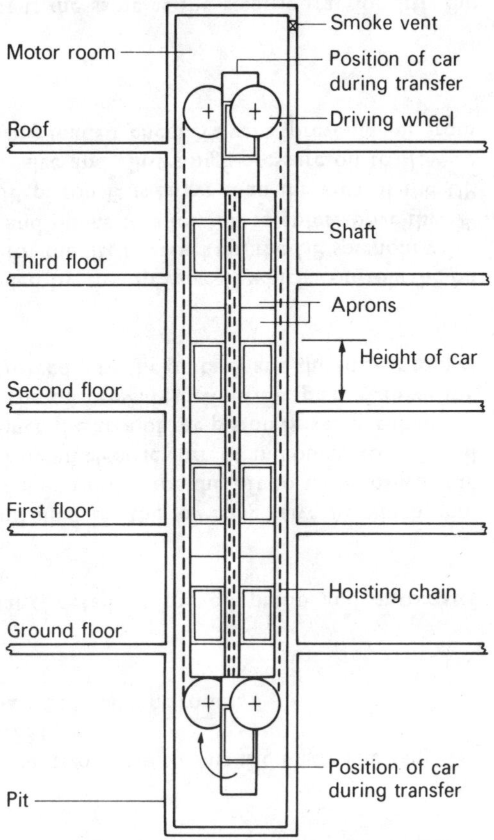
Smoke vent
Motor room
Position of car during transfer
Driving wheel
Roof
Shaft
Third floor
Aprons
Height of car
Second floor
First floor
Hoisting chain
Ground floor
Position of car during transfer
Pit
Fig.11.19 Elevation of paternoster

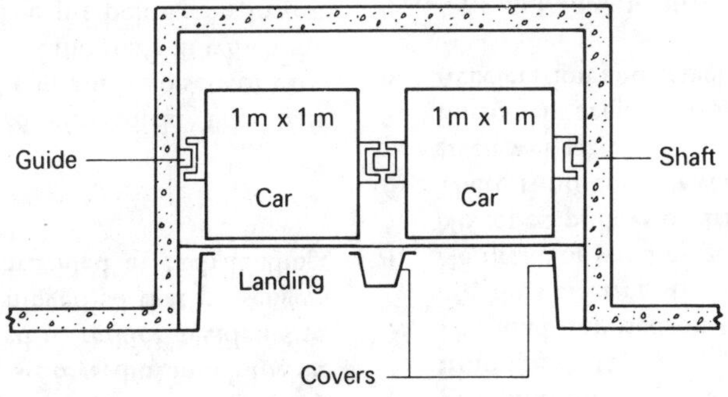
1 m x 1 m
Car
1 m x 1 m
Car
Guide
Shaft
Landing
Covers
Fig.11.20 Plan of shaft at floor level

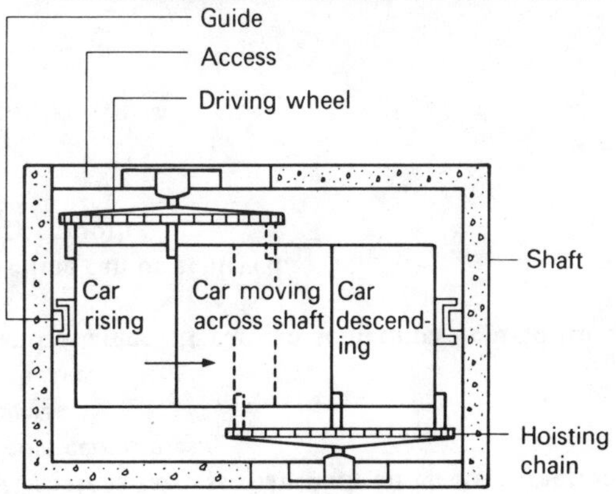
Guide
Access
Driving wheel
Shaft
Car rising
Car moving across shaft
Car descending
Hoisting chain
Fig.11.21 Plan of shaft at changeover

The absence of doors saves on capital costs, and since there are no locking devices or door gears the maintenance cost of the lift is greatly reduced. The manufacturers commonly claim that about 650 persons per hour can be carried by a paternoster with the cars moving at about 0.4 m/s. Paternosters are usually considered suitable for travel distances of up to seven storeys, but much higher installations are being used. Safety devices are provided to reduce accidents to passengers using the lift, for example landing sills are hinged so that passengers cannot be trapped and emergency stop buttons are provided at each landing, close to the cars.

Construction

A reinforced concrete or brick shaft is required 225 mm thick. The shaft measures about 2.6 m wide x 1.6 m front to back and contains two sets of cars, guides and chains. A motor room is required at the top of the shaft in which the motor, gearbox and sprocket wheels are mounted. The pit below the lowest floor served varies between 4.5 m and 5.8 m in depth. The headroom from the level of the top floor served to the ceiling of the motor room is at least 7 m, and the entrances to the cars at all floors are at least 2.6 m high and 950 mm wide.

Figure 11.19 shows the elevation of a paternoster installation and Figs. 11.20 and 11.21 show plans of the lift shaft, showing cars at different positions.

Advantages of a paternoster are as follows:

1. Absence of control gear and doors, and continuous running of the motor reduces maintenance costs.
2. Reduction in waiting time.
3. Lower capital costs.
4. High carrying capacity.

However, a paternoster is rather noisy and the risk of fire spread in buildings is increased.

Oil-hydraulic lifts

The older types of hydraulic lifts are operated by water from a high-pressure water main with a centralised pumping station and a good number are still in use today. However, the capital and maintenance costs of high-pressure water mains are high, and the modern lift uses oil pressure from a self-contained power pack driven by an electric motor.

The oil-hydraulic lift is most suitable where moderate car speeds and fairly short travel are acceptable; they are particularly suitable for goods lifts and for lifts in hospitals and old people's homes. The car speed ranges between 0.12 m/s and 1 m/s and the maximum travel is usually 21 m. The machine room is usually on the lowest level served, but it can be remote from the lift shaft, provided the oil-pipe length is not excessive.

All the lift loads are carried by the ram directly to the ground, thus simplifying the structural design of the shaft. The construction of the shaft is therefore cheaper and its design is normally decided by the degree of fire resistance required.

The simplicity of operation of the oil-hydraulic lift reduces maintenance costs and the power pack can be sited below the staircase, thus saving in space.

Advantages

1. The power pack is at low level and does not require an overhead machine room, thus eliminating the unsightly rooftop structure.
2. The machine room is relatively small and can be located at some distance from the shaft.
3. The load imposed on the lift shaft is far less than with an electric traction lift, thus offering structural cost economies.
4. No brake or winding gear necessary.
5. No ropes, pulleys or driving sheave.
6. There is no counterweight and a larger lift car can sometimes be used for a given well size.
7. Extremely accurate floor levelling can be achieved.
8. Acceleration and travel is very smooth.

Types of oil-hydraulic lifts

There are two types of hydraulic lifts:

1. *Direct acting:*

(*a*) Ram under the car which requires a lined borehole (see Fig. 11.22).
(*b*) Rams at the sides of the car located in the lift shaft which may not require a borehole.

2. *Suspended:* This requires one or two rams to suit the load. The rams are located in the lift shaft (see Fig. 11.23).

Figure 11.24 shows plans of shaft and machine room.

Operation of oil-hydraulic lift

Figure 11.25 shows a diagrammatic detail of the oil pump and automatic controller, which operates as follows:

Downward direction: This is controlled by the lowering valve A, which controls the oil returning to the oil tank. In order for the lift to travel down, the lowering solenoid valve is energised by an electric current and opens to allow oil to bypass the lowering piston B. Since the area of the piston B is larger than the lowering valve A, the reduction in the oil pressure behind the piston allows the lowering valve to open. Oil is thus forced into the oil tank and the lift car moves downwards.

Upward direction: This is controlled by the up valve C which controls the oil returning to the oil tank. In order for the lift to travel up, the UP solenoid valve is energised by an electric current and opens to allow oil to enter above the UP piston D. Since the area of the UP piston D is larger than the area of the UP valve C, the oil pressure closes the valve and allows high-pressure oil to flow to the ram and lift the car. The spring-loaded check valve E prevents oil from flowing back along pipe F.

Control equipment

The control of the oil-hydraulic lift is the same as the electric traction lift, the

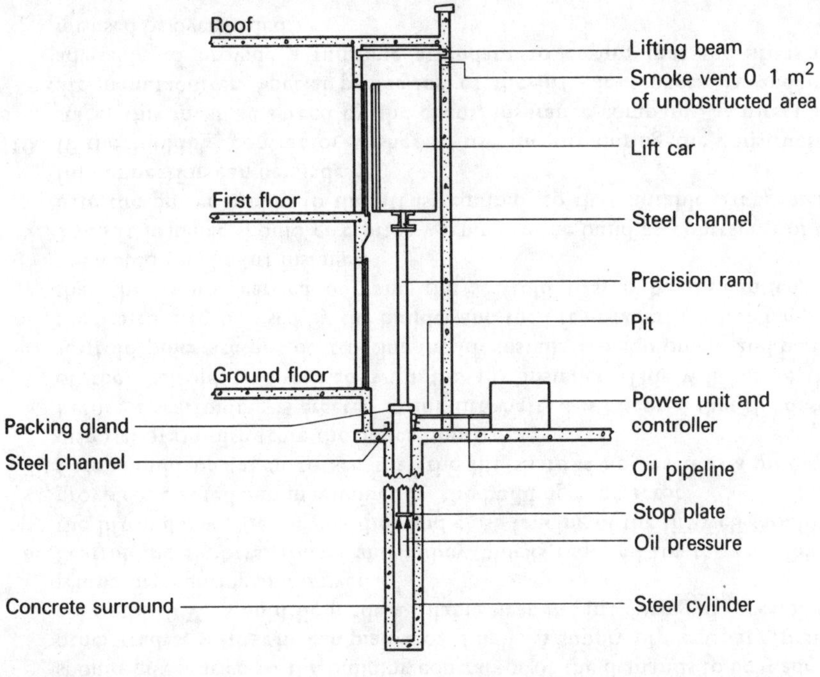
Roof
Lifting beam
Smoke vent 0 1 m^2 of unobstructed area
Lift car
First floor
Steel channel
Precision ram
Pit
Ground floor
Power unit and controller
Packing gland
Steel channel
Oil pipeline
Stop plate
Oil pressure
Concrete surround
Steel cylinder

Fig.11.22 Ram sited below lift car

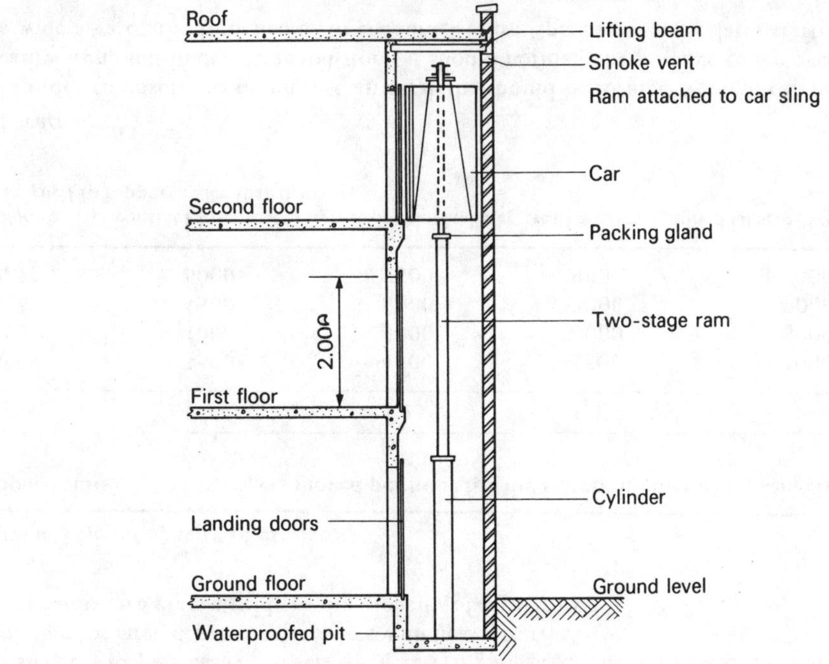
Roof
Lifting beam
Smoke vent
Ram attached to car sling
Car
Second floor
Packing gland
2.000
Two-stage ram
First floor
Cylinder
Landing doors
Ground floor
Ground level
Waterproofed pit

Fig.11.23 Vertical section of oil-hydraulic lift installation with rams on each side of lift car

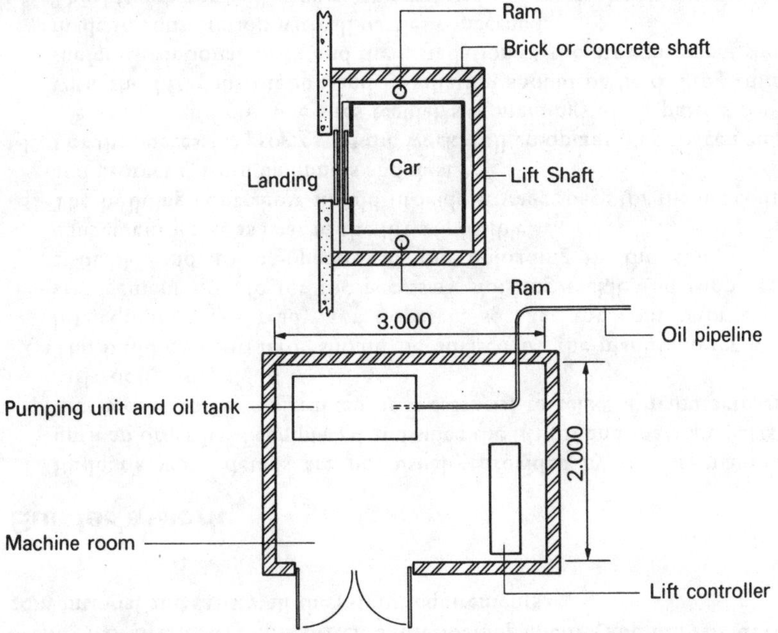
Ram
Brick or concrete shaft
Car
Landing
Lift Shaft
Ram
3.000
Oil pipeline
Pumping unit and oil tank
2.000
Machine room
Lift controller

Fig.11.24 Plans of shaft and machine room

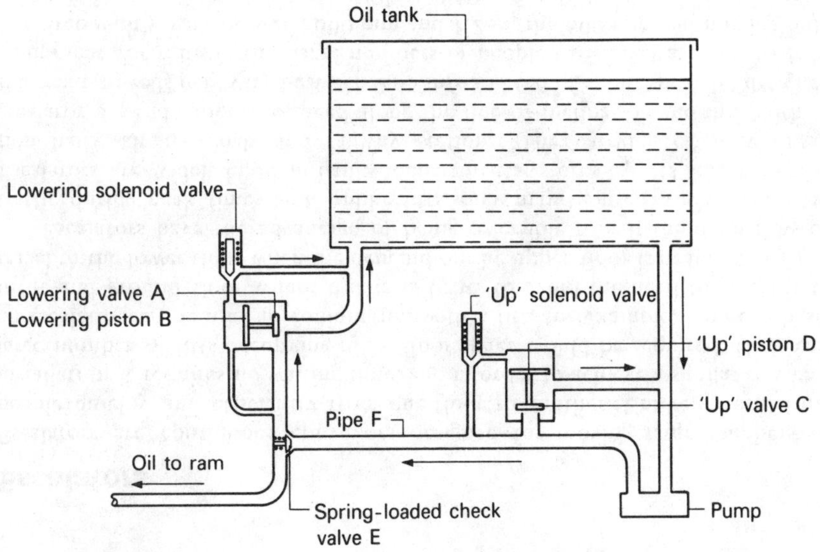
Oil tank
Lowering solenoid valve
Lowering valve A
Lowering piston B
'Up' solenoid valve
'Up' piston D
'Up' valve C
Pipe F
Oil to ram
Spring-loaded check valve E
Pump

Fig.11.25 Oil pump and automatic controller

push-button control panels being linked to the electric hydraulic system which can be designed to suit various requirements of speeds and loads.

Pressure relief valves are incorporated to safeguard the power pack in the event of an overload. Limit switches interlocking circuits and fail-safe devices are conventional and satisfy all the safety requirements.

Builder's work

1. Builder's work details are not usually provided by the lift manufacturer until an order has been placed and since the lift pit and shaft are parts of the first works to be constructed, it is essential to select a manufacturer at an early design stage.
2. The building contractor should be advised by the manufacturer when the lift equipment is ready for dispatch, so that the contractor can make arrangement on site for the necessary hoisting tackle and supports to be available and to co-operate in the unloading of the equipment at a convenient point as near to the lift as possible.
3. The building contractor should provide storage space for the lift equipment and protect it from dampness and dust.
4. For the successful progress of the work, full co-operation between all parties is essential, and on large sites regular site meetings of all parties concerned with the lift construction and installation should be held. Programmes for the constructional work and the installation of the lift equipment should be made in consultation with all parties concerned.
5. The lift erector will require the services of joiners, bricklayers, concretors and electricians as the work proceeds, and it is essential that the lift erector should give notice to the building contractor of the demands to be made on other trades, so that he can plan accordingly. A supply of electricity, preferably at 110 V, should be made available near the lift during its erection for temporary lighting and power.
6. Scaffolding timbers, rollers and pulley blocks required for the erection of the lift and also for the guarding and close fencing of the lift well should be provided, erected and maintained by the building contractor.
7. Care should be taken to see that the lift shaft is not used as a means of disposal of rubbish from the upper floors.
8. Before a scaffolding is erected in the lift shaft, it is essential that the design of the scaffold is discussed with the lift installer. This will ensure that scaffold poles are not obstructing the lift installation equipment and permit the scaffold to be used by the builder and the lift installer. It often happens that this is not carried out and the scaffold has to be dismantled a d re-erected for the lift installer.
9. The lift installer should give prior warning to the building contractor of the date the power supply to the lift is required, so that suitable rrangements for connection can be made.
10. If the building contractor wishes to use the lift during the construction stage, this must be agreed by the client, insurance company, architect and lift manufacturer. Special protection of the lift car is necessary and it is advisable to provide a full-time attendant to ensure that the lift is not misused or overloaded.
11. The lift installation should be thoroughly tested and examined before handing over and it is advisable for the client to enter into a maintenance contract with the manufacturer. The cost of the maintenance contract should be investigated when the lift manufacturers are being selected.

Escalators

Escalators are continuous conveyors designed for moving large numbers of people quickly and efficiently from one floor to another. Unlike a normal lift installation it requires no waiting time, and in order to achieve a similar service a large number of lifts occupying more floor space would be required. However, an escalator can be used in conjunction with a lift, for example, between basement and ground floor where traffic is light, to avoid the need for the lift to travel to the lower floor when the demand on the upper floor is heavy.

Escalators have the advantage of being reversible to suit the main flow of traffic during peak times and, unlike lifts, they may be used when stationary. Escalators are widely used in banks, departmental stores, sports stadia, exhibition halls, air terminals and railway stations. The carrying capacity of an escalator depends upon the speed along the line of inclination and the width of the tread. Speeds may vary between 0.45 m/s and 0.6 m/s, but specially designed equipment for transporting large numbers of people can travel at over 0.7 m/s.

Step widths vary between 600 mm and 1.2 m, the 600 mm width being only suitable for small installations, while a 810 mm width is suitable for small departmental stores and banks. A 1 m step width will allow two people to stand side by side or to pass on the step, while a 1.2 m width is normally used for air terminals and underground railway stations to allow adequate room for passengers to pass easily – even when carrying luggage. The 1.2 m width is also suitable for large departmental stores with heavy traffic.

Table 11.8 gives a guide to the capacity of escalators.

Table 11.8 Capacity of escalators

Speed (m/s)	Passengers moved per hour for the stated number of passengers per step			
	1	1.25	1.5	2
0.45	3500	4500	5500	7000
0.5	4000	5000	6000	8000
0.6	4500	6000	7000	9000
0.75	6000	7500	9000	12 000

Note: The contract load is 290 kg/m^2 of the total step tread area, but individual steps are designed to support twice that figure.

Location

In order to ensure maximum use an escalator should be located where it can be easily seen, and in departmental stores it should normally be possible to see over a wide area of the floors so as to encourage sales. Up and down escalator traffic

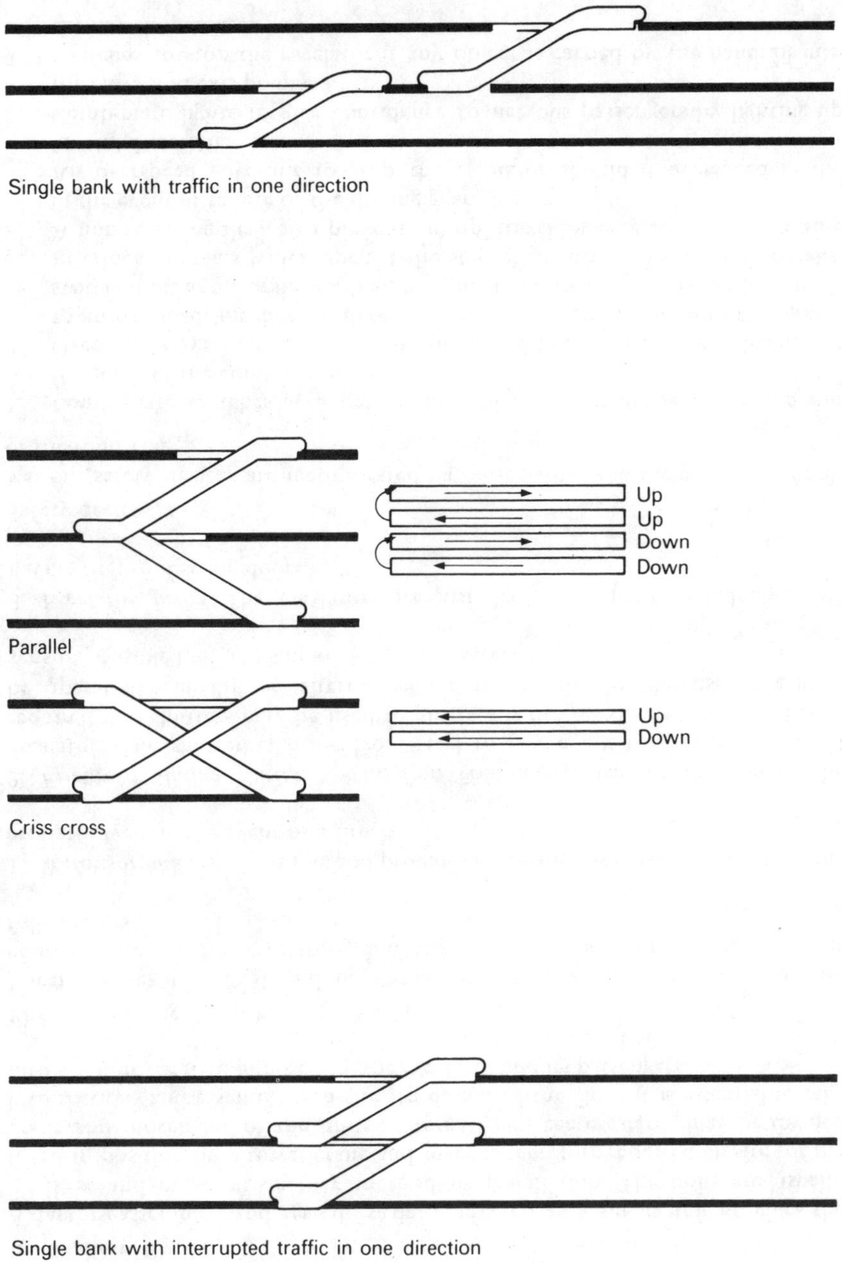
Single bank with traffic in one direction
Up
Up
Down
Down
Parallel
Up
Down
Criss cross
Single bank with interrupted traffic in one direction

Fig.11.26 Escalator arrangements

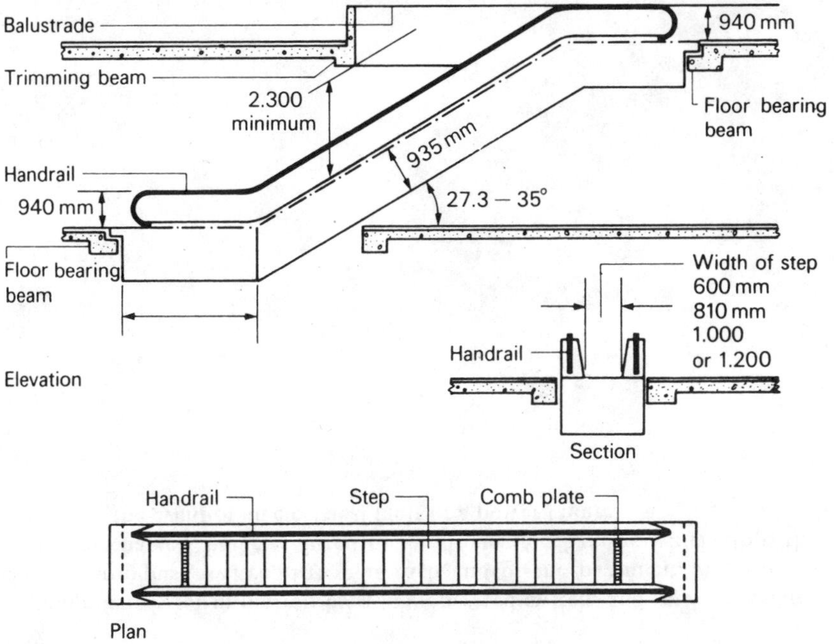
Balustrade
940 mm
Trimming beam
2.300
minimum
Floor bearing
beam
935 mm
Handrail
940 mm
27.3 – 35°
Floor bearing
beam
Width of step
600 mm
810 mm
1.000
or 1.200
Handrail
Elevation
Section
Handrail
Step
Comb plate
Plan

Fig.11.27 Details of escatator

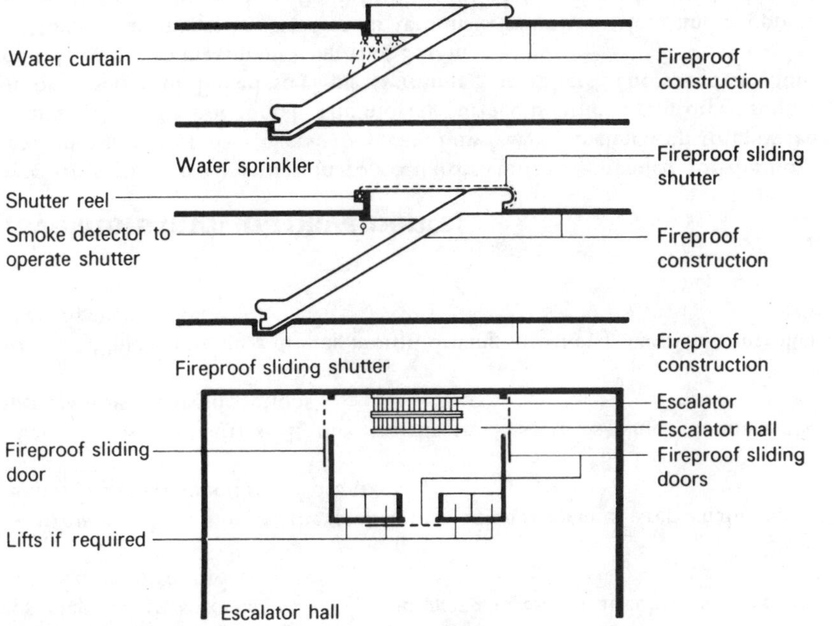
Water curtain
Fireproof
construction
Water sprinkler
Fireproof sliding
shutter
Shutter reel
Smoke detector to
operate shutter
Fireproof
construction
Fireproof
construction
Fireproof sliding shutter
Escalator
Escalator hall
Fireproof sliding
door
Fireproof sliding
doors
Lifts if required
Escalator hall

Fig.11.28 Fire control

should be kept distinct and should not be tapped or fed into from confined spaces, such as cul-de-sacs.

Arrangements

Various arrangements may be used for escalators depending upon the standard of service required and the cost of installation (see Fig. 11.26).

Installation (Fig. 11.27)

A factory-assembled and tested escalator erected as a single unit provides the quickest and most satisfactory method of installation. The units are usually lifted in position by a tower crane and space is therefore required on site for the crane and movement of the units. Alternatively, the escalator may be divided into sections and assembled on site. In design of the floor it is essential to take into account the load imposed by the escalator and its passengers.

Fire control (Fig. 11.28)

Local fire regulations should be consulted with regards to the type of fire control required for escalators and various methods used are described as follows:

1. *Water sprinklers:* This method provides a continuous curtain of water in the escalator well, in the event of a fire.

2. *Fireproof sliding shutter:* This can completely seal off the top of the escalator. The shutter is made from steel to give a 2 h fire resistance and if required two shutters may be used to give a 4 h fire resistance. The shutter can be operated manually or, alternatively, automatically by the use of a heat-sensitive fusible link or a smoke-detector device.

3. *Escalator hall:* The escalators are installed in a fire-protected enclosure having fireproof swing doors.

Safety devices

Various safety devices are incorporated with escalators and include most, or all of the following:

1. Comb plate switches actuated by any object caught between the step and the teeth of the comb plate.
2. Overload relays that trip if the motor should take an excessive current due to an overload, mechanical defect or any other cause. The power supply is switched off and a brake applied, bringing the escalator to a smooth stop.
3. Interlock contacts which open if the step chain stretches unduly or breaks.
4. A non-reversing device to prevent an 'up' travelling escalator from reversing in the event of failure of the driving gear.
5. An overspeed governor to stop the escalator should it overspeed in the 'down' direction.
6. Comb-plate lights to give confidence to nervous passengers by lighting up the entry and exit points.
7. Switches to stop the escalator if any object is carried by the handrail into the newel openings.

Construction

An escalator consists of a load-bearing steel truss structure which supports the steps, step rollers, sprocket motor, worm gear and electrical controls. There are three sections as follows:

1. *Bottom:* Which houses the step return idler sprockets, step chain safety switches and curved sections of the track.

2. *Centre:* Which carries all the straight track sections which connect the upper and lower curved sections.

3. *Top:* Which contains a driving motor, driving sprockets, electrical controller and emergency brake.

Travelators (moving pavements)

These are similar in construction to escalators, but are intended for the horizontal movement of passengers; they can, however, be inclined up to between 12° and 15° to the horizontal. The moving surface is either a reinforced rubber belt or a series of linked steel plates running on rollers. The speed is about 0.6–1.33 m/s with maximum lengths of 350 m.

Moving pavements are used at air terminals, railway stations and shopping centres; they can be used by the infirm, or by people with wheeled baskets or perambulators. Widths of the moving surface vary from 600 mm to 1 m and are flanked on both sides by balustrades incorporating handrails. The 600 mm width can carry 5000–6000 persons per hour.

Safety

The equipment is set in motion by a key-operated switch, and safety measures include a stop-push switch, fitted at each end of the pavement. A powerful electrical-mechanical brake is fitted to the driving mechanism, which stops the pavement in the event of an electrical fault or a power failure.

Chapter 12

Service ducts

Planning

Service ducts require careful planning and should be considered at an early stage in the design of a building. The purpose of a service duct is to conceal the services and to facilitate inspection, repair and alterations. A service duct also helps to reduce noise and protects the services from damage.

Entry into building

The point of entry of the services should, wherever possible, be on the side of the building nearest to the road in which the mains are installed. Information should be obtained from the local drainage, water, gas, electricity and telephone authorities, in order to ascertain the positions of the various mains relative to the building.

Pipe ducts should be laid during construction of the building for entry of the services. Pitch fibre, clay or plastic pipes are suitable materials for the pipe ducts. Figure 12.1(*a*) shows the method of installing a duct for small-diameter flexible pipes or cables, and Fig. 12.1(*b*) shows the method of installing a duct for larger pipes or cables or for smaller pipes of rigid material, such as galvanised steel. A pit is required which may have a brick lining, as shown.

The annular space between the duct and the pipe or cable should be sealed with a non-hardening plastic material to prevent the passage of water, gas or vermin.

Ducts for small pipes or cables

These may be formed in the floor or wall, or on the surface of the wall as shown in Figs. 12.2 and 12.3. The size of the ducts depends upon the outside diameter of the pipe or cable and the number of services installed. The outside diameter of any pipe or cable inside the small ducts shown, does not normally exceed 64 mm.

Large horizontal ducts

A floor trench with continuous covers may be used, as shown in Fig. 12.4. However, for larger pipes and cables it may be necessary to increase the depth of the trench and provide access openings at intervals, as shown in Fig. 12.5.

Floor trenches may be used internally at ground floor level or externally below ground. When used externally, the concrete cover slabs should be bedded and jointed together with lime mortar and should overlap the trench walls so that surface water may be directed clear of the wall.

For larger buildings, a crawlway or a sub-way may be required. Figure 12.6 shows a crawlway inside a building and Fig. 12.7 a crawlway in open ground. Crawlways provide easier means of access for inspection or repair of services than floor trenches.

Figure 12.8 shows a sub-way inside a building and Fig. 12.9 a sub-way in open ground. The internal height should be not less than 2 m to permit a man to walk through. Since the sub-way is used for very large pipes and cables, a pipe rack or strutted brackets are required. The main access to a sub-way will usually be gained from the boiler or plant room, but additional access points are usually required at changes of direction or junctions.

Construction

Floor trenches, crawlways and sub-ways should have concrete bases, but the sides may be concrete or brick. Expansion joints will be required on long runs, and in wet ground asphalt tanking may be necessary. The internal surfaces should be reasonably smooth and lime-washed. They should be ventilated preferably to the open air. The base should be laid to falls with a shallow channel provided to convey water, from whatever source, to a soakaway or drain. The connection to a soakaway or drain should be through a sealed gulley, to guard against the entry of vermin, sewer gas or back-flow. The position of the sealed gulley should be clearly marked on the side of the duct and the cover should only be removed when it is necessary to drain the duct. Wherever possible the ducts should be straight on plan to save the bending of the various pipes and cables.

Horizontal service ducts are often found in the voids of false ceilings. Figure 12.10 shows the method of installing services in the void of a false ceiling inside a corridor. It is essential to fill the holes around the various branch supplies when they pass through the walls. The filling must be of incombustible material for the full thickness of the wall.

Access to the services can be gained by removing the ceiling panels, which are either clipped or screwed in position. Figure 12.11 shows methods of supporting pipes. For heating pipes, a roller support may be required which provides for their movement due to expansion and contraction. Services may be installed inside the ceiling void of a ribbed, reinforced concrete floor and it is possible to segregate the various services as shown in Fig. 12.12.

Table 12.1 gives the distances between the pipe supports.

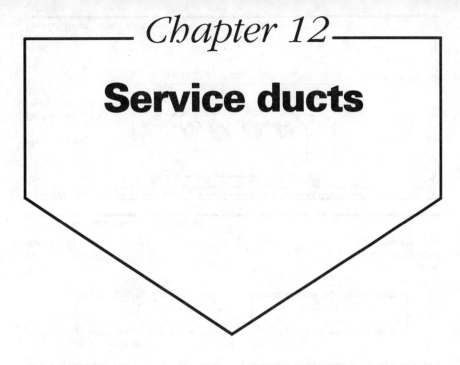

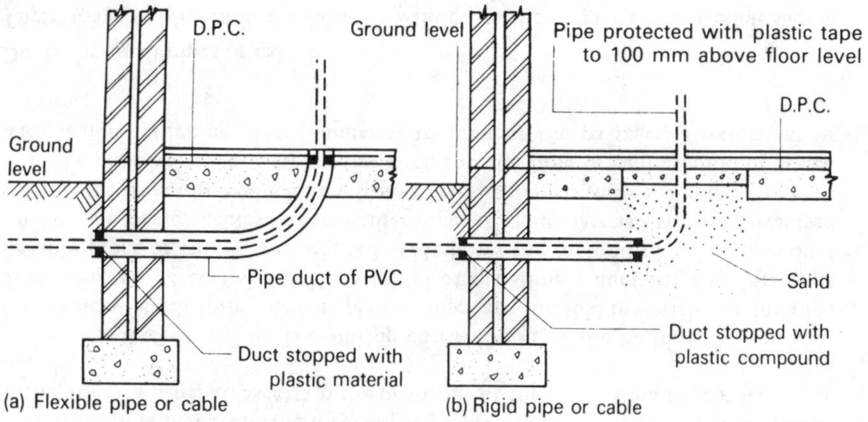
D.P.C.
Ground level
Pipe protected with plastic tape to 100 mm above floor level
D.P.C.
Ground level
Pipe duct of PVC
Sand
Duct stopped with plastic material
Duct stopped with plastic compound
(a) Flexible pipe or cable
(b) Rigid pipe or cable

Fig.12.1 Entry of services into building

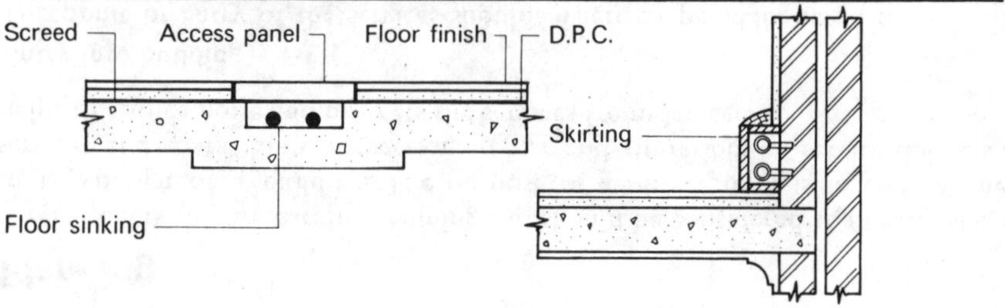
Screed
Access panel
Floor finish
D.P.C.
Skirting
Floor sinking

Fig.12.2 Horizontal ducts for small pipes

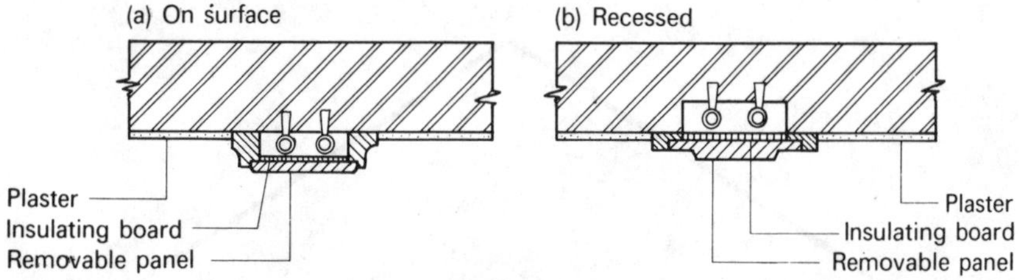
(a) On surface
(b) Recessed
Plaster
Insulating board
Removable panel
Plaster
Insulating board
Removable panel

Fig.12.3 Vertical ducts for small pipes

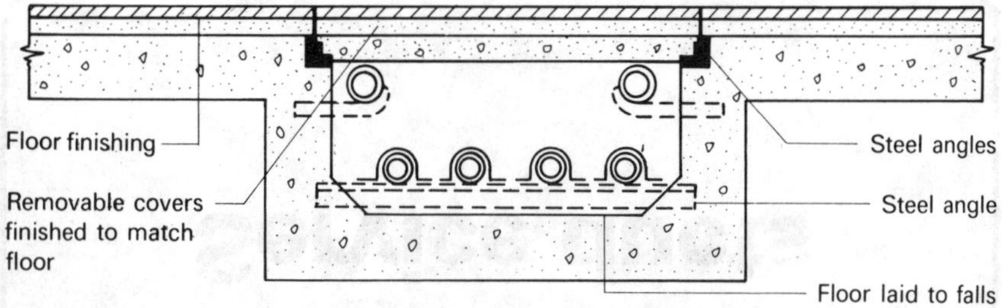
Floor finishing
Steel angles
Removable covers finished to match floor
Steel angle
Floor laid to falls

Fig.12.4 Floor trench with removable top for multiple pipes

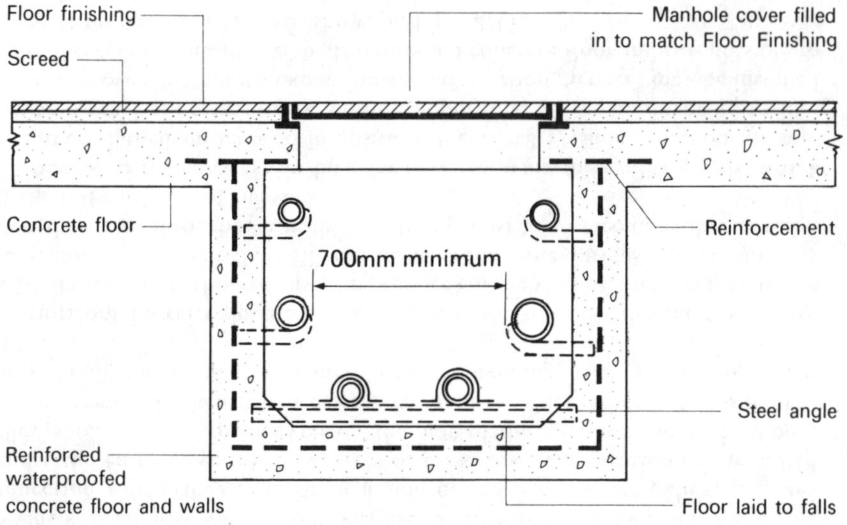
Floor finishing
Manhole cover filled in to match Floor Finishing
Screed
Concrete floor
Reinforcement
700mm minimum
Steel angle
Reinforced waterproofed concrete floor and walls
Floor laid to falls

Fig.12.5 Floor trench with access opening for multiple pipes

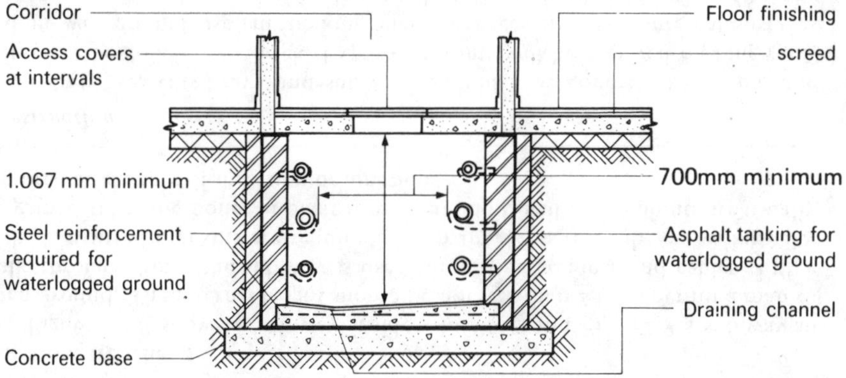
Corridor
Floor finishing
Access covers at intervals
screed
1.067 mm minimum
700mm minimum
Steel reinforcement required for waterlogged ground
Asphalt tanking for waterlogged ground
Draining channel
Concrete base

Fig.12.6 Crawlway inside a building

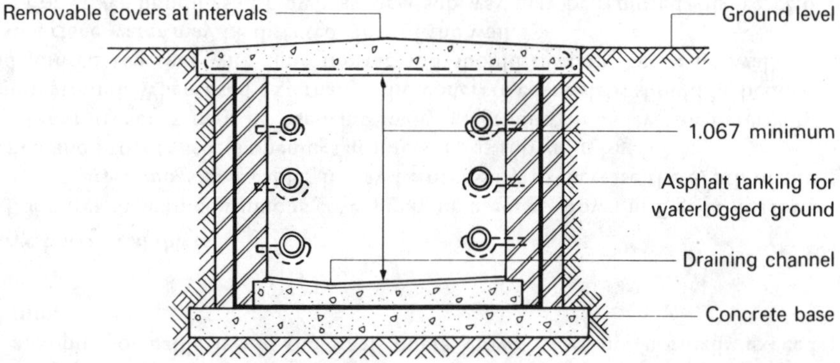
Removable covers at intervals
Ground level
1.067 minimum
Asphalt tanking for waterlogged ground
Draining channel
Concrete base

Fig.12.7 Crawlway in open ground

Table 12.1 Distances between pipe supports

Type of piping	Internal diameter of pipe (mm)	Interval of horizontal run (m)	Interval of vertical run (m)
Lead	All sizes	0.610	0.914
Copper (light gauge)	13	1.300	1.800
	19	1.800	2.400
	25	1.800	2.400
	32	2.400	3.000
	38	2.400	3.000
	51	2.700	3.000
	64	3.000	3.600
	76	3.000	3.600
	102	3.000	3.600
	152	3.600	4.200
Mild steel and heavy-gauge copper	13	1.800	2.400
	19	2.400	3.000
	25	2.400	3.000
	32	2.700	3.000
	38	3.000	3.600
	51	3.000	3.600
	64	3.600	4.500
	76	3.600	4.500
	102	4.000	4.500
	152	4.500	5.500
Cast or spun iron	51	1.800	1.800
	76	2.700	2.700
	102	2.700	2.700
	152	3.600	3.600
Polythene	Up to 25	12 x outside diameter	24 x outside diameter
	Over 25	8 x outside diameter	24 x outside diameter
PVC	19–25	0.914	1.800
	51–76	1.300	2.100
	89–127	1.500	2.400
	152	1.800	3.000

Large vertical ducts

These usually lead from the main horizontal ducts – such as floor trenches, crawlways or sub-ways – to the horizontal ducts on the various floors in the form of ceiling voids or floor ducts. Vertical ducts should extend throughout the full height of the building, the number required depending upon the practicability of accommodating the various services inside the same duct. The number and their spacing should allow for convenient lateral distribution of the services on each floor.

Figure 12.13 shows a recessed vertical duct and Fig. 12.14 a partly recessed vertical duct. Both these ducts are normally used for medium-sized pipes and cables up to 100 mm in diameter.

Vertical ducts for large buildings require space to house pipes and cables up to 150 mm in diameter. Figure 12.15 shows a built-out duct for large buildings with a separate duct for the electrical cables, and Fig. 12.16 a walk-in duct for large buildings.

If maintenance work on services has to be carried out without means of access from the corridors, a walk-in type of duct extending the full height of the building – provided with ladders and with access from the basement or the ground floor – will be required. Openings through the fire barriers at each floor level must be provided with a fire-resisting door which will close automatically in the event of a fire. However, it should be possible for any workman inside the duct to open the doors in order to escape through an access door on the upper or ground floor.

Fire risk

The risk of smoke or fire spread in a building may be influenced by the formation of service ducts and it is therefore essential that the following precautions are taken:

1. All service ducts, with the exception of casings and chases of small dimensions, should be constructed wholly of incombustible material such as brick, concrete or insulation board. The duct must have a fire resistance of an hour or more.
2. There must be no openings in the duct other than for any one or more of the following:
   (*a*) a hole for a pipe which shall be as small as possible and fire-stopped around the pipe;
   (*b*) an opening fitted with a door which has a fire resistance of not less than half an hour. The door must be fitted with an automatic self-closing device which operates in the presence of smoke.
3. Where ducts pass through walls, floors or partitions, the space around the duct must be filled in to the full thickness of the structure with non-combustible material, so that the fire resistance at the point at which the duct passes through it is not less than that of the surrounding structure.
4. Suitable fire-resisting barriers, fillings or doors should be provided inside the ducts where these pass through walls, floors or partitions, so as to reduce the risk of spread of fire or smoke through the building. The barriers, fillings or doors should consist of non-corrosive material having a fire resistance of not less than that of the floor, wall or partition through which the duct passes. The barriers or fillings should be constructed with due consideration given to the possibility of having to remove and replace them for extensions or repairs to the services.
5. Ducts which house services from a boiler room should be sealed at the boiler-room end.
6. If the service duct serves as, or contains, a ventilating duct, the duct must be fitted internally with automatic fire shutters at such intervals and in such

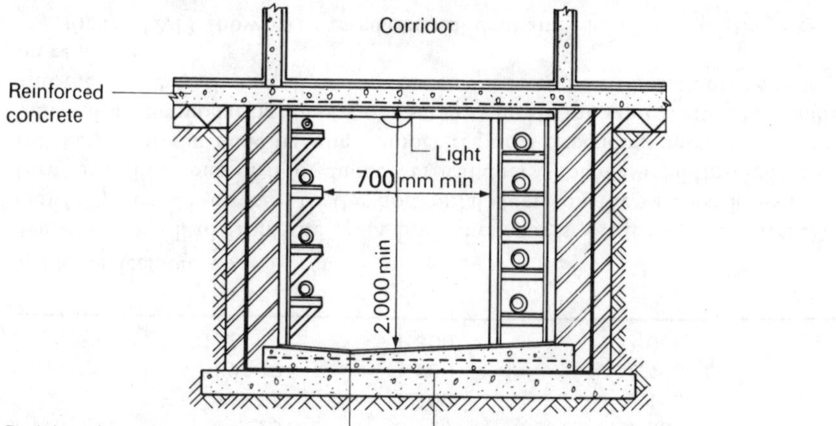
Corridor
Reinforced concrete
Light
700 mm min
2.000 min
Draining channel
Asphalt tanking for waterlogged ground

Fig.12.8 Subway inside building

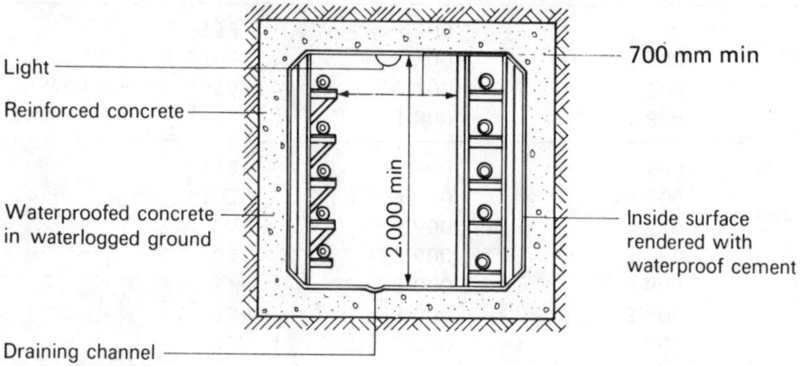
Light
700 mm min
Reinforced concrete
2.000 min
Waterproofed concrete in waterlogged ground
Inside surface rendered with waterproof cement
Draining channel

Fig.12.9 Subway in open ground

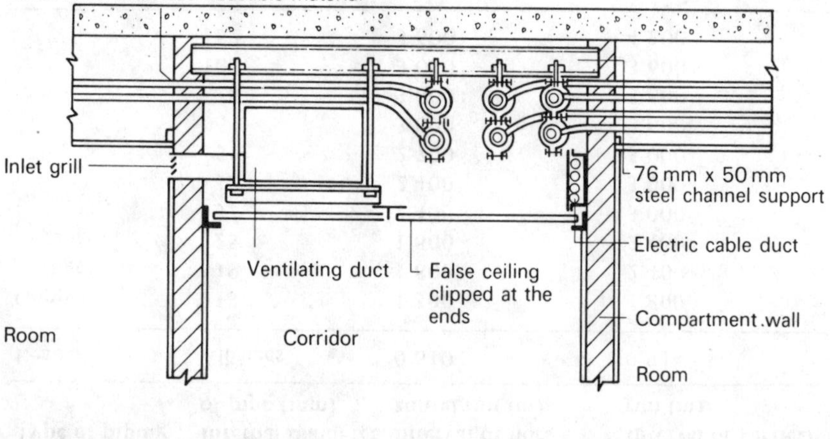
Holes filled with incombustible material
Inlet grill
76 mm x 50 mm steel channel support
Electric cable duct
Ventilating duct
False ceiling clipped at the ends
Compartment wall
Room
Corridor
Room

Fig.12.10 Installation of services in ceiling void

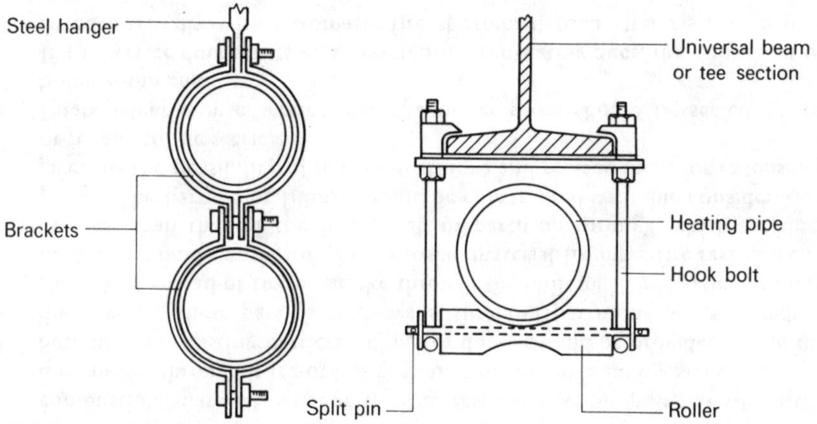
Steel hanger
Universal beam or tee section
Brackets
Heating pipe
Hook bolt
Split pin
Roller

Fig.12.11 Method of supporting pipes

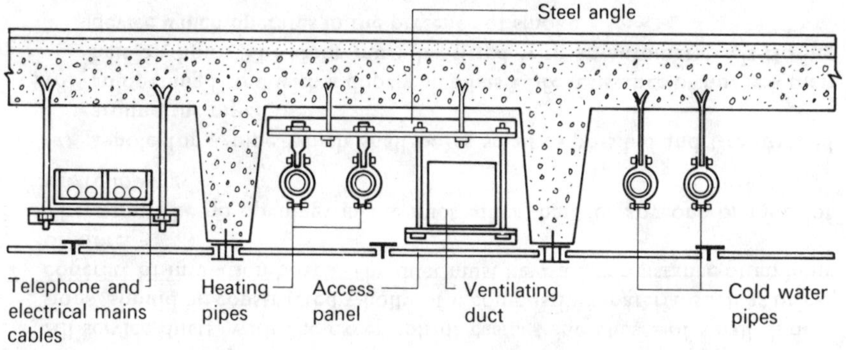
Steel angle
Telephone and electrical mains cables
Heating pipes
Access panel
Ventilating duct
Cold water pipes

Fig.12.12 Services installed in ceiling void of ribbed reinforced concrete floor

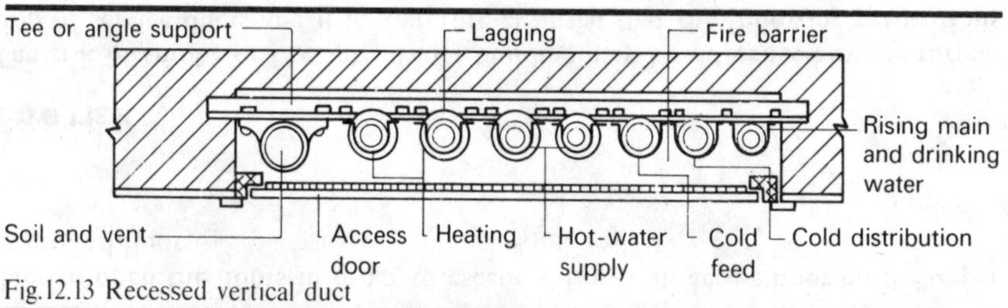
Tee or angle support
Lagging
Fire barrier
Rising main and drinking water
Soil and vent
Access door
Heating
Hot-water supply
Cold feed
Cold distribution

Fig.12.13 Recessed vertical duct

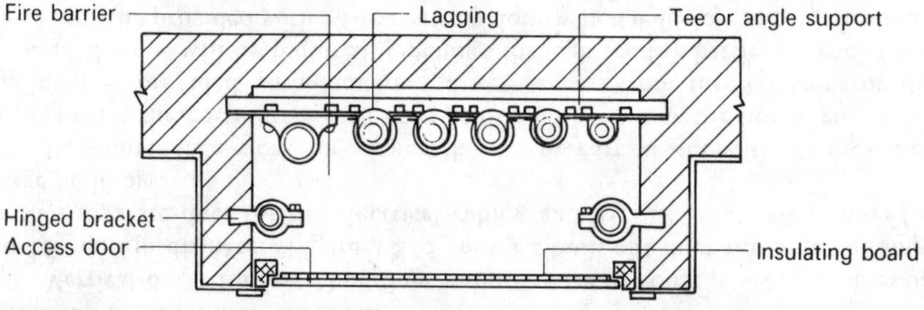
Fire barrier
Lagging
Tee or angle support
Hinged bracket
Access door
Insulating board

Fig.12.14 Partly recessed vertical duct

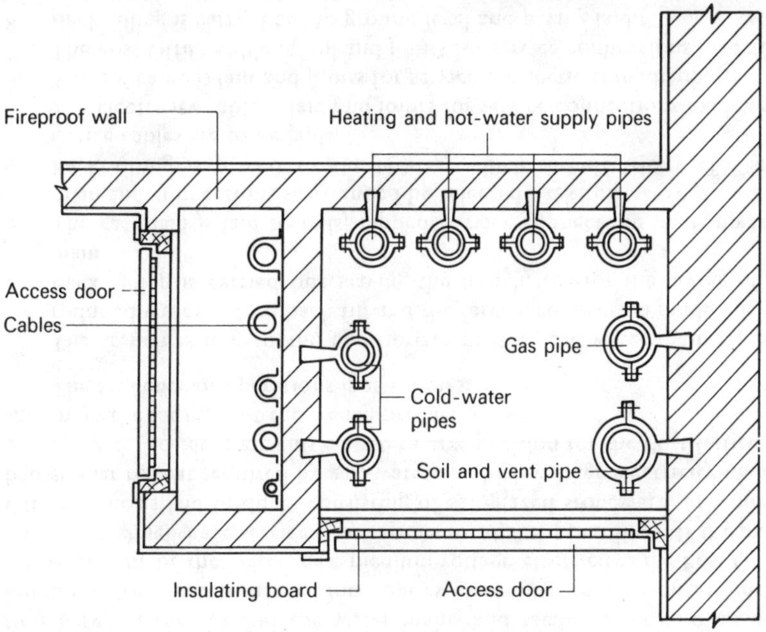
Fireproof wall
Heating and hot-water supply pipes
Access door
Cables
Gas pipe
Cold-water
pipes
Soil and vent pipe
Insulating board
Access door

Fig.12.15 Built-out vertical duct for large buildings with separate duct for electric cables

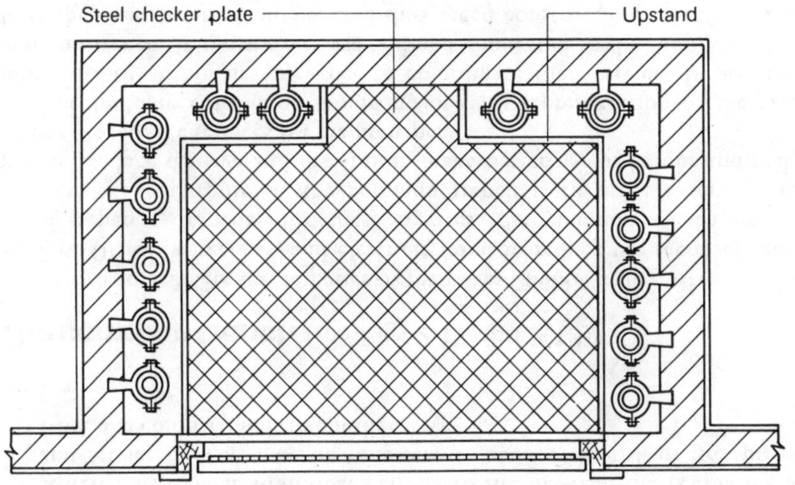
Steel checker plate
Upstand

Fig.12.16 Walk-in vertical duct for large buildings

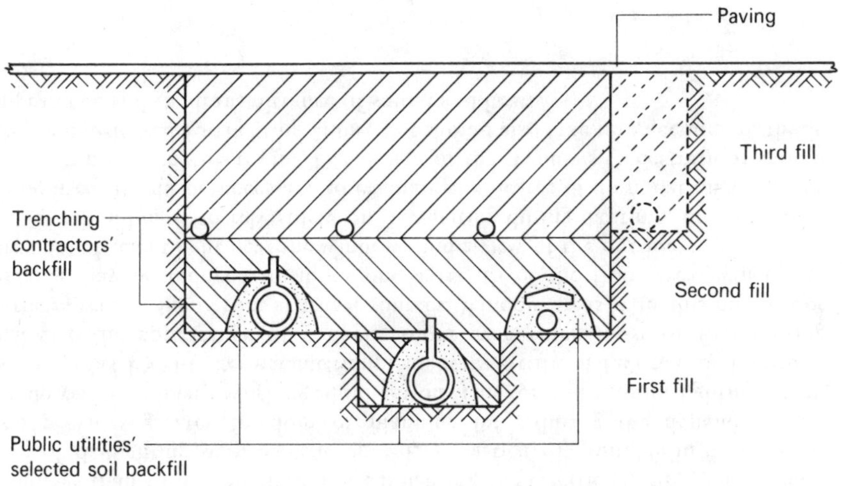
Paving
Third fill
Trenching
contractors'
backfill
Second fill
First fill
Public utilities'
selected soil backfill

Fig.12.17 Cross section through common trench showing sequence of back filling

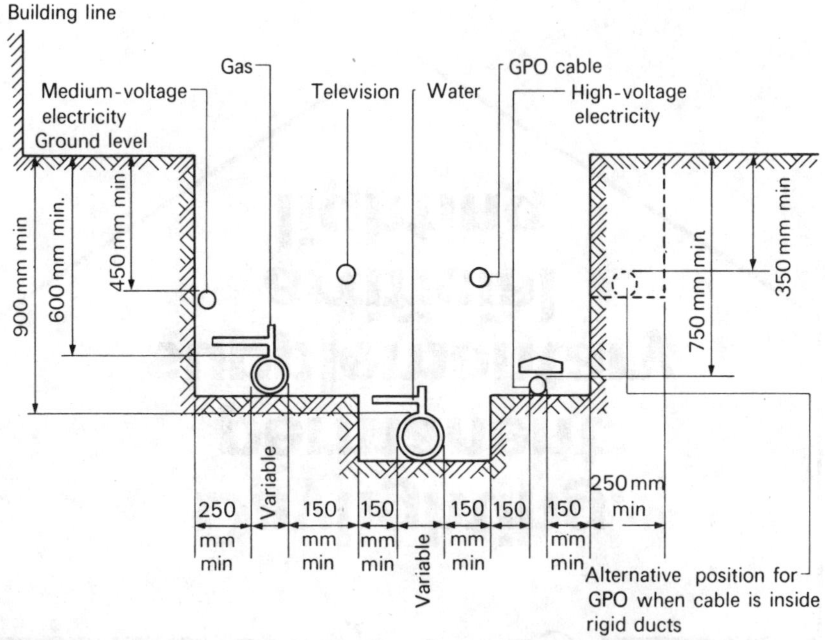
Building line
Gas
GPO cable
Medium-voltage
electricity
Television
Water
High-voltage
electricity
Ground level
900 mm min
600 mm min.
450 mm min
750 mm min
350 mm min
250
mm
min
Variable
150
mm
min
150
mm
min
Variable
150
mm
min
150
150
mm
min
250 mm
min
Alternative position for
GPO when cable is inside
rigid ducts

Fig.12.18 Cross section through common trench showing positions of services.

positions as may be necessary to reduce as far as is practicable the risk of fire spreading from a compartment to any other compartment.

The ventilating duct must not be constructed of, or lined with, any material which substantially increases the risk of fire spreading from a compartment to any other compartment. The service duct must also be constructed with such additional barriers to fire between the ventilating duct and service duct as may be necessary to reduce the risk of fire spreading from one compartment to another.

The common trench (see Figs 2.17 and 2.18)

In order to co-ordinate the activities of the various public utility services, such as the Post Office, water, gas, and electricity, and to reduce the inconvenience of having separate trenches for these services, the common trench has been developed. The purpose of the common trench is to provide services more quickly, reduce damage and disruption, make better use of labour and make it easier to locate the services for maintenance.

The building contractor should make early representation to the various public utility authorities in order to co-ordinate their services by use of the common trench. If the services are laid and connected to the supply at an early date, they can be used on site and thus speed construction. Once the services have been laid and the trench back filled, it is easier for the contractor to plan for the storage of materials and erection of site huts without the problem of leaving space for the laying of the services at a later date.

Trench profile

A stepped trench is recommended, as this gives vertical and horizontal segregation between the gas and the water mains and facilitates cross-overs at major junctions. The basic cross-section applies essentially to distribution mains and takes account of the water, gas, medium-voltage electricity, the Post Office and TV, all of which have service connections to adjacent buildings. If the main Post Office cable is laid in ducts consisting of salt-glazed stoneware, an undisturbed bed similar to that required by gas, water and high-voltage electricity mains must be provided. To facilitate this, an alternative position for the Post Office cable is shown on the far right of the trench profile.

The sequence of operations is as follows:

1. The trench is dug to the appropriate profile, the water main tapped, surrounded with selected back fill and the service connection made up.
2. Back filling is carried out leaving the trench free for the laying of the gas main.
3. The gas main is laid to falls, tapped, service connections made up and both main and connections surrounded by selected back fill.
4. Back filling is placed to the level at which the electricity, TV and Post Office cables are to be laid.
5. The electricity cable is laid and joints for service connections made up.
6. The TV cable is laid and joints for service connections made up.
7. The Post Office cable is laid and joints for service connections made up.
8. Back filling is carried out to ground level and paving laid.

Chapter 13

Daylighting, permanent supplementary artificial lighting

Daylighting

There is a tendency to give less attention to the daylighting of rooms in modern buildings than to a combination of daylighting and electric lighting. The integration of daylighting with electric lighting is particularly suitable in deep rooms and provides greater freedom of planning by reducing the dependence on windows for ordinary working illumination. This form of interior lighting which is known as PSALI (Permanent Supplementary Artificial Lighting of Interiors) will give the appearance of a daylit room even when most of the working surfaces receive light from artificial sources. Window areas with this method of lighting may be reduced to about one-sixteenth of the floor area, which will provide sufficient light near the windows and a view of the outside.

If a building is to rely solely on daylighting during daytime, or if PSALI is to be used, it will be necessary to evaluate the degree of illumination at various strategic points in the room. The design of the windows (fenestration) should not be considered for lighting alone, but should also take into account aesthetic appearance and the transmittance of sound and heat.

Daylight factor

The daylight received inside a building can be conveniently expressed as a ratio of the total daylight simultaneously available out of doors from the whole unobstructed sky. The ratio is usually expressed as a percentage and is known as the daylight factor.

$$\text{Daylight factor (percentage)} = \frac{\text{Instantaneous illumination indoors}}{\text{Simultaneously occurring illumination outdoors}} \times 100$$

The daylight factor for any given situation, depends upon the following:

1. The sky conditions.
2. The size, shape and position of the windows.
3. The effect of obstructions.
4. The reflection of light from external and internal surfaces.

The factor includes the light received in the building from the following sources:

Sky component

This is the ratio of that part of the daylight illumination at a point on a given plane, which is received directly from the sky of assumed or known luminance distribution, to the illumination on a horizontal plane due to an unobstructed hemisphere of this sky. Direct sunlight is excluded for both values of illumination.

Externally reflected component

This is the ratio of that part of the daylight illumination at a point on a given plane, which is received directly or indirectly by a sky of assumed or known luminance distribution, to the illumination on a horizontal plane due to an unobstructed hemisphere of this sky. Contributions of direct sunlight to the luminances of external reflecting surfaces and to the illumination of the comparison plane are excluded.

Internally reflected component

This is the ratio of that part of the daylight at a point on a given plane, which is received from internal reflecting surfaces (the sky being of assumed or known luminance distribution) to the illumination on a horizontal plane due to an unobstructed hemisphere of this sky. Contributions of direct sunlight to the luminance of internal reflecting surfaces and to the illumination of the comparison plane are excluded.

Figure 13.1 illustrates the three components that make up the daylight factor.

Amounts of daylight

The amount of light required for a given visual task depends upon the size of the relevant detail and the brightness contrast between the detail and its background. The ease and safety of performing the visual task increases with more illumination. The length of time during which keen concentration is required must be taken into account and rooms occupied by older people, or people with defective vision, require more daylight. The recommended daylight factors for various buildings is given in Table 13.1.

Table 13.1 (BSCP 31A: 1964)

Building type	Recommended daylight factor (%)	Recommendations
Dwellings:		
Kitchen	2	Over at least 50% of floor area (min. approx. 4.5 m^2)
Living room	1	Over at least 50% of floor area (min. approx. 7.0 m^2)
Bedroom	0.5	Over at least 75% of floor area (min. approx. 5.5 m^2)
Schools	2	Over all teaching areas and kitchens
Hospitals	1	Over all ward areas
Offices:		
General	1	With side lighting at a penetration of approx. 3.75 m^2
General	2	With top light, over whole area
Drawing	6	On drawing boards
Drawing	2	Over remainder of working area
Typing and computing	4	Over whole working area
Laboratories	3–6	Depending on dominance of side or top lighting
Factories	5	General recommendation
Art galleries	6	Maximum on walls or screens where no special problems of fading
Churches	1	General, over whole area
Churches	1.5–2	In sanctuary area
Public buildings	1	Depending upon function, the recommendation may exceed 1%, but a minimum value of 1% is generally desirable

Flow of light in buildings

The direction of light (its vector quality) will appear to change as light moves across the room (see Fig. 13.2). The light received at the point furthest from the window relies more on internal and external reflected light than upon direct daylight, resulting in a nearer horizontal direction of light.

Distribution of daylight

The distribution of daylight in a room is the variation of the daylight factor between one point and another on the surfaces of that room. The distribution of daylight in an interior can be shown conveniently on plan by means of equal daylight-factor contours, and in vertical section by drawing a curve showing the values of the corresponding daylight factors at the appropriate reference points. The pattern of the contours and the shape of the curves will indicate the nature of the distribution.

Figure 13.3 shows a plan of a room with contours of equal daylight factor and Fig. 13.4 shows the effect of window design on the daylight factor.

The CIE standard overcast sky

In countries where clear skies are predominant, the direct rays from the sun may

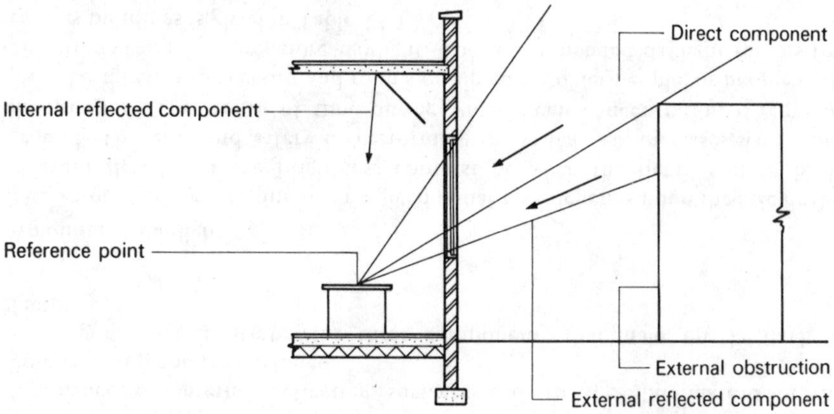
Direct component
Internal reflected component
Reference point
External obstruction
External reflected component

Fig.13.1 Components of the daylight factor

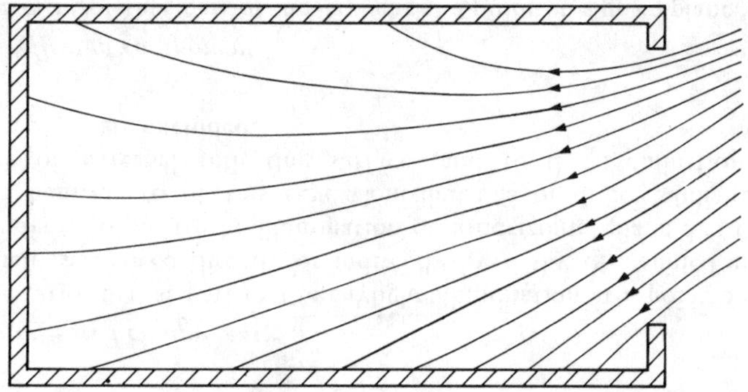

Fig.13.2 Vertical section of a room showing vectorial flow of light due to daylight

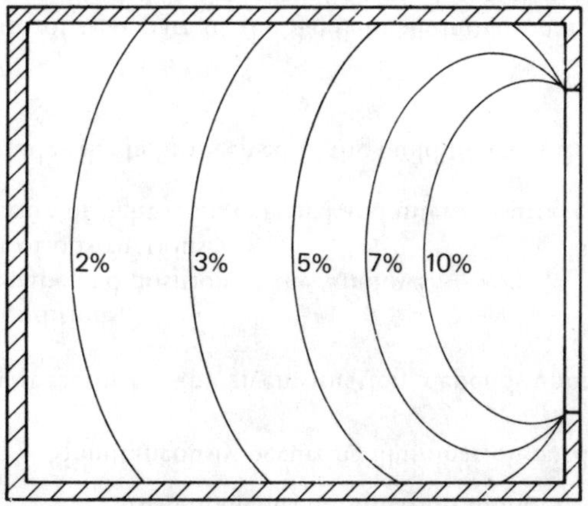
2%
3%
5%
7%
10%

Fig.13.3 Plan of a room showing contours of equal daylight factor

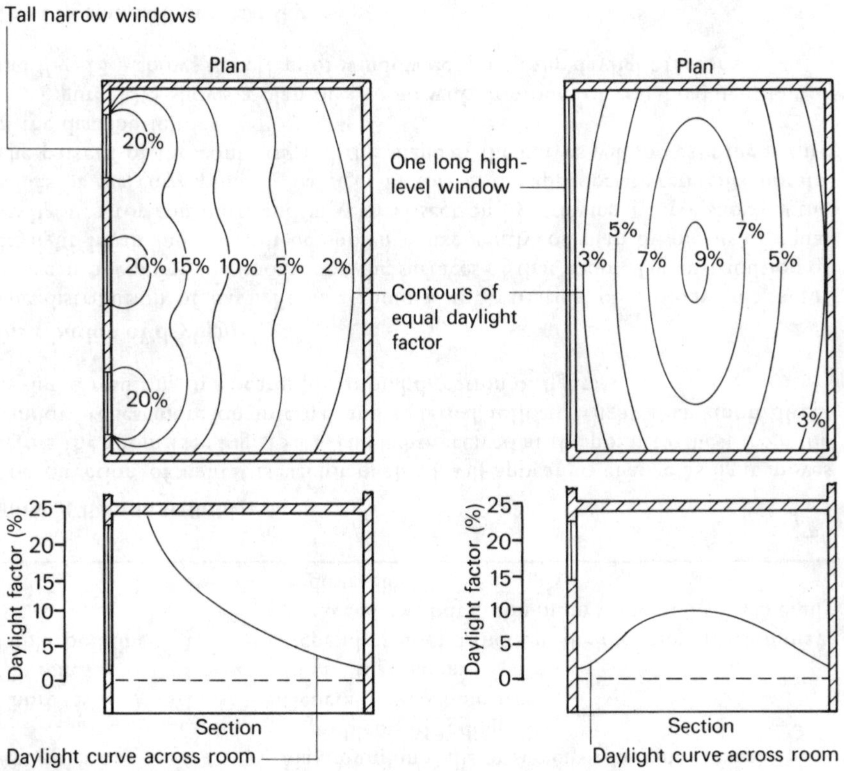
Tall narrow windows
Plan
Plan
20%
One long high-level window
5%
7%
20% 15% 10% 5% 2%
3% 7% 9% 5%
Contours of equal daylight factor
20%
3%
Daylight factor (%)
25
20
15
10
5
0
Daylight factor (%)
25
20
15
10
5
0
Section
Section
Daylight curve across room
Daylight curve across room

Fig.13.4 The effect of window design on daylight factor

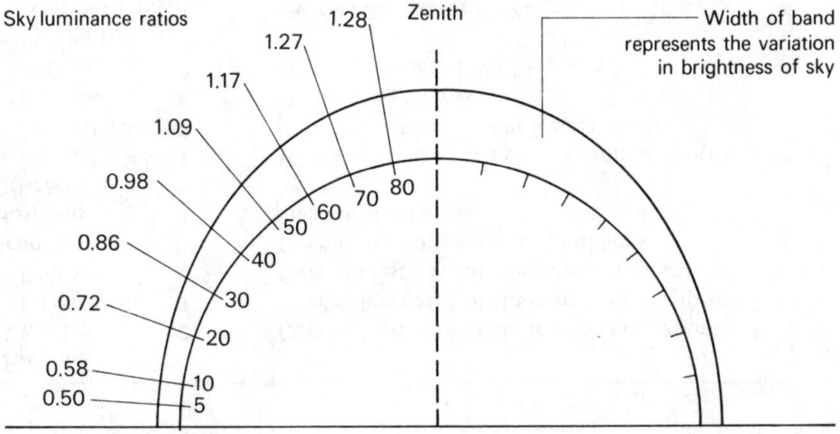
Sky luminance ratios
Zenith
Width of band represents the variation in brightness of sky
1.28
1.27
1.17
1.09
0.98
0.86
0.72
0.58
0.50
80
70
60
50
40
30
20
10
5

Fig.13.5 Luminance distribution of densely overcast sky compared with average luminance taken as unity

be taken into account for daylighting analysis. In Britain and many other countries the sky is often covered with clouds and the daylighting varies greatly from day to day. For daylighting analysis, therefore, these countries use the Commission Internationale de l'Enclairage standard overcast sky, which is a completely overcast sky for which the ratio of its luminance at an altitude θ above the horizon at the zenith is assumed to be

$$\frac{1 + 2 \sin \theta}{3}$$

The total unobstructed illumination on a horizontal plane at ground level using the CIE standard overcast sky is in the region of 5000 lx (lux), but this value is exceeded for about 85 per cent of the normal working time throughout the year. The luminance of the sky varies regularly from the horizon to the zenith, and is about three times as bright at the zenith as at the horizon (see Fig. 13.5).

Evaluation of the daylight factor

The daylight factor can be predicted at the design stage either by actually measuring the daylight received in a model of the proposed building, or more frequently by calculation using drawings and other necessary data.

BRE daylight protractors

These comprise of a set of circular protractors which compute the sky component and enable the external reflected component to be found. They are designed for use with either a uniform sky or a CIE standard overcast sky, and for various types of windows.

The use of protractors is given in Table 13.2.

Table 13.2 Conditions of application of BRE daylight protractors

Protractor	Type of sky	Slope of glazing
1	Uniform	Vertical
2	CIE	Vertical
3	Uniform	Horizontal
4	CIE	Horizontal
5	Uniform	30°
6	CIE	30°
7	Uniform	60°
8	CIE	60°
9	Uniform	Unglazed
10	CIE	Unglazed

The protractors are designed for use with scaled drawings of the proposed buildings. Care should be taken that the drawings show the thickness of window walls and include internal and external projections such as sills, blind boxes and overhangs.

A large-scale drawing helps in the accurate location of sight lines to the operative edges of the window openings. However, in some cases it is possible to use a smaller scale of the entire room from which to obtain a first estimate of the probable daylighting levels, and a large-scale drawing of the windows from which to assess the effective area of the glazing.

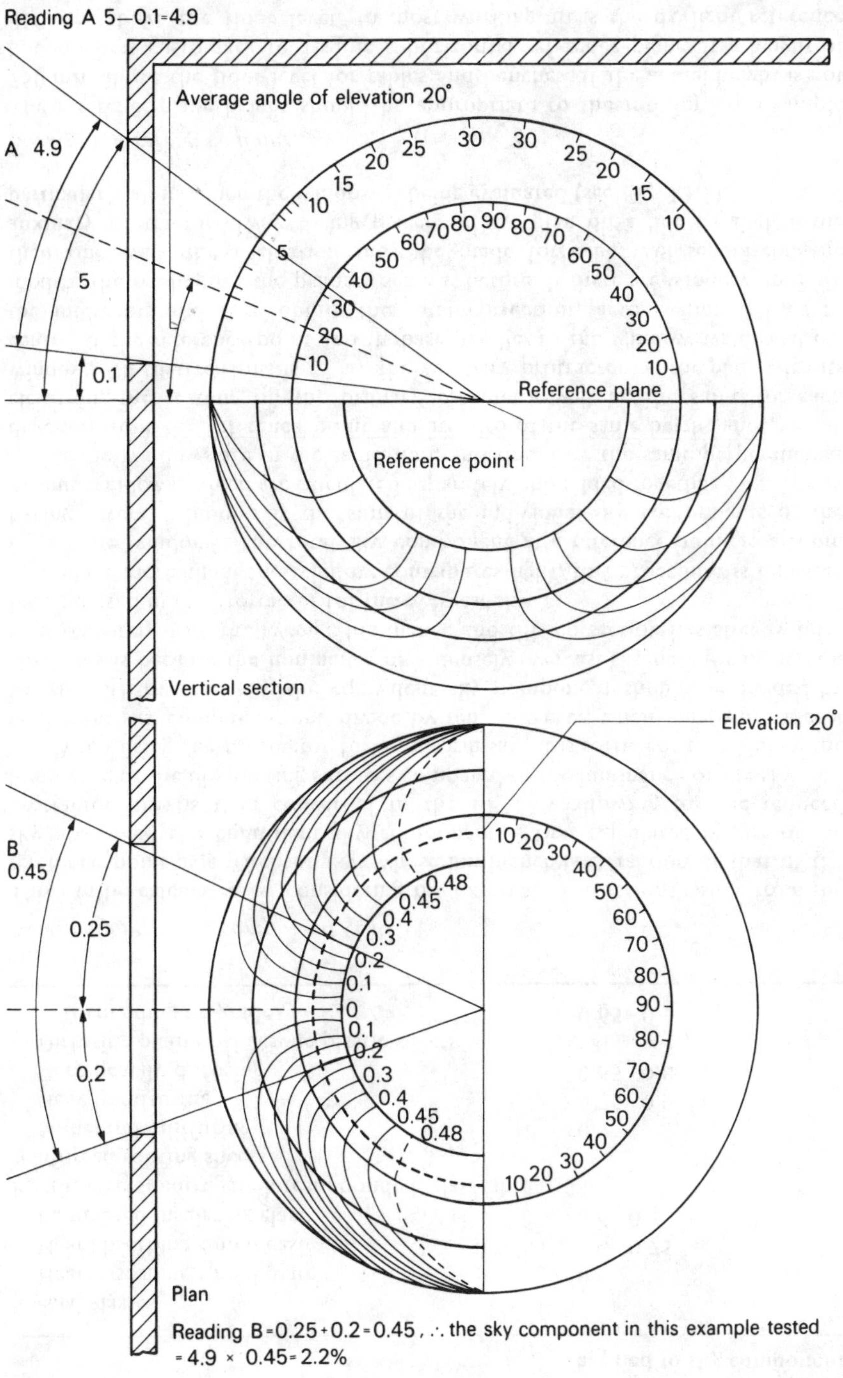

Reading A 5—0.1=4.9
Average angle of elevation 20°
A 4.9
5
0.1
Reference plane
Reference point
Vertical section
Elevation 20°
B
0.45
0.25
0.2
Plan
Reading B=0.25+0.2=0.45, ∴ the sky component in this example tested =4.9 × 0.45=2.2%

Fig.13.6 Use of BRE daylight protractor

Using the protractor

First, sight lines must be drawn from the reference point in the interior to the edges of the windows or the visible sky, as shown in Fig. 13.6.

The centre of the appropriate protractor – and for vertical glazing, protractor No. 2 for CIE overcast sky is used – is laid over the reference point on the sectional drawing with the centre line of the protractor in the reference plane.

The intercepts of the sight lines to the upper and lower edges of the window, or obstructions, are read off on the outer sky component scale. This gives the sky component for an infinitely long window and must be corrected as follows.

The protractor is laid on the plan drawing with its centre at the reference point, with its centre line parallel to the window and with the correction factor scales in the lower half of the protractor directed towards the window. A semi-circle on the correction factor scale corresponding to the average angle of elevation of the window is estimated, and the intercepts with the sight lines to the edges of the window, or obstructions, are read off on the curves.

The value of the sky component, for the infinitely long window (found previously) is then multiplied by this correction factor to obtain the corrected sky component for the window of infinite length.

Note: The protractors incorporate glazing losses corresponding to a single sheet of ordinary clear glass. An additional correction is necessary if more than one sheet of glass is used, if some form of glazing other than clear sheet or plate glass is used or if the glazing is expected to become dirty.

It is usually sufficient to reduce the sky component by a factor to allow for the effects of dirt or patterned non-diffusing glass, but special glasses such as diffusing glass or prismatic glass require special treatment.

Table 13.3 gives the conversion factors to allow for the reduced light transmission of some typical glazing materials.

Table 13.3

Glazing material	Conversion factor to be applied to sky component
Transparent glasses	
Flat drawn sheet	1.0
Polished plate	1.0
Polished wired	0.95
Patterned and diffusing glasses	
Rolled	0.95
Rough cast	0.95
Wired cast	0.9
Cathedral	1.0
Hammered	1.0
Arctic	0.95
Reeded	0.95
Small morocco	0.9

Table 13.3 continued on p. 136

Table 13.3 – *continued*

Glazing material	Conversion factor to be applied to sky component
Special glasses	
Heat-absorbing tinted plate	0.9
Heat-absorbing tinted cast	0.6–0.75
Laminated insulating glass	0.6–0.7
Plastic sheets: corrugated, resin-bonded, glass-fibre reinforced roofing sheets	
Moderately diffusing	0.9
Heavily diffusing	0.75–0.9
Very heavily diffusing	0.65–0.8
Diffusing plain opal acrylic plastic sheets (depending on grade)	0.65–0.9

Externally reflected component (ERC)

This can be calculated by considering the external obstructions visible from the reference point as a patch of sky, whose luminance is a fraction of that of the sky obscured. The equivalent sky component is first calculated by use of the protractor; this is then converted to the ERC by allowing for the reduced luminance of the obstructing surfaces compared to the luminance of the sky.

When using the protractor for a uniform sky, it is sufficient to calculate the equivalent sky component and divide by ten. However, when using a protractor for the CIE overcast sky the equivalent sky component should be divided by five. This is because the luminance of a densely overcast sky near the horizon is approximately half the average luminance and this correction has already been incorporated in the protractor for the overcast sky.

When more than one window contributes light, this procedure is repeated for all the windows. The total sky component at a reference point in a room having several windows is the sum of the individual sky components of the various windows, which are calculated separately and added together.

If the windows are in the same wall, and therefore the same perpendicular distance from the reference point and are also of the same height and average elevation, the reading on the primary protractor will be the same for each window. All that remains is to lay the auxiliary protractor on the plan, with its centre at the reference point and its base parallel to the window wall, read off the intercepts and then obtain the total correction factor which is used to modify the reading on the primary scale as before. If there are windows in more than one wall, the evaluation must be made for each wall separately, the auxiliary protractor always being placed on the plan on a line parallel to the particular wall in which the window is being evaluated (see Fig. 13.7).

Location of reference point

The reference-plane height should be appropriate to the interior, for example 750 mm above the floor level for tables and benches. If the actual height is not known, it is customary to assume a horizontal reference plane at a height of 850 mm above the floor level. In most working areas the daylight reference

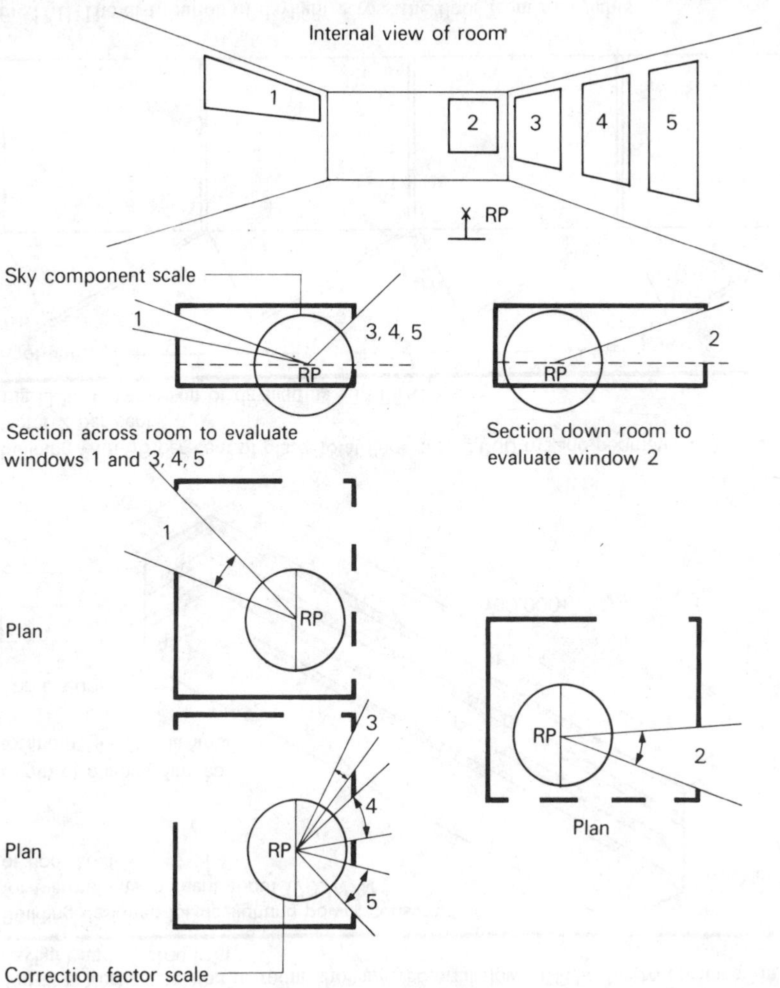
Internal view of room
1
2
3
4
5
RP
Sky component scale
1
3, 4, 5
RP
RP
2
Section across room to evaluate windows 1 and 3, 4, 5
Section down room to evaluate window 2
1
Plan
RP
RP
2
Plan
3
4
Plan
RP
5
Correction factor scale

Fig.13.7 Use of BRE daylight protractor to evaluate sky component for several windows

points all lie on a single horizontal plane, such as the plane of desks in classrooms or offices, or work benches or machines in factories.

For an example in evaluating the sky component for a room having several windows, Fig. 13.7 shows a room with windows in three walls:

Window 1. A high clerestorey window in one wall.
Window 2. A window in an adjacent wall, the sill of which is the reference plane.
Windows 3, 4 and 5. Windows in the opposite wall of the same height as window 2.

The procedure for evaluating the total sky component at reference point (RP) is as follows:

1. Draw sight lines on the section of the room through windows 1 and 3 (or 4 or 5) from RP.
2. Place protractor No. 2 with its centre at the RP. Window 1 is then evaluated on the primary scale and the value of the primary sky component noted. Window 3 (or 4 or 5) is also evaluated with the protractor in the same position and the value of the primary sky component noted.
3. Note the average angles of elevation of window 1 and window 3.
4. Draw sight lines on the section of the room through window 2.
5. Use protractor No. 2 to evaluate window 2 as above and note the value of the primary sky component.

   Note: Because the perpendicular distances from the reference point to windows 2 and 3 are different, the evaluation obtained for window 3 will not suffice for window 2, even though the two windows have the same sill and head height.

6. Note the average angle of elevation of window 2.
7. Draw sight lines, on the plan of the room, from the RP to each window.
8. Place protractor No. 2 with its centre line parallel to window 1 and with the auxiliary scales directed towards window 1. Obtain the correction factor for window 1 and multiply the primary sky component for window 1 by this value.
9. Rotate the protractor so that its centre line is parallel to windows 3, 4 and 5 and the auxiliary protractor directed towards these windows. Obtain the individual correction factors for windows 3, 4 and 5, add these together and multiply the primary sky component for windows 3 (or 4 or 5) by this value.
10. Rotate the protractor so that its centre line is parallel to window 2 and the auxiliary scales directed towards window 2. Obtain the correction factor for window 2, ensuring that the correct angle of elevation is used, and multiply the primary sky component for window 2 by this value.
11. Summate the three sky component values obtained above for the three window walls to determine the total sky component at the RP.

Glare from daylight

In general, glare will be experienced when parts of a room are excessively bright in relation to the remainder. If a room is uniformly bright, the eye adapts to the general level of brightness. Glare can be described in two ways: disability glare which reduces visual efficiency, and discomfort glare which causes physical discomfort.

A typical case of disability glare occurs when objects in a room are seen silhouetted against or close to a window giving a view of an area of bright sky. The sensation of discomfort glare, unlike disability glare which operates at once, only becomes noticeable after a period of sustained, exacting work. The sensation of discomfort depends principally on the relative brightness and the apparent size and position of the brightest surfaces in the field of view.

Glare control may not be required continuously and glare may only occur temporarily, for example on bright days in summer or during late afternoons in rooms facing west. In such cases, temporary control by use of blinds or curtains may be used and, where possible, working positions should be arranged so that unscreened windows are not in the line of sight of occupants when attention is directed to the task.

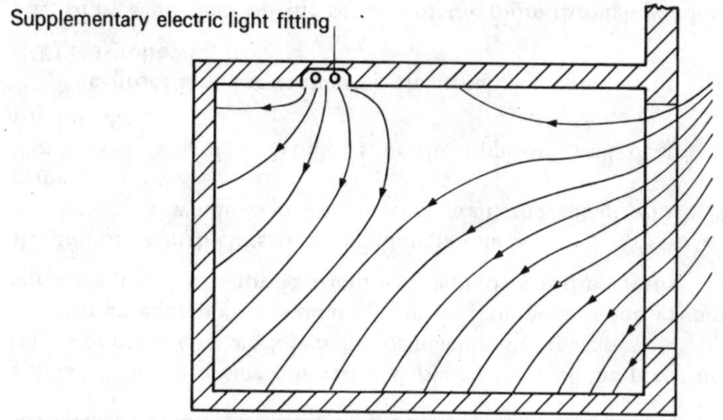
Supplementary electric light fitting

Fig.13.8 Vertical section of room showing vectorial flow of light due to combination of daylight and electric light

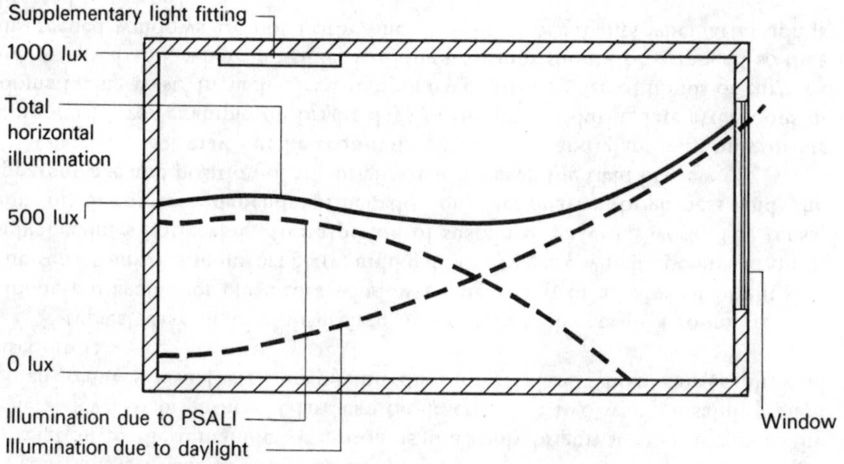
Supplementary light fitting
1000 lux
Total
horizontal
illumination
500 lux
0 lux
Illumination due to PSALI
Illumination due to daylight
Window

Fig.13.9 Effect of PSALI in a room

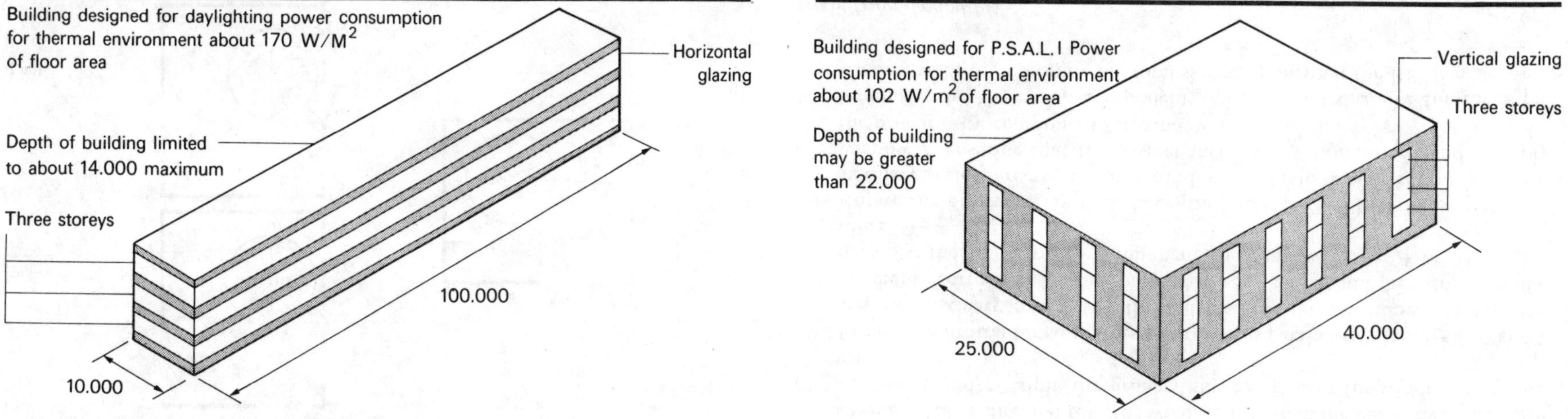
Building designed for daylighting power consumption for thermal environment about 170 W/M²
of floor area
Horizontal
glazing
Depth of building limited
to about 14.000 maximum
Three storeys
100.000
10.000
Building designed for P.S.A.L.I Power consumption for thermal environment about 102 W/m² of floor area
Vertical glazing
Three storeys
Depth of building
may be greater
than 22.000
25.000
40.000

Building with 66 per cent of glass, total floor area 3.000 m^2 and perimeter 220m, daylight factor 2 per cent

Building with 20 per cent of glass, total floor area 3000 m^2 and perimeter 130 m illumination level 1000 lux

Fig.13.10 Comparison of daylighting and PSALI

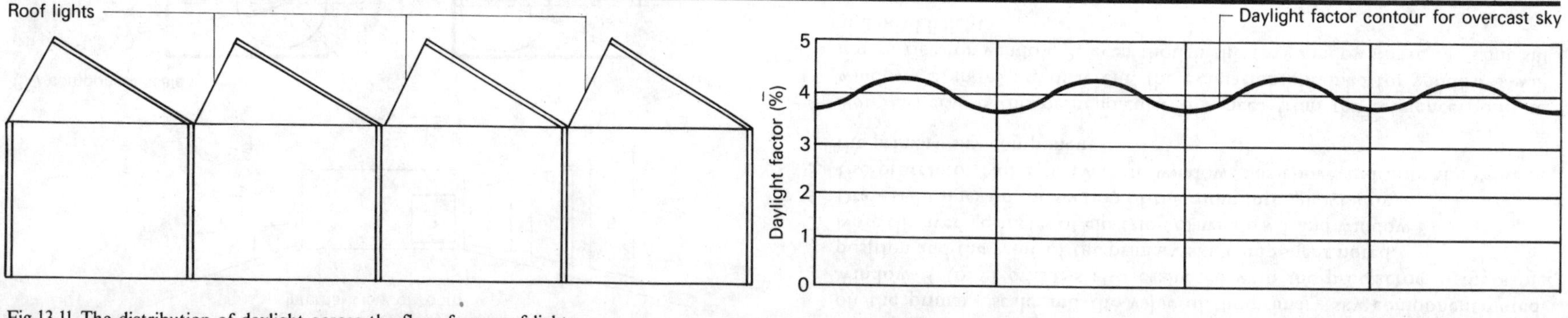
Roof lights
Daylight factor contour for overcast sky
Daylight factor (%)
5
4
3
2
1
0

Fig.13.11 The distribution of daylight across the floor from roof lights

The internally reflected component

Having calculated the amount of light reaching a reference point – first from the sky directly, and secondly the reflection from surfaces outside the window – the next stage is to calculate the light reaching the reference point after inter-reflection and reflection from surfaces inside the room.

Three methods of calculation have been devised which are sufficiently accurate for general use:

1. The inter-reflection formula.
2. Tables based on the formula.
3. Nomograms based on a modified form of the formula.

Table 13.4 gives the minimum internally reflected component of the daylight factor. The table was formulated by the BRE to allow a rapid assessment of the minimum internally reflected component of the daylight factor where certain limitations on the size of room and scheme of decoration can be accepted. The table was designed primarily for rooms about 40 m^2 in floor area and 3 m ceiling height, with a window on one side extending from a sill height of 900 mm to the ceiling. However, by means of simple conversion factors the internally reflected component can be calculated for rooms from 10 m^2 to 90 m^2 floor area, with ceiling heights ranging from 2.5 m to 4 m (see Table 13.5).

Table 13.4 The minimum internally reflected component of daylight factor (per cent)

Window area as percentage of floor area	Floor reflection factor											
	10%				20%				40%			
	Average wall reflection factor (excluding window)											
	20%	40%	60%	80%	20%	40%	60%	80%	20%	40%	60%	80%
2	—	—	0.1	0.2	—	0.1	0.1	0.2	—	0.1	0.2	0.2
5	0.1	0.1	0.2	0.4	0.1	0.2	0.3	0.5	0.1	0.2	0.4	0.6
7	0.1	0.2	0.3	0.5	0.1	0.2	0.4	0.6	0.2	0.3	0.6	0.8
10	0.1	0.2	0.4	0.7	0.2	0.3	0.6	0.9	0.3	0.5	0.8	1.2
15	0.2	0.4	0.6	1.0	0.2	0.5	0.8	1.3	0.4	0.7	1.1	1.7
20	0.2	0.5	0.8	1.4	0.3	0.6	1.1	1.7	0.5	0.9	1.5	2.3
25	0.3	0.6	1.0	1.7	0.4	0.8	1.3	2.0	0.6	1.1	1.8	2.8
30	0.3	0.7	1.2	2.0	0.5	0.9	1.5	2.4	0.8	1.3	2.1	3.3
35	0.4	0.8	1.4	2.3	0.5	1.0	1.8	2.8	0.9	1.5	2.4	3.8
40	0.5	0.9	1.6	2.6	0.6	1.2	2.0	3.1	1.0	1.7	2.7	4.2
45	0.5	1.0	1.8	2.9	0.7	1.3	2.2	3.4	1.2	1.9	3.0	4.6
50	0.6	1.1	1.9	3.1	0.8	1.4	2.3	3.7	1.3	2.1	3.2	4.9

Table 13.5 Conversion factors for rooms whose floor area corresponds to 10–90 m^2

Floor area	Wall reflection factor (%)			
	20%	40%	60%	80%
10 m^2	0.6	0.7	0.8	0.9
90 m^2	1.4	1.2	1.0	0.9

Changes in the height of the sill and of the window head will not as a rule affect the estimate of the internally reflected component, neither will the shape of the room, so long as the ratio of length to width does not exceed 2 : 1. The ceiling reflection factor in the table is assumed to be 70 per cent, but other values can be allowed for by means of conversion factors (see Table 13.6).

Table 13.6 Conversion factors for rooms with a ceiling reflection factor of other than 70 per cent

Ceiling reflection factors (%)	Conversion factor
40	0.7
50	0.8
60	0.9
80	1.1

Light losses due to glazing bars

The BRE protractor allows for full glazing openings and the following reduction should be made for glazing bars (see Table 13.7).

Table 13.7 Light losses due to glazing bars

For large-paned windows	Reduction factor
All-metal windows	0.8
Metal windows in wood frames	0.75
All-wood windows	0.65–0.70

Maintenance factors

Maintenance factors recommended by the Illuminating Engineering Society (IES) that allow for dirt on the windows and room surfaces are given in Tables 13.8 and 13.9.

Table 13.8 Maintenance factors to be applied to the total daylight factor or to each of its three components to allow for dirt on glazing

Location of building	Inclination of glazing	Maintenance factor	
		Non-industrial or clean industrial work	Dirty industrial work
Non-industrial or clean industrial area	Vertical	0.9	0.8
	Sloping	0.8	0.7
	Horizontal	0.7	0.6
Dirty industrial area	Vertical	0.8	0.7
	Sloping	0.7	0.6
	Horizontal	0.6	0.5

Table 13.9 Maintenance factor to be applied to the internally reflected component of daylight to allow for dirt on room surfaces

Location of building	Maintenance factor	
	Non-industrial or clean industrial work	Dirty industrial work
Non-industrial or clean industrial area	0.9	0.7
Dirty industrial area	0.8	0.6

Example 13.1 *Find the minimum internally reflected component of the daylight factor for a room measuring 8 m x 5 m x 3 m high, having a window in one wall and with an area of 8 m². The floor has an average reflection factor of 10 per cent and the walls and ceiling average reflection factors of 40 per cent and 70 per cent, respectively.*

$$\text{Window area as a percentage of floor area} = \frac{8}{40} \times \frac{100}{1} = 20 \text{ per cent}$$

Referring to Table 13.4, the minimum internally reflected component = 0.5 per cent.

Assuming that the building is in a non-industrial area this value can be corrected for the maintenance factor. Therefore

IRC = 0.5 x 0.9 = 0.45 per cent

Final computation of daylight factor

Using the example given in Fig. 13.6 and assuming that the external reflected component is 0.15 per cent and the internal reflected component is 0.6 per cent, the following conditions must now be considered. These are, that the building is in a non-industrial area with vertical single glazing, all-metal framed.

The uncorrected factors are

Sky component = 2.2 per cent
ERC = 0.15 per cent
IRC = 0.6 per cent

First correct the IRC for the maintenance factor. Therefore
IRC = 0.6 x 0.9 = 0.54 per cent

Therefore

total daylight factor = 2.2 + 0.15 + 0.54
= 2.89 per cent

This will represent the minimum value if the sky component and external reflection component are taken at the furthest point from the windows.

Correction for glazing bars = 0.8

Correction for dirt on windows = 0.9

These factors affect the total light entering the room and the total correction is 0.8 x 0.9.

Corrected minimum daylight factor = 2.89 x 0.8 x 0.9
= 2.08 per cent

Integration of daylight with electric light (PSALI)

As mentioned previously, this form of interior lighting is particularly suitable in deep rooms in multi-storey buildings, or where daylight is partially obstructed by adjacent buildings or trees. It also gives greater freedom in solving internal planning problems by reducing dependence on glazing for ordinary working illumination.

Experimental studies carried out by the BRE have shown that the form of a supplementary lighting installation depends on the daylight distribution in the interior and may differ according to whether a deep room is lit by one window at one end, or by windows on more than one wall, or whether a room is top-lit or side-lit. But whatever the design solution may be, the basic principles to be applied must be the same, that is, supplementary illumination must be related to the available daylight and the walls, ceiling and other surfaces must be so illuminated that there is complete integration with the daylight character of the room.

The colour of the fluorescent lighting must blend acceptably with the daylight it supplements. This applies to the colour rendering of the room surfaces illuminated by artificial light and to the colour appearance of the light sources themselves. Colour rendering is an important factor in PSALI design, and though many fluorescent lamps that look like daylight are poor in this respect, good colour rendering can be obtained from lamps with a wide range of source colours.

For certain selected interiors supplementary illumination in the range 300–1000 lx has been recommended, and fluorescent lamps are available that can provide high levels of illumination of a colour comparable with daylight, making the use of artificial light throughout the day a practicable proposition. The supplementary illumination required depends on the average daylight factor (per cent) in the area to be supplemented. A guide to the level of supplementary illumination is given in Table 13.10.

Table 13.10 Level of supplementary illumination needed on average daylight

Average daylight factor in area to be supplemented (%)	Illumination to be provided (lx)
0.5	250
1.0	500
1.5	750
2.0	1000
3.0	Full supplementary lighting is unlikely to be required

PSALI after dark

The designing of an integrated scheme involves two different lighting systems:

1. A system for daylight hours in which the electric lighting level is considered along with the daylight factor.
2. A system for after dark in which the lighting level is considered as a total electric light system which is usually lower than the electric lighting level required during daylight hours.

When daylight fades, the supplementary lighting will begin to dominate the room and this is the time when the after-dark lighting system is brought into use. However, it is inadvisable to make a sudden change from a high level of supplementary lighting to a lower level of total electric lighting: the change should be made in stages while the daylight fades. The change can be made automatically by the use of photo-receptors activated by the sky brightness.

Advantages of PSALI

Daylight constantly varies in quality and quantity, and buildings designed to use it as a primary light source must have large areas of glazing and must also be restricted in depth. Large areas of glazing subjected to solar radiation leads to a greenhouse effect with an attendant rise in temperature. Thermal comfort is not possible inside glass cladding and this leads to the use of blinds, special glasses or complex controls.

Figure 13.10 shows two buildings each having a total floor area of 3000 m^2; the buildings have 66 per cent and 20 per cent, respectively, of glass area to wall area. The building having the large area of glass requires substantially more energy, since what little is saved on lighting is more than lost in the heating and cooling load, due to the poor thermal properties of the glass. The building with 20 per cent of glass area is also usually cheaper to construct, saves on window cleaning and reduces glare and noise from outside.

Roof lights (see Fig. 13.11)

These are often used in single-storey factories and provide an even spread of light over the working area; higher daylight factors are obtained but these are usually drastically reduced by overhead equipment and poor maintenance of the windows.

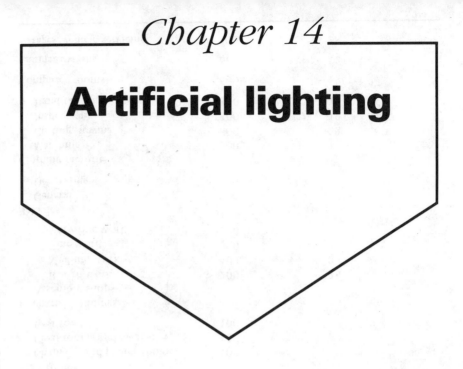

Chapter 14

Artificial lighting

Principles

Artificial light is constant and controllable, whereas daylight is continually varying with the time of year, time of day and weather conditions. The amount of light needed is directly related to the nature of the task, and the higher the level of illumination the better can small details be seen and the quicker the task carried out without mistakes.

Factors which affect the choice of illumination level include the size of the object, the degree of contrast which assists vision and whether or not the eyes or the object have movement. The reflective characteristics of the main surfaces, furnishings and content of a room contribute to the illumination and may have a marked effect on visual comfort, and therefore should be considered at an early stage in the lighting design. Glare may also be caused by the reflection of the light sources from polished surfaces such as table-tops and floors: matt surfaces are generally preferable where there is a risk of this occurring, and for the same reason matt finishes may be used for walls and ceilings.

Although contrast in luminance between illuminated surfaces in a working interior in excess of a ratio of 12 : 1 are generally to be avoided, complete lack of contrast is also undesirable as it can lead to a monotonous, lifeless effect. In order to create an interior that has a lively and stimulating appearance, variety in the visual scene is required with contrasts in light and dark, highlight and shadow – as well as variations in texture and colour.

An appropriate distribution of light can also help to concentrate attention on specific parts of the interior. The eye is attracted involuntarily to that part of the field of view which appears the brightest or has the strongest local brightness contrast. For example, in a library it helps the reader if the page of the book he is reading appears rather brighter than the surface of the desk, and in turn the desk surface appears brighter than the floor below it.

Design of lighting

It is essential that the decision about the method of lighting is taken at an early stage in the design of the building and the architect should consult the lighting engineer and others concerned during the conceptual stage. The first step is to establish the general requirements for the artificial lighting in terms of the main visual tasks to be carried out in the building (sometimes referred to as the 'work lighting'). The next step is to determine the lighting requirements in terms of revealing the form of the building and helping to create the right character of the interior (sometimes referred to as the 'building lighting'). The possibility of changes in the use of the building should be considered at this stage.

The artificial lighting should be considered as part of the design of the interior environment as a whole and in addition to its relationship to daylighting, it may need to be related to the thermal and acoustic requirements (this design process is often termed 'integrated environmental design'). The proportion of the total cost of the building to be allocated to the artificial lighting installation should be decided as early as possible, and cross-checks should be made as the design proceeds. The client will be concerned that the right balance is struck between capital and running costs.

The architect and the lighting engineer should be able to consider more detailed aspects of the lighting design under the following headings:

1. The extent to which artificial lighting will be used alone, or to supplement the daylighting.
2. The illuminances required for lighting specific visual tasks.
3. The required luminance distribution throughout the interior.
4. The evaluation of discomfort glare in terms of the whole visual environment.
5. The directional characteristics of the lighting required to give the desired modelling effects and to reveal form and texture.
6. The main features of the colour schemes of the building interior in terms of hue, chroma and colour rendering.

In addition, practical considerations such as siting, weight and access for maintenance of light fittings and electrical control gear should be considered at an early design stage. On completion of the building the client should be given drawings showing the layout of the artificial lighting installations and clear instructions for its operation and maintenance.

The recommended minimum service values of illumination and the limiting value of glare index are obtainable from the Code of the IES and the Lighting Industry Federation Ltd.

Table 14.1 gives a summary of the more common minimum values.

Table 14.1 Recommended minimum service values of illumination and limiting service values of glare index. Extracted from the Lighting Industry Federation handbook, Interior Lighting Design

	Lux (lm/m^2)	Limiting glare index
General		
Corridors, passageways	100	22
Lifts	200	
Lift lobbies	200	22
Stairs, escalators	100	
Entrance halls	200	
Kitchens (general)	200	22
Kitchens (food preparation)	400	
Food stores	200	
Medical		
Consulting rooms and First-aid cubicles	400	
Rest rooms	50	
Treatment rooms	400	
Medical stores	200	25
Staff restaurants		
Canteens, cafeteria	200	22
Dining rooms (general)		
Counters	400	
Staff rooms		
Changing and locker rooms	100	
Cloakrooms and lavatories	100	
Rest rooms	100	
Industrial buildings		
Assembly shops		
Rough work	200	28
Medium work	400	25
Fine work	900	22
Very fine work	2000	16
Bakeries		
General	200	25
Decorating	400	22
Clothing factories		
Matching	600	19
Cutting, sewing	600	22
Inspection	1300	19
Hand tailoring	1300	19
Computer rooms	600	19
Cabinet making	600	22
Garages, vehicle servicing	200	28

	Lux (lm/m²)	Limiting glare index
Laundries (general)	200	25
Printing (general)	400	25
Textile mills		
Spinning	600	25
Weaving	900	19
Inspection	1300	19
Hairdressing salons		
General	200	22
At chairs	600	
Supermarkets		
General	600	25
Display	900	
Cash desk	600	
Hospitals		
Corridors	100–200	
Wards (general)	100	22
Bed heads	30–50	
Reading	150	
Laboratories	400	19
Operating theatres (general)	400	
Operating area	Special lighting	
Anaesthetic room	300	
Recovery, intensive care rooms	30–50	
Consulting rooms	200	16
Homes		
Living room (general)	100	
Reading	200	
Sewing	600	
Study	400	
Bedrooms	200	
Kitchens	200	
Bathrooms	100	
Halls, landings	100	
Stairs	50–100	
Garages	50	
Hotels		
Entrance halls	200	
Reception	400	
Cloakrooms	100	
Dining rooms	100	
Lounges	200	
Bedrooms	100	
Kitchens	200	

Table 14.1 – *continued*

	Lux (lm/m²)	Limiting glare index
Offices		
General	400	19
Print room	200	22
Drawing	400	16
Records	400	22
Shops		
Circulation areas	200	22
Counters	600	
Cash desks	400	
Churches		
General	50–100	16
Pulpit	200	
Altar	400	
Vestries	200	
Libraries		
Reading rooms	200	19
Reading tables	600	16
Counters	600	
Museums, art galleries		
General	200	16
Displays	Special lighting	
Schools		
General	200	16
Examination rooms	300	16
Classrooms	300	16
Lecture theatres	200	16
Art rooms	600	16
Laboratories	400	16
Needlework rooms	600	16

Table 14.2 gives the recommended minimum service values of illuminance for different classes of visual task.

Reflection factors

Only part of the light emitted from the lamps is received as useful illumination on the working plane. Walls, ceilings, floors, furnishings and the light fittings themselves absorb a considerable amount of light and it becomes necessary to know what proportion of the light that escapes from the fittings is reflected from the surrounding surfaces and contributes to the illumination at the required level.

Table 14.3 shows the percentage of incident of light which is reflected from coloured matt surfaces based on the British Standard colours for ready-mixed paints (BS 381). This percentage figure is known as the 'reflection factor' of the surface. White enamel, whether synthetic or porcelain, reflects from 60 per cent to 80 per cent of light.

Table 14.2 Recommended minimum service values of illuminance (CP3: Part 2: 1973, Chapter 1)

Nature of work	Illuminance (lx)
Rough or routine work; large detail; medium to light material of good contrast	200
Ordinary work usually involving workers' inspection; medium detail and contrast	400
Fairly critical work; fairly small detail or poor contrast	600
Fine or critical work; very small detail; very poor contrast or very dark material	900
Very fine, exacting work	2000
Minute work	3000

Table 14.3

Reflection factors	(%)	Reflection factors	(%)
White	75–88	Golden yellow	62
Light stone	53	Orange	36
Middle stone	37	Eau-de-nil	48
Light buff	60	Sea green	38
Middle buff	43	Sage green	20
Deep buff	34	Sky blue	47
French grey	45	Turquoise	27
Quaker grey	35	Peacock blue	16
Battleship grey (light)	44	Light brown	30
Battleship grey (dark)	26	Middle brown	20
Pale cream	73	Salmon pink	42
Deep cream	70	Post Office red	21
Lemon	70	Crimson	13

When the type of light fitting, room index and reflection factors are known, the utilisation factor may be found from tables obtained from the following:

1. The IES.
2. Manufacturers' technical catalogues.
3. Lighting Industry Federation Ltd handbook, *Interior Lighting Design*.

British Zonal (BZ) classification

Different countries have different techniques for the optical performance of light fittings. In the BZ system, the intensity distribution of fittings are classified as one of the ten types labelled BZ1 to BZ10, which are mathematically derived curves. These are based on ten different polar curves which cover the performance of most types of light fittings in use.

The BZ classification relates to the intensity of light distribution in the lower hemisphere only (see Fig. 14.1).

Figure 14.2 shows the direction of light output for the classifications. The maximum ratio of spacing to mounting height of the fittings may be related to the classification (see Table 14.4).

Table 14.4 Ratio of spacing to mounting heights of fittings

BZ classification	Maximum spacing/mounting height ratio
1 and 2	1 : 1
3 and 4	1.25 : 1
5 to 10	1.5 : 1

Care must be taken in the selection of light fittings to avoid too little light on the ceilings which will give the rooms a gloomy appearance, even though there is adequate light on the working plane.

The Bodmann ratio of 10 : 4 : 3 for task, immediate surroundings and walls and ceilings has been generally accepted as a criteria for the distribution of light for most types of interiors. The illumination ratios derived from Bodmann's recommendations coincide almost exactly with BZ3 fittings, which emit about 10 per cent of their light output upwards on to the ceiling. Fittings of this kind are often used in offices and school classrooms because they are known to be an agreeable and economical means of complying with the limiting glare indices given in the IES Code for these rooms.

In normal circumstances, the direction of view does not rise much above the horizontal, so that the likelihood of glare from lighting fittings within 35° of the vertical may be largely ignored (see Fig. 14.3). Light emitted between 35° to the vertical and the horizontal is more likely to give rise to glare. Methods of glare control depend largely on restricting luminance within this zone, while ensuring that the background against which the fittings are seen (normally the ceiling) is light enough.

However, if the ceiling is too light it may distract attention from other objects, while the zone between 35° to the vertical and horizontal is also important for lighting walls. The more severe the measures taken to restrict glare, the more the risk of having an over-bright ceiling and dark walls.

Glare index

The IES gives acceptable maximum values for the glare index in many situations. These have been arrived at by field surveys of successful installations, and tables are available which enables the value of the glare index to be determined for an installation consisting of a regular pattern of fittings, mounted or suspended from the ceiling, given only the room dimensions, reflectances of surfaces and details of the lighting fittings.

Work at the BRE established a relationship between direct glare discomfort and the four factors which most affect it. These are as follows:

1. Brightness of the source (B_s).
2. The apparent size of the source (w).
3. The brightness of the background (B_b).
4. An index (p) representing the position of the source in relation to the direction in which the eyes are looking.

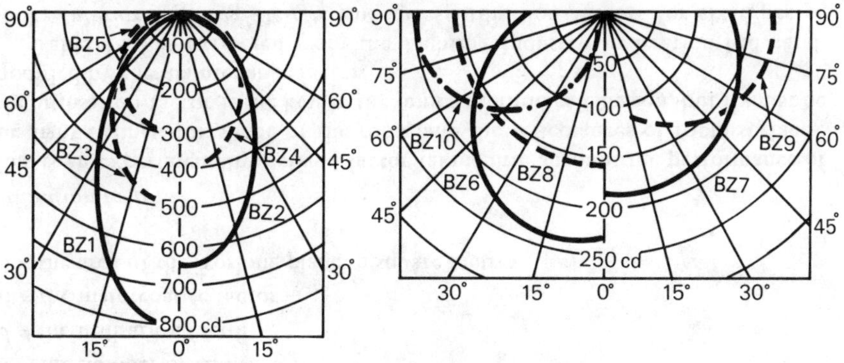
90°
60°
45°
30°
BZ5
BZ3
BZ1
BZ4
BZ2
100
200
300
400
500
600
700
800 cd
15°
0°
15°
90°
60°
45°
30°
90°
75°
60°
45°
BZ10
BZ6
BZ8
BZ7
BZ9
50
100
150
200
250 cd
30°
15°
0°
15°
30°
90°
75°
60°
45°
BZ polar curves 1 - 5
BZ polar curves 6 — 10

Fig.14.1 Polar curves in the BZ classification

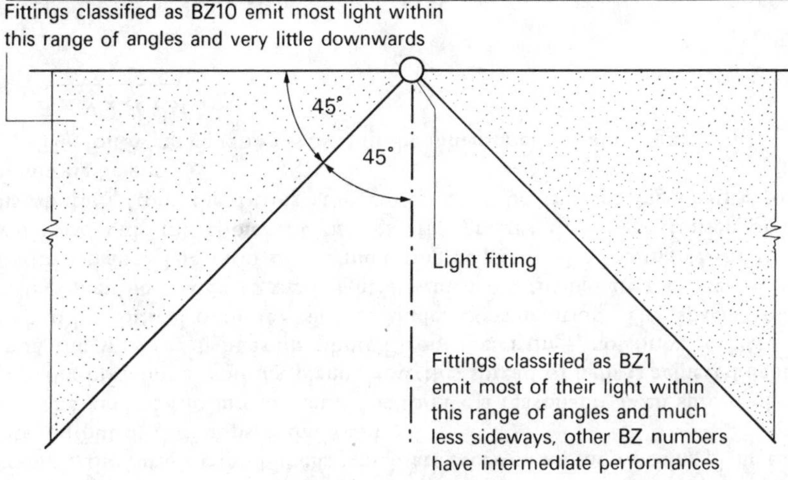
Fittings classified as BZ10 emit most light within this range of angles and very little downwards
45°
45°
Light fitting
Fittings classified as BZ1 emit most of their light within this range of angles and much less sideways, other BZ numbers have intermediate performances

Fig.14.2 Diagram illustrating BZ classification of fittings

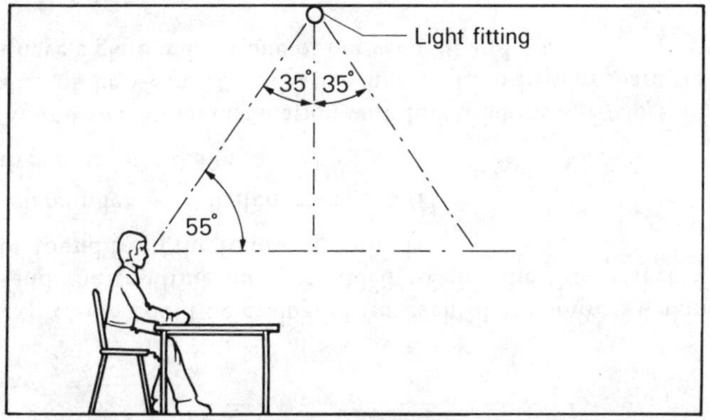
Light fitting
35°
35°
55°

Fig.14.3 Light emitted within 35° of the vertical is not perceived from normal angles of view and is therefore unlikely to contribute to a sensation of glare

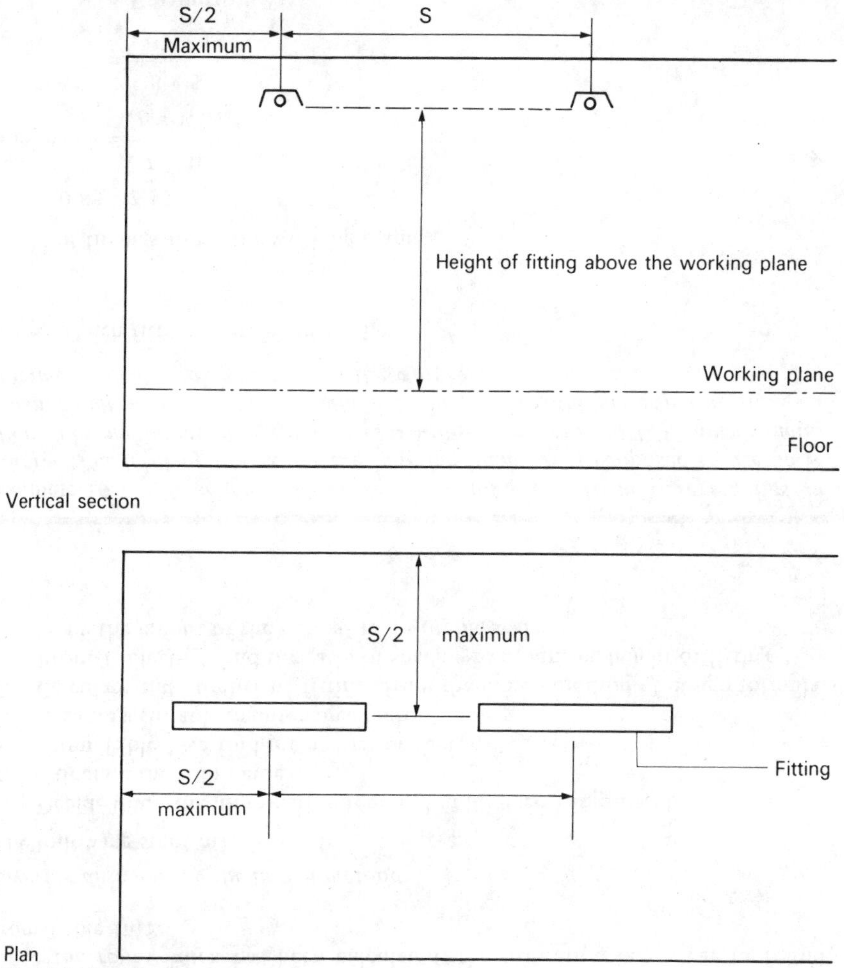
S/2
Maximum
S
Height of fitting above the working plane
Working plane
Floor
Vertical section
S/2
maximum
Fitting
S/2
maximum
Plan

Fig.14.4 Method of spacing fluorescent fitting

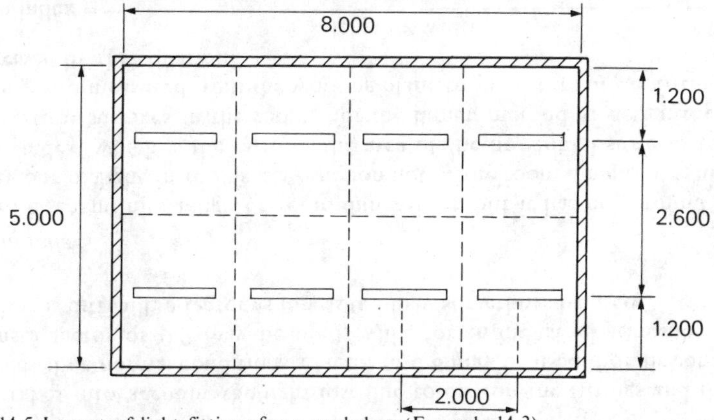
8.000
1.200
5.000
2.600
1.200
2.000

Fig. 14.5 Layout of light fittings for a workshop (Example 14.3)

Knowing the above factors, a glare constant may be obtained from the following:

$$\text{glare constant } G = \frac{B_s^{1.6} \times w^{0.8}}{B_b \times p^{1.6}}$$

The above expression could be evaluated for each of the sources making up the installation and the resulting number added to give the total effect. The glare index may be found from the following formula:

$$\text{glare index} = 10 \log[\text{constant} \times \Sigma G]$$

In SI units the constant = 0.478

It has been found that in an installation with low brightness, fittings might have an index value of between 10 and 16, while a room lit with bare fluorescent lamps might have a glare index value of between 28 and 30.

Lumen method of design

The lumen method is the most widely used approach to the systematic design of electric lighting. The method depends essentially on the utilisation factor, for example the ratio of the lumens which are received on the working plane to the total output of the lamps in the room.

The aim of the lumen method is to give a reasonably even spread of light over the horizontal working plane. How this spread of light is achieved depends upon the way the light is distributed from the fittings, not only in relation to each other but also to the surface under consideration. The spacing of the fittings may be related to the height at which the fittings are mounted over the working plane. The ratio of the mounting height to the spacing of the fittings will vary with the choice of fittings: the greater the concentration of light distribution from the fitting, the closer must be the spacing relative to the mounting height.

The following equation is used in the lumen method of design:

$$E = \frac{F \times N \times U \times M}{A}$$

where

E = the average horizontal illumination at the working plane in lux
F = the lamp lighting design lumens
N = the number of lamps
U = the utilisation factor
M = the maintenance factor
A = the area of the working plane in square metres

Utilisation factor

This is an experimentally derived factor taking into account the performance of the light fittings, the shape of the room and the reflectances of room surfaces. The utilisation factor is the key to the lumen method calculation, but it must be modified by the maintenance factor.

Table 14.5 is extracted from the Lighting Industry Federation Ltd handbook, *Interior Lighting Design*, and gives utilisation factors for six types of fittings.

Maintenance factor

This takes into account the light lost due to dirt on the fittings and the room surfaces. For normal conditions a factor of 0.8 may be used. For air-conditioned rooms a factor of 0.9 may be used, while for an industrial atmosphere where cleaning is difficult, a factor as low as 0.5 may sometimes be used.

Room index

There is an infinite range of room dimensions, but it has been found that the behaviour of light in rooms is a function not of the room dimensions, but of the room index, which is the ratio of the area of the horizontal surfaces to that of the vertical surfaces in the room. For the lumen method of design the vertical surfaces are measured from the working plane to the centre of the fitting. This is expressed by the equation:

$$\text{room index} = \frac{\text{length} \times \text{width}}{\text{length} + \text{width (height of fitting above the working plane)}}$$

Once the room index has been calculated, the utilisation factor can be found from Table 14.5.

Lighting design using the lumen method

The following steps may be used:

1. Decide upon the illumination required in lux (see Table 14.1).
2. Calculate the room index.
3. From Table 14.5 find the utilisation factor.
4. Assume a suitable maintenance factor.
5. Calculate the number of fittings from the lumen method of design formula.
6. From Table 14.4 find the ratio of spacing to mounting height of fitting.
7. Draw the layout of the fittings to a suitable scale.

Example 14.1. *A school laboratory 15 m long and 10 m wide requires an illumination level of 400 lx on the working plane. It is proposed to use 65 W floor. The manufacturer's table gives a utilisation factor of 0.45 and a maintenance factor of 0.8 may be assumed. If the flux is 8000 lm per fitting, design a suitable electric lighting system for the workshop.*

Note: Each fitting contains two tubes.

Height of fittings above the working plane:

$$3 - 0.85 = 2.150$$

$$\text{room index} = \frac{L \times W}{(L + W)H}$$

$$= \frac{8 \times 5}{13 \times 2.15}$$

$$= 1.4 \text{ approx.}$$

Description of fitting and typical downward light output ratio (%)	Typical outline	Basic downward LOR (%)	Reflectance (%) Ceiling / Walls / Room index	70 / 50	70 / 30	70 / 10	50 / 50	50 / 30	50 / 10	30 / 50	30 / 30	30 / 10
(F) Suspended opaque-sided fitting, upward and downward light, diffuser or louvre beneath (40–50)	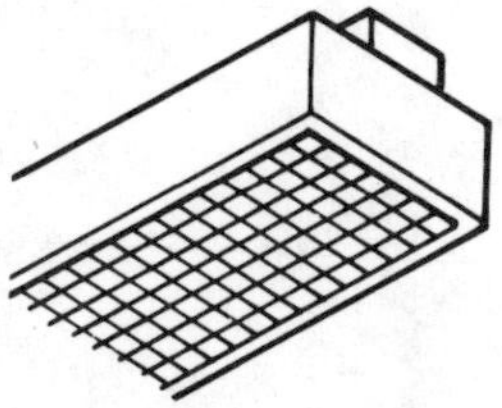	45	0.6	0.28	0.24	0.20	0.26	0.22	0.19	0.24	0.20	0.19
			0.8	0.36	0.30	0.28	0.33	0.29	0.26	0.31	0.27	0.24
			1.0	0.41	0.36	0.32	0.37	0.33	0.30	0.34	0.30	0.27
			1.25	0.45	0.41	0.36	0.41	0.37	0.34	0.37	0.33	0.30
			1.5	0.49	0.45	0.40	0.44	0.40	0.37	0.39	0.35	0.33
			2.0	0.55	0.50	0.46	0.48	0.45	0.42	0.42	0.39	0.37
			2.5	0.58	0.53	0.50	0.51	0.48	0.45	0.45	0.42	0.40
			3.0	0.60	0.56	0.53	0.53	0.50	0.48	0.47	0.44	0.42
			4.0	0.63	0.59	0.57	0.55	0.53	0.51	0.48	0.46	0.44
			5.0	0.65	0.62	0.60	0.57	0.55	0.53	0.50	0.48	0.46
(F) Plastic trough-louvred (45–55)	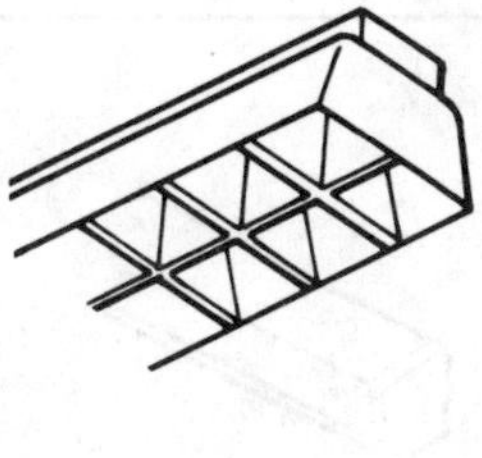	50	0.6	0.26	0.22	0.19	0.25	0.21	0.19	0.24	0.20	0.18
			0.8	0.34	0.29	0.26	0.32	0.28	0.25	0.31	0.27	0.24
			1.0	0.39	0.34	0.30	0.36	0.32	0.29	0.34	0.31	0.28
			1.25	0.43	0.38	0.34	0.39	0.36	0.33	0.37	0.34	0.31
			1.5	0.46	0.41	0.37	0.42	0.39	0.36	0.30	0.36	0.33
			2.0	0.50	0.46	0.43	0.43	0.42	0.40	0.43	0.39	0.37
			2.5	0.53	0.49	0.46	0.49	0.46	0.43	0.45	0.42	0.40
			3.0	0.55	0.51	0.49	0.51	0.48	0.46	0.47	0.45	0.43
			4.0	0.58	0.54	0.52	0.53	0.51	0.49	0.48	0.47	0.45
			5.0	0.60	0.57	0.55	0.55	0.53	0.51	0.50	0.48	0.47
(F) Plastic trough, unlouvred (60–70)	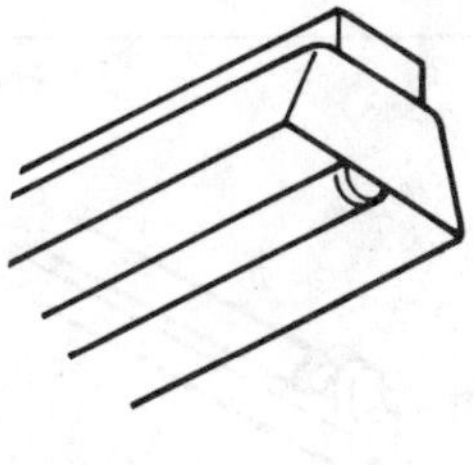	70	0.6	0.33	0.28	0.25	0.32	0.28	0.25	0.31	0.27	0.25
			0.8	0.42	0.37	0.33	0.41	0.36	0.33	0.40	0.36	0.33
			1.0	0.48	0.43	0.38	0.46	0.42	0.38	0.45	0.42	0.38
			1.25	0.52	0.47	0.43	0.50	0.46	0.42	0.40	0.45	0.42
			1.5	0.56	0.51	0.47	0.54	0.50	0.46	0.52	0.48	0.45
			2.0	0.62	0.56	0.53	0.58	0.55	0.51	0.56	0.52	0.50
			2.5	0.65	0.60	0.57	0.61	0.58	0.55	0.50	0.56	0.53
			3.0	0.67	0.63	0.60	0.64	0.61	0.58	0.62	0.59	0.56
			4.0	0.70	0.66	0.64	0.67	0.64	0.61	0.64	0.62	0.59
			5.0	0.73	0.69	0.67	0.69	0.67	0.64	0.66	0.64	0.62
(T) Near-spherical diffuser, open beneath (50)		50	0.6	0.28	0.22	0.18	0.25	0.20	0.17	0.22	0.18	0.16
			0.8	0.39	0.30	0.26	0.33	0.28	0.23	0.27	0.25	0.22
			1.0	0.43	0.36	0.32	0.38	0.34	0.29	0.31	0.29	0.26
			1.25	0.48	0.41	0.37	0.42	0.38	0.33	0.34	0.32	0.29
			1.5	0.52	0.46	0.41	0.46	0.41	0.37	0.37	0.35	0.32
			2.0	0.58	0.52	0.47	0.50	0.46	0.43	0.42	0.39	0.36
			2.5	0.62	0.56	0.52	0.54	0.50	0.47	0.45	0.42	0.40
			3.0	0.65	0.60	0.56	0.57	0.53	0.50	0.48	0.45	0.43
			4.0	0.68	0.64	0.61	0.60	0.56	0.54	0.51	0.48	0.46
			5.0	0.71	0.68	0.65	0.62	0.59	0.57	0.53	0.50	0.48

Table 14.5 Utilisation factors – *continued*

Description of fitting and typical downward light output ratio (%)	Typical outline	Basic downward LOR (%)	Ceiling	70			50			30		
			Walls	50	30	10	50	30	10	50	30	10
			Room index									
(F) Bare lamp on ceiling (F) Batten fitting (60–70)		65	0.6	0.29	0.24	0.19	0.27	0.22	0.19	0.24	0.21	0.19
			0.8	0.37	0.31	0.27	0.35	0.30	0.25	0.31	0.28	0.24
			1.0	0.44	0.37	0.33	0.40	0.35	0.31	0.35	0.32	0.29
			1.25	0.49	0.42	0.38	0.45	0.40	0.36	0.39	0.36	0.33
			1.5	0.54	0.47	0.42	0.50	0.44	0.40	0.43	0.40	0.37
			2.0	0.60	0.52	0.49	0.54	0.49	0.45	0.48	0.44	0.41
			2.5	0.64	0.57	0.53	0.57	0.53	0.49	0.52	0.48	0.45
			3.0	0.67	0.61	0.57	0.60	0.57	0.53	0.56	0.52	0.49
			4.0	0.71	0.66	0.62	0.64	0.61	0.57	0.59	0.55	0.52
			5.0	0.74	0.70	0.66	0.68	0.64	0.61	0.62	0.58	0.54
(F) Enclosed plastic diffuser (45–55)	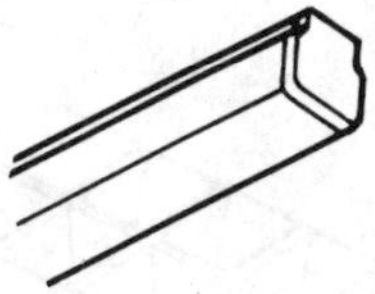	50	0.6	0.27	0.21	0.18	0.24	0.20	0.18	0.22	0.19	0.17
			0.8	0.34	0.29	0.26	0.32	0.28	0.25	0.29	0.26	0.24
			1.0	0.40	0.35	0.31	0.37	0.33	0.30	0.33	0.30	0.28
			1.25	0.44	0.39	0.35	0.40	0.36	0.33	0.36	0.33	0.31
			1.5	0.47	0.42	0.38	0.43	0.39	0.36	0.38	0.35	0.33
			2.0	0.52	0.47	0.44	0.47	0.44	0.41	0.41	0.39	0.37
			2.5	0.55	0.51	0.48	0.50	0.47	0.44	0.44	0.42	0.40
			3.0	0.58	0.54	0.51	0.52	0.49	0.47	0.47	0.45	0.43
			4.0	0.61	0.57	0.54	0.55	0.52	0.50	0.49	0.47	0.45
			5.0	0.63	0.59	0.57	0.57	0.55	0.53	0.51	0.49	0.47

Note: **In the first column of the table: (F) denotes a fitting for fluorescent lamp(s); (T) denotes a fitting for tungsten filament lamp.**

Using the lumen method of design equation

$$N = \frac{E \times A}{F \times U \times M}$$

$$N = \frac{500 \times 40}{8000 \times 0.45 \times 0.8}$$

N = 6.94 approx.

Seven fittings would be required, which does not permit regular spacing. Six fittings would be acceptable in terms of spacing, but the illumination would be proportionally reduced. Eight fittings would therefore be used which would give a higher illumination value of 576 lx.

Spacing

Fittings have a BZ number of 4 have a ratio of spacing to mounting height of 1.25 and this gives a maximum spacing of 1.25 x 2.15 = 2.7 m approx., centre to centre of the fittings. The distances of the fittings from the wall should not exceed half of the normal spacing and should be less if there is a working surface near the wall. The maximum distance from the centre of the fittings to the wall is therefore 1.35 m. Figure 14.4 shows the method of spacing fluorescent fittings for symmetrical distribution. Figure 14.5 shows a method of spacing of the fittings for the factory workshop in Example 14.3.

A glare check may be carried out on the installation by the use of a circular calculator or tables in the IES Technical Report No. 10 and the handbook, *Interior Lighting Design*, published by the Lighting Industry Federation Ltd.

Lamps and light fittings

Tungsten filament lamps (Fig. 14.6)

Tungsten filament lamps operate on the principle of passing an electric current through a fine tungsten wire filament and thus raising its temperature to incandescence, i.e. the wire gives out light.

fluorescent light fittings with a rated output of 4300 lm each. Assuming a maintenance factor of 0.8 and a utilisation factor of 0.5, calculate the number of light fittings required.

$$E = \frac{F \times N \times U \times M}{A}$$

by transposition,

$$N = \frac{E \times A}{F \times U \times M}$$

$$N = \frac{400 \times 15 \times 10}{4300 \times 0.5 \times 0.8}$$

$N = 34.8$ approx.

$N = 35$ fittings

Example 14.2 *A general office of dimensions 16 m x 10 m x 4 m high (floor to false ceiling) has a white ceiling and light-coloured walls with reflection factors of 70 per cent and 50 per cent, respectively. The working plane of the office is 1 m above the floor level. It is proposed to use 70 W fluorescent light fittings with a rated output of 5000 lm. If the illumination level is 400 lx determine the number of fittings required.*

$$\text{Room index} = \frac{16 \times 10}{3(16 + 10)}$$

$$= 2$$

From Table 14.4 the utilisation factor for the fluorescent fitting is 0.62. Assuming normal atmospheric conditions, the maintenance factor may be taken as 0.8.

Using the lumen method of design formula:

$$E = \frac{F \times N \times U \times M}{A}$$

by transposition,

$$N = \frac{E \times A}{F \times U \times M}$$

$$N = \frac{400 \times 16 \times 10}{5000 \times 0.62 \times 0.8}$$

$$N = 25.8$$

Number of fittings required = 26

Example 14.3 *A workshop is to be lit to a design level of 500 lx, using two 1.8 m long fluorescent tubes having a BZ4 classification. The room dimensions are 8 m x 5 m x 3 m high and the reflectances of the ceiling and walls are 70 per cent and 30 per cent respectively. The working plane is to be 850 mm above the*

In the early type of lamps the filament was sealed within an evacuated glass bulb. The vacuum of early lamps has been replaced by an inert gas, the pressure of which tends to reduce evaporation from the filament and to make it possible to operate the lamp at a higher temperature, thus giving an increased light output without shortening the life of the lamp.

The light output and life of a lamp depends upon the operating temperature; the higher the temperature the greater the efficacy in lumens per watt, but the lamp life is shorter. The best balance between life and output is that which gives the highest number of lumen-hours per unit cost.

Lamps for general lighting service (GLS) should comply with the requirements of BS 161, which lists the available range of lamp ratings from 25 W to 1500 W (see Table 14.6). The GLS lamps have a rated life of 1000 h. Some of the lower wattage lamps are available with either coiled-coil or single-coil filaments, the former having a higher luminous efficacy. The BS 161 lamp gives a good balance between life and light output.

Pear- and mushroom-shaped bulbs are available in domestic ratings with an internal coating of white powder for greater diffusion. Lamps with internal reflectors are also available for display and other purposes where directional effects are required.

Table 14.6 Tungsten lamps

Watts	Finish	Standard cap*	Lighting design lumens at 240 V
25	Pearl	bc	200
40	Pearl	bc	390 (coiled-coil)
60	Pearl	bc	665 (coiled-coil)
100	Pearl	bc	1260 (coiled-coil)
150	Pearl	bc	1960
200	Clear	es	2720
300	Clear	ges	4300
500	Clear	ges	7700
750	Clear	ges	12 400
1000	Clear	ges	17 300
1500	Clear	ges	27 500

* bc = bayonet cap, es = Edison screw cap, ges = giant Edison screw cap.

Lamp failure

The most likely cause of premature lamp failure is vibration. If mechanical movement brings two adjacent turns of the coiled filament into contact while the lamp is in use, they will weld together, shorten the path of current and so cause the filament to run too hot and fail prematurely.

Heat from lamps

All light sources produce heat and every unit of electrical energy put into an interior lighting system appears as heat. For corresponding values of installed flux, the high efficacies of fluorescent lamps mean, in comparison with tungsten filament lamps, that a lower wattage is required and therefore less heat is produced.

Most of the energy from tungsten filament lamps appears as radiation, while the proportion of radiation from fluorescent lamps is much smaller and most of the heat is removed by conduction and convection. It is usually difficult to control the heat from filament lamps, and an installation designed to provide 200 lx will probably produce a noticeable amount of radiant heat. The light from lamps eventually becomes heat as it is absorbed by the room surfaces.

Discharge lamps

These operate on the principle of passing an electric current through a gas or vapour, thereby producing a luminous arc. Control gear to initiate the discharge and to stabilise it is essential for all discharge lamps, this gear being normally located within or near the lamp.

The lamps take a short time, according to type, to run-up to full light output from being switched on, i.e. high-pressure mercury may take up to 5 min, low-pressure sodium up to 15 min and high-pressure sodium up to 2 min.

Figures 14.7 and 14.8 show the details of mercury-vapour and sodium-vapour lamps.

Mercury discharge lamps

A high-pressure mercury lamp has a compact discharge tube within an outer glass envelope, which may have an internal coating of fluorescent powder. Some mercury lamps have internal reflectors to modify the light distribution. The cost of the lamps and the control gear is fairly high compared with that of tungsten filament lamps, but this can be offset by reduced cost arising from their high luminous efficacy and long life (rated at 7500 h).

The colour rendering of plain mercury lamps is poor, thus restricting their use to industrial interiors and external lighting, unless they are mixed with other sources of light. A better spectral distribution is obtained by the use on the inside of the outer envelope, of fluorescent powders and recent improvements to these powders make the lamps suitable for many commercial applications.

Sodium discharge lamps

The high pressure sodium (SON) lamp has become the high efficacy light source for a wide range of applications. Its golden glow can be found in factories, offices, leisure centres and streets – anywhere where colour is not critically important. The high pressure sodium lamp has an efficacy of 67 to 121 lumens per watt and a lamp life of 6000 to 12 000 hours depending on construction.

There are four main versions of high pressure sodium lamp:

SON The standard SON has poor colour rendering but high efficacy and is generally accepted as a replacement for any colour-corrected mercury lamp. There are two main circuits in use, one utilises an electronic igniter to ensure rapid starting while the other, cheaper circuit, relies on a device within the lamp such as a bi-metallic switch or an auxilliary electrode. The main disadvantage of this cheaper circuit is that it can take some time to relight the lamp following a short supply interruption. When lighting a large area therefore, it is good practice to install lamps having an electronic igniter circuit installed at strategic points to ensure a rapid return of some of the lighting. Igniter circuits usually ensure restriking in less than one minute and circuits are available that will guarantee an instantaneous strike.

A recent development to further assist in rapid relighting is a SON lamp with two arc tubes. If the lamp is extinguished by a brief supply interruption the second arc, being cooler, will relight promptly. Some types of SON lamps can be used as plug-in replacements for high pressure mercury discharge lamps.

SONP Although the standard SON lamp has a very high efficacy, some manufacturers produce an even high output lamp such as the 'SONPLUS'. This is claimed to match all the features of the SON, such as colour rendering and life, but provides about 20 per cent more light. The lamp requires an external igniter and is not available as a plug-in replacement, as is available with the SON lamp.

SONDL The SONDL (high-pressure sodium deluxe) lamp has a warm colour appearance but its colour rendering is about equal to the white flourescent lamp. This has made the lamp suitable for supermarkets, sports halls, swimming pools, gymnasia, banks, offices and department stores etc.

SONW The SONW which is also known as 'White SON' is a low wattage lamp with colour properties very similar to a filament lamp but provides between two or three times the lumens per watt. Its main application has been in display lighting as an alternative to filament lamps.

Low pressure sodium lamps

The low-pressure sodium lamp gives an orange coloured light used predominantly for security, street and motor-way lighting. Whilst it has no colour rendering (being monochromatic in nature) it has the highest efficacy of any lamp available. It is quite unacceptable, however, where any reasonable standard of colour rendering is required.

The low-pressure sodium lamp has an efficacy of 70 to 175 lumens per watt and a lamp life of 6000 to 12 000 hours, depending on lamp construction.

Note: When calculating the installed efficacy of sodium discharge lamps the power consumption of the associated control gear should be inducled in the estimate of luminous efficacy.

Tubular fluorescent lamps (see Fig. 14.9)

Hot-cathode type

The common fluorescent lamp is a low-pressure mercury-vapour discharge lamp which utilises the ultraviolet radiation from the discharge to excite a coating on the inside of the lamp, so that it fluoresces. The colour of the light emitted depends on the chemical composition of the internal coating. The choice of colour of the light source is necessary in order to achieve good colour rendering.

Table 14.7 gives the type of colour rendering of the lamps for various uses. The life of the lamp is rated at 5000–7500 h.

Table 14.7 Colour rendering effects of tubular fluorescent lamps

Correlated colour temperature (K)	Tube designation	Colour rendering
2700	Softone	Similar to tungsten light and blends well with it
3000	De luxe warm white	Used for homes, hotels, restaurants

3600	De luxe natural	Good on all colours, reds are emphasised. Used on food displays, especially meat
3900	Colour 34	Similar to daylight with some sunlight
4000	Kolor-rite	Similar to daylight with some sunlight
4050	Natural	A good general-purpose colour for offices, shops and lighting pictures
4200	Trucolour 27	For critical colour matching, especially in hospitals and art galleries
5000	Graphic A 47	For critical colour matching in graphic arts. Also used for general lighting
6500	Northlight	Similar to north sky daylight
6500	Colour matching	Cool appearance, colour matching only to reasonable accuracy
6500	Artificial daylight	For critical colour matching

Cold-cathode type

These are similar to hot-cathode fluorescent lamps, but they operate on a higher voltage and require a transformer but no starting gear or ballast. The capital cost of the lamps and equipment is usually higher than that of hot-cathode lamps and equipment to give the same light output. The main advantage is the longer life of the lamp rated at 20 000 h and this can be of benefit where replacement is difficult or costly, or both. A range of colours are available; efficacies are slightly below those of the hot-cathode type of comparable colour.

Advantages

The first advantage of tubular fluorescent lamps is their high efficacy. The 100 W coiled-coil tungsten lamp gives about 1260 lm (12.6 lm/W), whereas a fluorescent tube such as the 65 W warm white has a lighting design output of 4600 lm. The control gear of the fluorescent lamp consumes about 15 W so that the total efficacy is about 57 lm/W, which is more than four times per watt of the light output of the 100 W tungsten lamp. Another advantage of the tubular fluorescent lamp is reduction of radiant heat input to the building, when compared to tungsten lamps.

The fluorescent lamp provides a light which is compatible with daylight and this permits the use of permanent supplementary lighting of interiors. Their long life and high efficacy has made the lamps very popular for all types of buildings.

Electric control circuits

Control gear is necessary with all discharge lamps to start the discharge and to keep it steady once the lamp is in operation. The cost of this gear partly accounts for the generalisation that fluorescent lighting systems are more expensive to install than tungsten filament lamp lighting systems.

There are several circuits available, but the basic types are a starter switch circuit shown in Fig. 14.10 and a transformer circuit shown in Fig. 14.11.

Starter switch circuits are slightly cheaper, but the starter switch is a replaceable component and its life is 20 000 h or more. The transformer circuit is one of the simplest and most reliable circuits used, but the lamp may not strike at once; however, delay should not be more than a few seconds. Tube life to failure may be marginally greater with a transformer circuit. Table 14.8 gives details of tubular fluorescent lamps.

Table 14.8 Details of tubular fluorescent lamps

Nominal watts	Nominal length (mm)	Nominal tube diameter (mm)	Lighting design lumens (de luxe colour)
15	450	25 or 38	500
20	600	38	790
30	900	25 or 38	1300
40	600	38	1160
40	1200	38	2000
50	1500	25	2300
65	1500	38	3300
80	1500	38	3650
85	1800	38	4150
85	2400	38	4800
125	2400	38	6200

Note: Approximate values for lighting design lumens from high-efficacy lamps may be found by multiplying the above figures by 1.3.

Oval fluorescent tubes

In order to give sufficient light output, tubular fluorescent tubes for commercial buildings may have diameters of 100 mm or more. These tubes look very cumbersome and are sometimes referred to as 'drainpipe lighting tubes'. In order to provide the equivalent light output of these tubes but with the slim appearance of smaller diameter tubes, oval tubes have been developed.

The oval tube system is now very popular for commercial buildings and the majority of commercial lighting manufacturers now regard it as one of their more important lamp products.

Compact fluorescent lamps (Fig. 14.20)

These new, energy efficient lamps use compact fluorescent tubes and can be used to replace the ordinary incandescent filament lamps in homes. The lamps perform far better, lasting over 8000 hours rather than the 1000 hours offered by incandescent filament lamps; they also use only 20–25 per cent of the energy. An electronic compact fluorescent lamp is the most efficient using only 20 per cent of the energy of the incandescent filament lamp (which is a massive 80 per cent saving).

Because the compact fluorescent lamps use less energy the demand for electricity from the power stations is less. In the long term the national energy requirement is reduced, and less carbon dioxide gas will be released into the

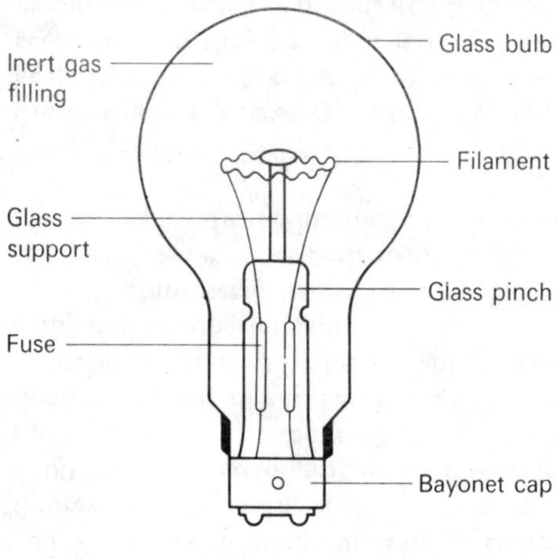
Inert gas filling
Glass bulb
Filament
Glass support
Glass pinch
Fuse
Bayonet cap

Fig.14.6 The general lighting service filament lamp

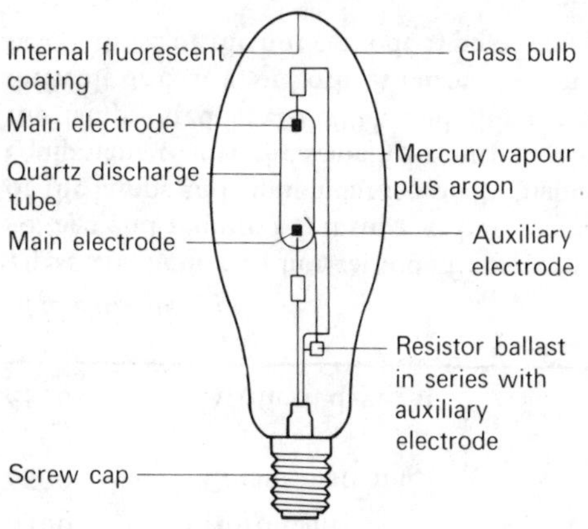
Internal fluorescent coating
Glass bulb
Main electrode
Quartz discharge tube
Mercury vapour plus argon
Main electrode
Auxiliary electrode
Resistor ballast in series with auxiliary electrode
Screw cap

Fig.14.7 Mercury-vapour discharge lamp

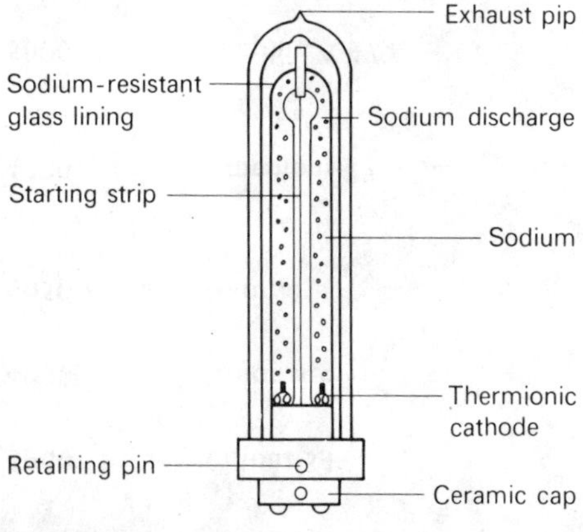
Exhaust pip
Sodium-resistant glass lining
Sodium discharge
Starting strip
Sodium
Thermionic cathode
Retaining pin
Ceramic cap

Fig.14.8 Sodium-vapour discharge lamp

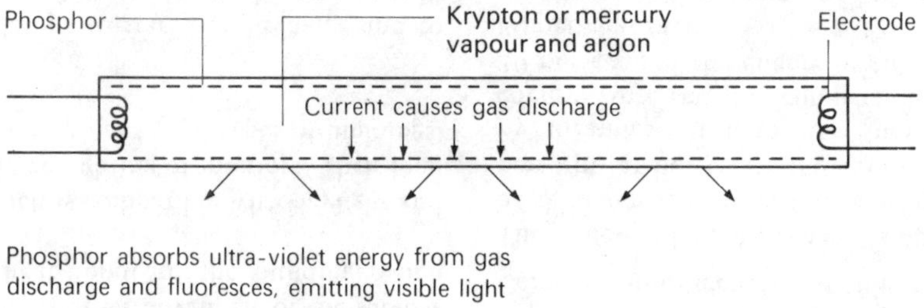
Phosphor
Krypton or mercury vapour and argon
Electrode
Current causes gas discharge
Phosphor absorbs ultra-violet energy from gas discharge and fluoresces, emitting visible light

Fig.14.9 The tubular fluorescent lamp

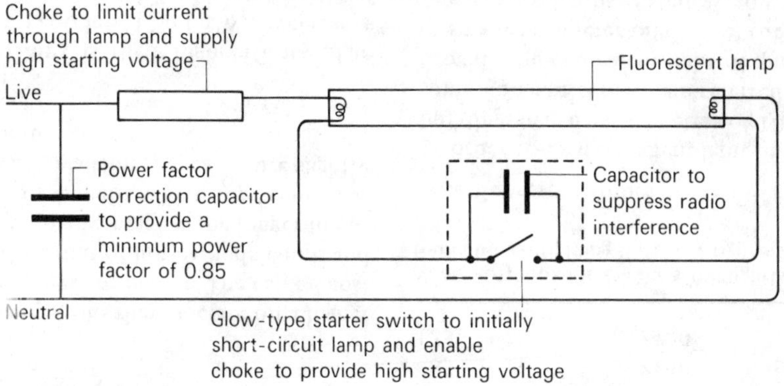
Choke to limit current through lamp and supply high starting voltage
Fluorescent lamp
Live
Power factor correction capacitor to provide a minimum power factor of 0.85
Capacitor to suppress radio interference
Neutral
Glow-type starter switch to initially short-circuit lamp and enable choke to provide high starting voltage

Fig.14.10 Glow-type starter switch circuit

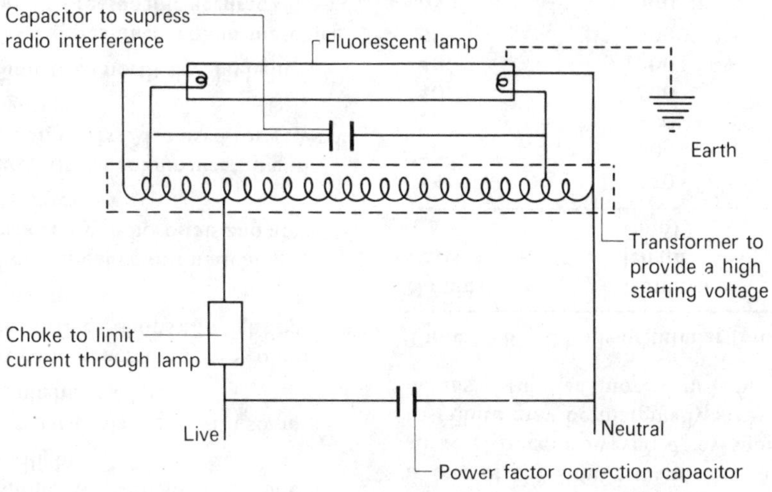
Capacitor to supress radio interference
Fluorescent lamp
Earth
Transformer to provide a high starting voltage
Choke to limit current through lamp
Live
Neutral
Power factor correction capacitor

Fig.14.11 Transformer start circuit

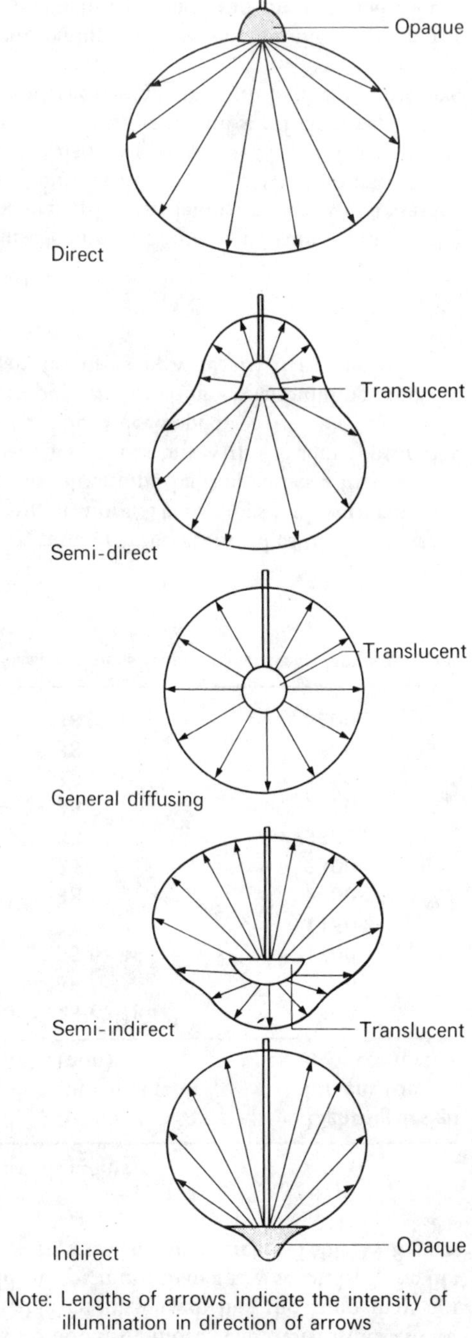
Opaque
Direct
Translucent
Semi-direct
Translucent
General diffusing
Semi-indirect
Translucent
Indirect
Opaque
Note: Lengths of arrows indicate the intensity of illumination in direction of arrows

Fig.14.12 Light distribution

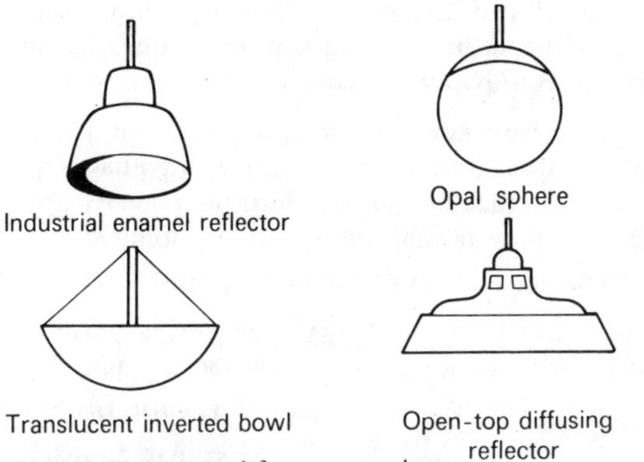
Industrial enamel reflector
Opal sphere
Translucent inverted bowl
Open-top diffusing reflector

Fig.14.13 Fittings used for tungsten lamps

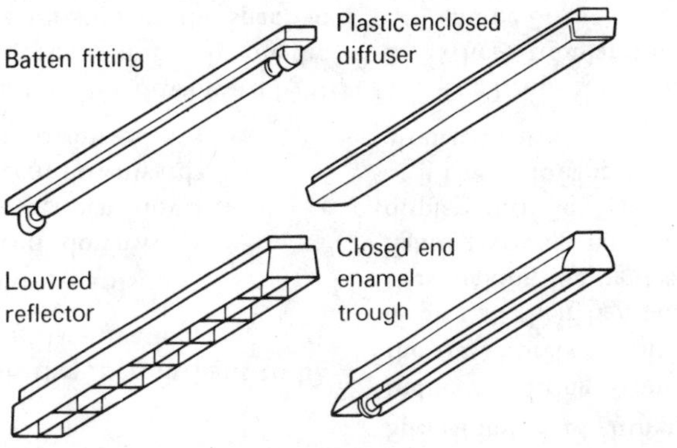
Batten fitting
Plastic enclosed diffuser
Louvred reflector
Closed end enamel trough

Fig.14.14 Fittings used for fluorescent lamps

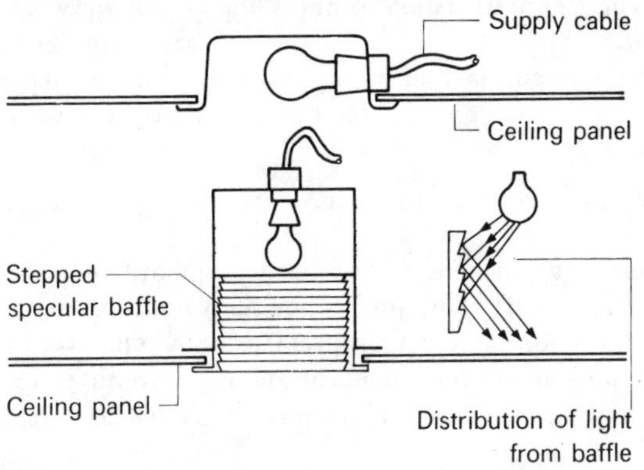
Supply cable
Ceiling panel
Stepped specular baffle
Ceiling panel
Distribution of light from baffle

Fig.14.15 Recessed fittings (black hole spotlights)

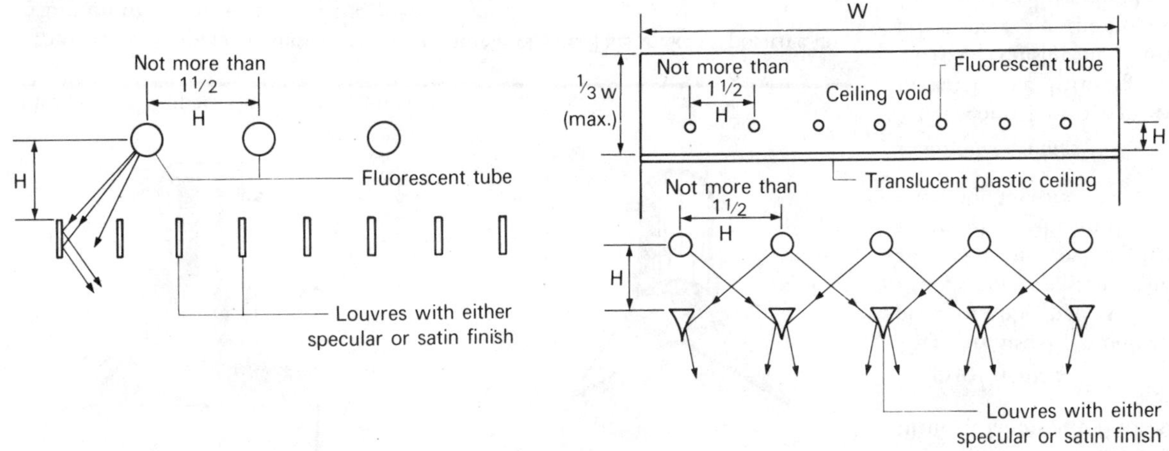
Not more than
1 1/2
H
H
Fluorescent tube
Louvres with either specular or satin finish
W
1/3 w (max.)
Not more than
1 1/2
H
Ceiling void
Fluorescent tube
H
Translucent plastic ceiling
Not more than
1 1/2
H
H
Louvres with either specular or satin finish

Fig.14.16 Luminous ceilings

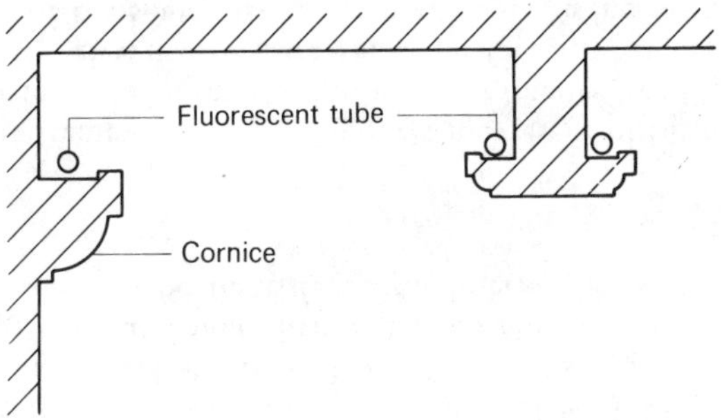
Fluorescent tube
Cornice

Fig 14.17 Special indirect lighting

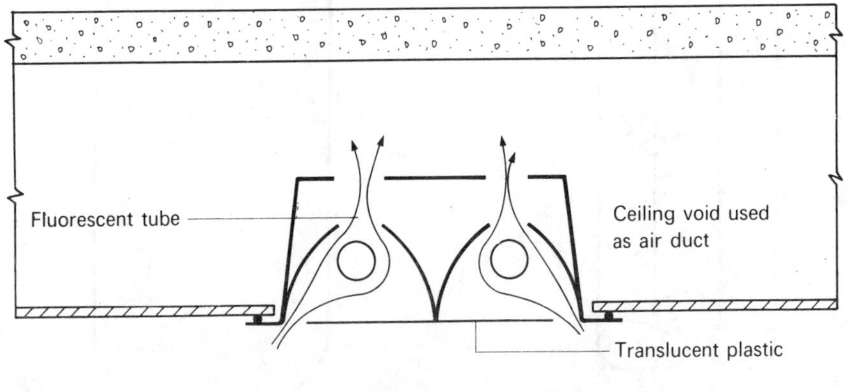
Fluorescent tube
Ceiling void used as air duct
Translucent plastic

Fig.14.18 Air-handling light fitting with translucent plastic diffuser

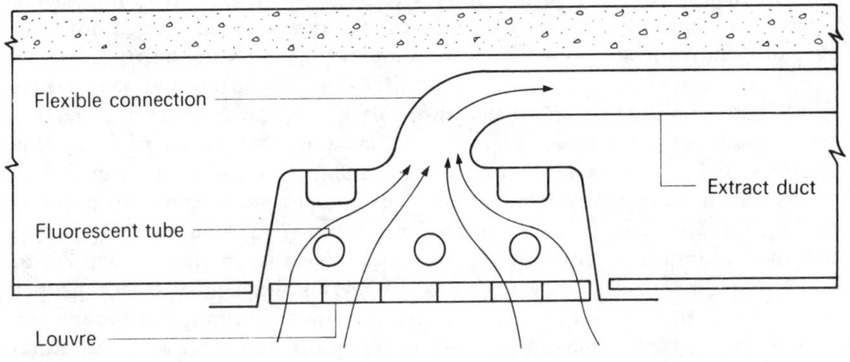
Flexible connection
Extract duct
Fluorescent tube
Louvre

Fig.14.19 Air-handling light fitting with louvre and extract duct

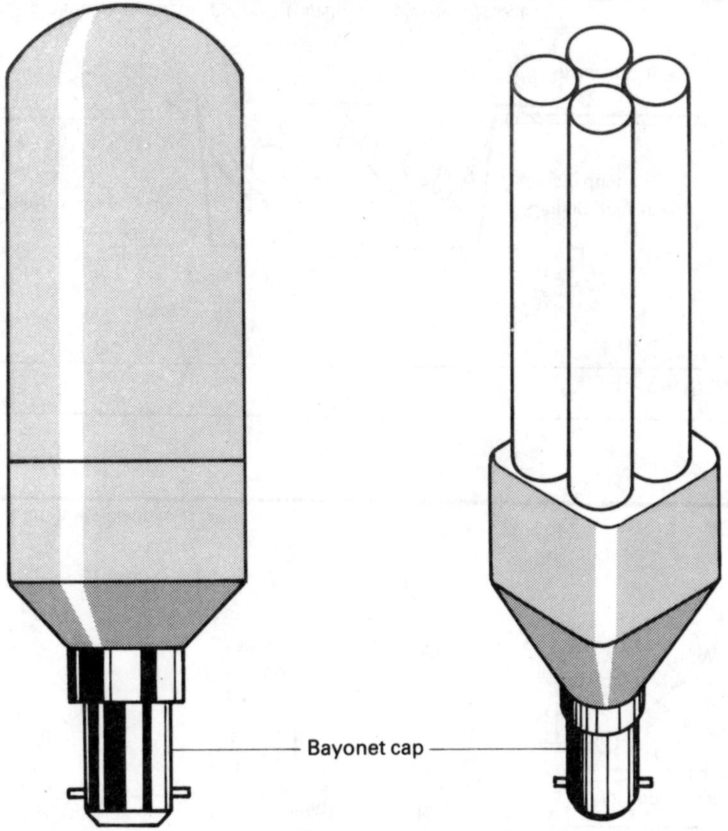
Bayonet cap

Fig. 14.20 Compact fluorescent lamps

atmosphere, helping to reduce the greenhouse effect. This makes a positive contribution to the protection of the environment.

Lighting fittings

Classification (Fig. 14.12)

Lighting fittings are classified by the manner in which they permit light to be emitted. CP3: Part 2: 1973, Chapter 1 lists the following classifications:

1. Direct: 0–10 per cent upwards, 100–90 per cent downwards.
2. Semi-direct: 10–40 per cent upwards, 90–60 per cent downwards.
3. General diffusing: 40–60 per cent upwards, 60–40 per cent downwards.
4. Semi-direct: 60–90 per cent upwards, 40–10 per cent downwards.
5. Indirect: 90–100 per cent upwards, 10–0 per cent downwards.

This system has now been extended to take account of how the downward light is distributed, as the final result in an installation depends not only on the distribution from the fitting but also upon the characteristics of the space in which it is used. Both these aspects are combined into the BZ method described earlier.

Design and selection of fittings

The efficiency of a fitting depends solely on the amount of absorption of light in transmission or reflection from the various components and a fitting that has many reflecting or diffusing surfaces will generally be less efficient than one with a small area of controlling surfaces. The way in which light is distributed by a fitting does not effect its overall efficiency, although it does effect the proportion of light reaching the working plane. If efficiency of light output is the only criterion on which the design of the installation is to be based, then a system that permits an unscreened light source is the answer, as no light is lost by absorption in the fitting. However, in general it is preferable to sacrifice some efficiency in order to achieve a better visual environment, as when bright sources are screened to avoid glare.

Light fittings may be divided into three categories: general fittings which are used for offices, schools and factories; decorative fittings which are used in homes, hotels and restaurants; and directional fittings used for display purposes in shops, departmental stores and art galleries. All light fittings must be strong enough to withstand handling and protect the lamp. They must be provided with means of suspending the lamp in space and access to the electrical gear.

Figures 14.13 and 14.14 show common types of fittings used for tungsten filament lamps and fluorescent lamps.

Recessed fittings

These are usually modular units which may be inserted in suspended ceilings in place of solid panels; they are usually fitted with fluorescent tubes. Recessed fittings for tungsten filament lamps are shown in Fig. 14.15. These have the advantage of providing lighting without visible fittings, but do not provide any appreciable light on the ceiling and may be unsatisfactory in interiors having low reflection factors.

Luminous ceilings

These consist of a lightweight framework which supports louvres or diffusing material. The diffusing material may be of translucent plastic or louvres (see Fig. 14.16). Translucent plastic may give rise to glare since the factors that may affect glare, i.e. the brightness of the source is increased as its size is increased. Luminous ceilings should not exceed the recommended luminance limits given in the IES Technical Report No. 10, *Evaluation of Discomfort Glare*.

Special indirect lighting (Fig. 14.17)

Indirect lighting from cornices, recessed coves and coffers may be used in theatres, hotels and living rooms of houses. The arrangement permits the lamps to give all their light above the horizontal and provides bright spots in the room, thus making the lighting scheme more interesting and stimulating. This form of lighting arrangement may be used to provide a decorative effect and also foi display purposes.

The following points should be taken into account for this type of lighting arrangement:

1. The lamps should not be fixed too close to the ceiling.
2. Where tubular fluorescent lamps are used they should be well maintained and be of the same colour and luminance.
3. Black streaks may form on walls and ceilings due to convection currents, and frequent redecoration of adjacent surfaces may be required.

4. The ceiling and upper wall surfaces should be of a matt finish and in a white or very pale colour, as they affect both the amount of light reflected and its colour.
5. Where continuous lines of light are required in cornices, etc., cold-cathode tubular fluorescent lamps can be specially made to suit the dimensions of the room.

Air-handling fittings

Combined heating, lighting and air-conditioning systems require special lighting fittings (see Figs. 14.18 and 14.19). The heat generated by the lighting equipment is extracted through the fitting and is transferred to the air-conditioning or warm-air unit. The air-conditioning or heating unit may thus be reduced in size. The heat generated by lighting equipment can be considerable, and a lighting load of 30–75 W/m^2 may be sufficient to provide all, or the major part, of the heating requirements of a well-insulated building in winter.

Air-handling lighting fittings allow air to be drawn past the lamps which reduces their operating temperature to a more efficient level – often increasing their light output by about 10 per cent. The air stream also helps to keep the lamps and fittings clear of dust, thus reducing the amount of cleaning.

In deep rooms, the heat from the lighting fittings in the central areas may be fed to the colder area around the perimeters, thus providing a more even distribution of heat throughout the rooms. Air-handling fittings also simplify the layout of the fittings in the ceiling.

Lighting controls

It is no use having the most efficient lighting system and equipment if it is not used efficiently. In public buildings, experience has shown that the occupants of the building are much better at switching lights on than they are at switching them off. To reduce lighting costs therefore it is necessary to make the control of the electric lighting as automatic as is practicable.

Time switches

The simplest fully automatic control system is the time-switch which can be set to switch the lights on and off at predetermined times. It is essential that the switch has a day omitting device so that the timeswitch does not operate the lights during weekends and holidays. Timeswitches may, of course, be used for security lighting.

Photo-electric cell

An alternative to the time-switch is the photo-electric cell which responds to the amount of light falling upon it. The photo-electric cell can be positioned at any point where the lighting level is critical. As the daylight fades the lights will be switched on and, subsequently, the photo-electric cell controls will switch off the lights when daylight increases.

Switching the lights on and off, however, can cause problems. If for instance the recommended task illuminance is 600 lux it might be assumed that the photo-electric cell controls should switch off the lights when 600 lux of daylight is available. This, however, will mean that the instant before the lights are switched off there will be 1200 lux of illuminance available: 600 lux of daylight and 600 lux from the electric lights. Suddenly reducing the available light from 1200 lux to 600 lux is likely to cause complaints from the building's occupants. A point must therefore be chosen, say 1000 lux or more of daylight, where the change is less noticeable. This means that some of the power used for the lighting is wasted while waiting for daylight to increase.

Dimmers

To maximise the savings in power, it is essential to start reducing the lighting as soon as the daylight starts to make a useful contribution. This can be achieved by using regulated high frequency ballasts controlled by a photo-electric cell. When daylight is sensed by the cell the brightness of the electric light can start to be reduced and eventually the light is switched off when the daylight contribution alone provides sufficient illuminance. Conversely, as daylight fades, the electricity supply to the light can gradually be increased and should not use the full load straight away.

There is one important requirement with these systems: that the dimmer must be capable of switching off entirely when the electric light is not required.

Whilst using regulated ballasts with photo-electric cells to react with daylight and either switch on or off the lights, the ballast can also be programmed to respond to mains borne signals. This means that the system is ideal for other methods of control. At the end of a working day, for instance, half the luminaires may be switched off automatically, the other half switching off, say, thirty minutes later. A few luminaires could be left on manual control for safety, and only switched off by caretakers or security staff when the building is empty.

In such systems, rearranging switching to suit a change in the office layout need only be a case of changing the signal to which the luminaire responds. Systems are now available where coded signals are sent via the telephone network to a microcomputer which then transmits the necessary codes to the luminaires. In this way the period of operation and the illuminance may be controlled remotely.

Presence detector

Controlling the lighting automatically will save costs but this saving can be lost if the lights are on when the building is empty. A presence detector which senses body heat or movement can be used to detect if there is a person or persons in the room and will switch the lights off if there is not. A luminaire is available having a high frequency and with its own built-in photo-electric cell and presence detector. If a person is close to the luminaire it adjusts its output and takes into account the available daylight, and if a person is absent for about ten minutes the luminaire switches off.

Depending upon the complexity of the installation and the amount of control that is required, some or all of these automatic controls may be used to provide the most economic operation.

Definition of terms

BZ system The British Zonal system. A system for classifying lighting fittings according to their distribution of downward light.

Candela (cd) The SI unit of luminance intensity.

Colour rendering A general expression for the effect of an illuminant on the colour appearance of the objects in conscious or sub-conscious comparison with their appearance under a reference illuminant.

Colour temperature Of a light source: the temperature of a full radiator which would emit radiation of substantially the same spectral distribution in the visible region as the radiation from the light source, and which would have the same colour.

Daylight factor At a given point inside a building. The ratio of the illumination measured on a horizontal plane at that point to that simultaneously existing on a horizontal plane under an unobstructed sky of uniform luminance. Light reflected from interior and exterior surfaces is included in the illumination at the point.

Directional lighting Lighting designed to illuminate predominantly from a preferred direction.

Discharge lamp A lamp in which light is produced by the passage of electricity through a metallic vapour or gas enclosed in a tube or bulb.

Efficacy Of a light source: the ratio of luminous flux emitted to power consumed by the source. In the case of an electric lamp it is expressed in lumens per watt.

Fluorescent lamp A tubular discharge lamp internally coated with a powder which fluoresces under the action of discharge, producing a shadowless white or coloured light.

General lighting Lighting designed to illuminate an area without special provision for local lighting.

Glare An uncomfortable effect on vision, due to an excessive range of luminances or extreme contrast.

Glare index A numerical expression of the potential glare from lighting.

Illumination At a point of a surface: the quotient of luminous flux incident on an infinitesimal element of surface containing the point under consideration, by the area of that element. Unit lux (lx).

Lightmeter A photocell, which when light falls on it, generates current to move a pointer over a graduated scale giving the value of illumination in lux. Usually a switch is fitted to make it a double-range instrument, say 0–500 and 0–2500 lx.

Light output ratio The ratio of light emitted by a fitting to that from the lamp, or lampit houses.

Localised lighting Lighting designed to illuminate an area, and at the same time to provide higher illumination over part of the area for local requirements.

Local lighting Lighting designed to illuminate a relatively small area.

Lumen The unit of luminous flux. Used in describing the total light emitted by a source or received by a surface.

Luminance A quantitative expression of brightness (cd/m^2).

Luminosity Apparent brightness. The attribute of visual perception in accordance with which an area appears to emit more or less light.

Luminous flux The flow of light, i.e. the rate at which light is emitted from a source, or arrives at a surface, or is reflected from it, measured in lumens. Unit lumen (lm).

Luminous intensity The strength of a light source in a given direction, measured in candelas. Symbol cd, unit candela.

Lux The unit of illumination. An illumination of one lumen per square metre. Abbreviation lx.

Maintenance factor An allowance for reduced light emission due to dirt on the fittings and room surfaces. A factor of 0.8 is applied for normal interiors.

Polar curve The distribution of light from a source indicated graphically.

Reflection factor The ratio of reflected luminous flux to the incident luminous flux. Where mixed reflection occurs, the total reflection factor can be divided into two parts: direct reflection factor and diffuse reflection factor.

Room index

$$\frac{\text{length} \times \text{width}}{(\text{length} + \text{width})\ \text{height of fitting above the working plane}}$$

Scalar illumination The average illumination on the surface of a small sphere within the space, i.e. the average illumination coming from all directions thus takes into account all reflection from walls and floor.

Spacing/height ratio The ratio of the distance between adjacent fittings to their height above the illuminated plane.

Utilisation factor An experimentally determined factor taking into account the shape of the room, reflection factors of surfaces, and the performance of light fittings.

Working plane The plane on which illumination measurements are made. Unless otherwise stated, the plane is assumed to be horizontal and 850 mm above the floor.

Appendix

SI units (International System of units)

In 1971 the Council of Ministers of the European Economic Community (EEC) decided to commit all member countries to amend their legislation in terms of SI units. The United Kingdom had already decided that SI units would become the primary system of measurement, and legislation is established in some twenty-five countries, including Germany, France, India, the USSR and Czechoslovakia. The system is also being considered in the USA where a 3-year study has been completed on behalf of the Department of Commerce.

Base units

The SI system is based on seven units.

Quantity	Unit	Symbol
Length	metre	m
Mass	kilogram	kg
Time	second	s
Electric current	ampere	A
Thermodynamic temperature	kelvin	K
Luminous intensity	candela	cd
Amount of substance	mole	mol

Note: For ordinary temperature and the difference between two temperatures, i.e. temperature interval, the degree Celsius (°C) is used.

Supplementary units

Quantity	Unit	Symbol
Plane angle	radian	rad
Solid angle	steradian	sr

The radian is the angle between two radii of a circle which cut off on the circumference an arc equal in length to the radius. The steradian is an angle which having its vertex in the centre of a sphere, cuts off an area of the surface of the sphere equal to that of a square having sides of length equal to the radius of the sphere.

Derived units

These are expressed algebraically in terms of base units and for supplementary units.

Quantity	Name of derived units	Symbol	Units involved
Frequency	hertz	Hz	1 Hz = 1 s^{-1} (1 cycle per second)
Force	newton	N	1 N = 1 kg m/s^2
Pressure and stress	pascal	Pa	1 Pa = 1 N/m^2
Work, energy, quantity of heat	joule	J	1 J = 1 N m
Power	watt	W	1 W = 1 J/s
Quantity of electricity	coulomb	C	1 C = 1 A s
Electric potential, potential difference, electromotive force	volt	V	1 V = 1 W/A
Electric capacitance	farad	F	1 F = 1 As/V
Electric resistance	ohm	Ω	1 Ω = 1 V/A
Electric conductance	siemens	S	1 S = 1 Ω^{-1}
Magnetic flux, flux of magnetic induction	weber	Wb	1 Wb = 1 V s
Magnetic flux density, magnetic induction	tesla	T	1 T = 1 Wb/m^2
Inductance	henry	H	1 H = 1 Vs/A
Luminous flux	lumen	lm	1 lm = 1 cd sr
Illuminance	lux	lx	1 lx = 1 lm/m^2

Multiples and sub-multiples of SI units

Factor		Prefix Name	Symbol
One billion	10^{12}	tera	T
One thousand million	10^{9}	giga	G
One million	10^{6}	mega	M
One thousand	10^{3}	kilo	k
One hundred	10^{2}	hecto	h
Ten	10	deca	da
One-tenth	10^{-1}	deci	d
One-hundredth	10^{-2}	centi	c
One-thousandth	10^{-3}	milli	m
One millionth	10^{-6}	micro	μ
One thousand millionth	10^{-9}	nano	n
One million millionth	10^{-12}	pico	p

Units for general use

Quantity	Unit	Symbol	Definition
Time	minute	min	1 min = 60 s
	hour	h	1 h = 60 min
	day	d	1 d = 24 h
Plane angle	degree	°	$1^\circ = (\pi/180)$ rad
	minute	′	$1' = (1/60)^\circ$
	second	″	$1'' = (1/60)'$
Volume	litre	l	1 l = 1 dm^2
Mass	tonne	t	1 t = 10^3 kg

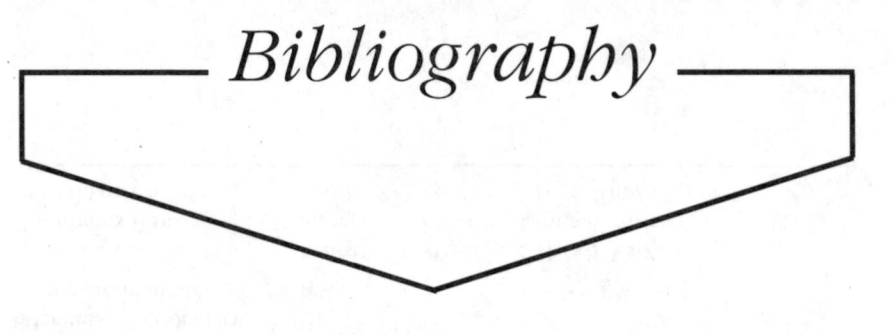

Bibliography

Relevant BS. British Standards Institution.

Relevant BSCP. British Standards Institution.

Building Regulations. HMSO.

Relevant BRE Digests. HMSO.

Electrical Regulations. 16th edition. The Institution of Electrical Engineers.

Fire Protection Handbook. Mather & Platt Ltd.

Electrical Services in Buildings. The Electricity Council.

Electrical Installations in Buildings. J. A. Crabtree & Co. Ltd.

D. C. Pritchard. *Lighting*. Longman Group Ltd.

Smoke Control in Covered Shopping Malls. The Fire Research Association.

Lifts. Hammond and Clampness.

Relevant manufacturers' catalogues contained in the Barbour Index and Building Products Index Libraries.

Index

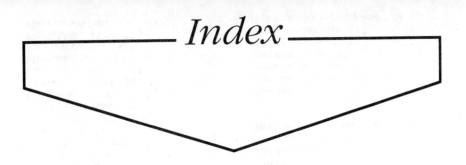